Tetrel Bonds

Tetrel Bonds

Editor

Steve Scheiner

MDPI • Basel • Beijing • Wuhan • Barcelona • Belgrade • Manchester • Tokyo • Cluj • Tianjin

Editor
Steve Scheiner
Utah State University
USA

Editorial Office
MDPI
St. Alban-Anlage 66
4052 Basel, Switzerland

This is a reprint of articles from the Special Issue published online in the open access journal *Molecules* (ISSN 1420-3049) (available at: https://www.mdpi.com/journal/molecules/special_issues/Tetrel_Bonds).

For citation purposes, cite each article independently as indicated on the article page online and as indicated below:

LastName, A.A.; LastName, B.B.; LastName, C.C. Article Title. *Journal Name* **Year**, *Volume Number*, Page Range.

ISBN 978-3-0365-0236-6 (Hbk)
ISBN 978-3-0365-0237-3 (PDF)

© 2023 by the authors. Articles in this book are Open Access and distributed under the Creative Commons Attribution (CC BY) license, which allows users to download, copy and build upon published articles, as long as the author and publisher are properly credited, which ensures maximum dissemination and a wider impact of our publications.

The book as a whole is distributed by MDPI under the terms and conditions of the Creative Commons license CC BY-NC-ND.

Contents

About the Editor vii

Preface to "Tetrel Bonds" ix

José Luis Casals-Sainz, Aurora Costales Castro, Evelio Francisco and Ángel Martín Pendás
Tetrel Interactions from an Interacting Quantum AtomsPerspective
Reprinted from: *Molecules* **2019**, *24*, 2204, doi:10.3390/molecules24122204 1

Guillermo Caballero-García, Gustavo Mondragón-Solórzano, Raúl Torres-Cadena,
Marco Díaz-García, Jacinto Sandoval-Lira and Joaquín Barroso-Flores
Calculation of $V_{S,max}$ and Its Use as a Descriptor for the Theoretical Calculation of pKa Values
for Carboxylic Acids
Reprinted from: *Molecules* **2019**, *24*, 79, doi:10.3390/molecules24010079 17

Daniel Sethio, Vytor Oliveira and Elfi Kraka
Quantitative Assessment of Tetrel Bonding Utilizing Vibrational Spectroscopy
Reprinted from: *Molecules* **2018**, *23*, 2763, doi:10.3390/molecules23112763 33

Ephrath Solel and Sebastian Kozuch
On the Power of Geometry over Tetrel Bonds
Reprinted from: *Molecules* **2018**, *23*, 2742, doi:10.3390/molecules23112742 55

Antonio Bauzá and Antonio Frontera
Tetrel Bonding Interactions in Perchlorinated Cyclopenta- and Cyclohexatetrelanes:
A Combined DFT and CSD Study
Reprinted from: *Molecules* **2018**, *23*, 1770, doi:10.3390/molecules23071770 63

Tiddo J. Mooibroek
Intermolecular Non-Covalent Carbon-Bonding Interactions with Methyl Groups: A CSD, PDB
and DFT Study
Reprinted from: *Molecules* **2019**, *24*, 3370, doi:10.3390/molecules24183370 75

Ekaterina Bartashevich, Yury Matveychuk and Vladimir Tsirelson
Identification of the Tetrel Bonds between Halide Anions and Carbon Atom of Methyl Groups
Using Electronic Criterion
Reprinted from: *Molecules* **2019**, *24*, 1083, doi:10.3390/molecules24061083 89

Raymond C. Trievel and Steve Scheiner
Crystallographic and Computational Characterization of Methyl Tetrel Bonding in
S-Adenosylmethionine-Dependent Methyltransferases
Reprinted from: *Molecules* **2018**, *23*, 2965, doi:10.3390/molecules23112965 103

Mehdi D. Esrafili and Parisasadat Mousavian
Strong Tetrel Bonds: Theoretical Aspects and Experimental Evidence
Reprinted from: *Molecules* **2018**, *23*, 2642, doi:10.3390/molecules23102642 121

Steve Scheiner
Tetrel Bonding as a Vehicle for Strong and Selective Anion Binding
Reprinted from: *Molecules* **2018**, *23*, 1147, doi:10.3390/molecules23051147 141

Ibon Alkorta and Anthony C. Legon
An Ab Initio Investigation of the Geometries and Binding Strengths of Tetrel-, Pnictogen-, and Chalcogen-Bonded Complexes of CO_2, N_2O, and CS_2 with Simple Lewis Bases: Some Generalizations
Reprinted from: *Molecules* **2018**, *23*, 2250, doi:10.3390/molecules23092250 . . . . **161**

Wenbo Dong, Qingzhong Li and Steve Scheiner
Comparative Strengths of Tetrel, Pnicogen, Chalcogen, and Halogen Bonds and Contributing Factors
Reprinted from: *Molecules* **2018**, *23*, 1681, doi:10.3390/molecules23071681 . . . . **177**

Janet E. Del Bene, José Elguero and Ibon Alkorta
Complexes of CO_2 with the Azoles: Tetrel Bonds, Hydrogen Bonds and Other Secondary Interactions
Reprinted from: *Molecules* **2018**, *23*, 906, doi:10.3390/molecules23040906 . . . . **195**

Wiktor Zierkiewicz, Mariusz Michalczyk and Steve Scheiner
Comparison between Tetrel Bonded Complexes Stabilized by σ and π Hole Interactions
Reprinted from: *Molecules* **2018**, *23*, 1416, doi:10.3390/molecules23061416 . . . . **209**

Sławomir J. Grabowski
Tetrel Bonds with π-Electrons Acting as Lewis Bases—Theoretical Results and Experimental Evidences
Reprinted from: *Molecules* **2018**, *23*, 1183, doi:10.3390/molecules23051183 . . . . **225**

About the Editor

Steve Scheiner

Steve Scheiner earned his B.S. at City College of New York in 1972, and his Ph.D. in 1976 from Harvard University applying quantum chemical methods to enzyme activity, under the tutelage of William Lipscomb. After two years as a Weizmann Postdoctoral Fellow at Ohio State University, he took a faculty position at Southern Illinois University, Carbondale, and then moved to Utah State University in 2000. His research has centered generally around molecular interactions of relevance to biological systems, with a particular focus on H-bonds and proton transfers, and more recently on weak noncovalent interactions. He has published more than 400 articles, authored one monograph, edited three volumes, and has contributed 25 chapters to various books. He is a regular reviewer of papers submitted to more than 150 different journals, at the rate of some 130 submissions each year.

Preface to "Tetrel Bonds"

After many decades of study of the H-bond which probed every facet of its origin and behavior, researchers realized there are a group of very similar interactions that had largely defied recognition. It was realized that the replacement of the bridging proton of the H-bond by a halogen atom led to a stabilizing interaction of strength comparable to the H-bond itself. The stability of this halogen bond was based in part on the anisotropic distribution of electron density around the halogen atom that allowed a positive region on an otherwise negatively charged atom. It was not long after that point that research turned to close parallels where the halogen atom was itself replaced by members of the chalcogen and pnicogen families, whose bonding properties were likewise parallel to those of halogen bonds. A recent addition to this pantheon is the tetrel bond, a family that encompasses Si, Ge, Sn, and the ubiquitous and immensely important C atom.

It is the tetrel bond that serves as the focus of this volume. Papers collected from a recent Special Issue of Molecules cover this topic from a range of perspectives. The first two papers concentrate on some of the theoretical aspects that contribute to the bond, providing a lens by which to understand it and how it might be used to predict certain chemical properties. Coverage next turns toward some of the structural and spectroscopic aspects of these bonds, from an understanding based on vibrational modes, to effects of geometrical distortions, and to specific lessons learned from analysis of crystal structures.

The C atom is of special importance in all of chemistry and biochemistry, and the next three papers focus attention on the tetrel bonds specifically centered on C. One common element of these papers is to alert the community to not only the widespread presence of tetrel bonds but also the importance of these bonds to methyl groups. Along with the acceptance that the presence of a negative charge on the electron donor will strengthen a H-bond, the next two contributions illustrate that the same phenomenon is true of tetrel bonds and describe how this bond enhancement can be applied to advantage. The last five papers offer a broad perspective, describing how tetrel bonds fit into the general picture of noncovalent bonds in general. This section also includes a visualization as to the roles played by π-systems, not only in the tetrel bond donor but also in the partner nucleophile.

Steve Scheiner
Editor

Article

Tetrel Interactions from an Interacting Quantum Atoms Perspective

José Luis Casals-Sainz, Aurora Costales Castro, Evelio Francisco and Ángel Martín Pendás *

Departamento de Química Física y Analítica, Julián Clavería, 6, Facultad de Química, Universidad de Oviedo, 33006 Oviedo, Spain; jluiscasalssainz@gmail.com (J.L.C.-S.); costalesmaria@uniovi.es (A.C.C.); evelio@uniovi.es (E.F.)
* Correspondence: ampendas@uniovi.es

Received: 17 May 2019; Accepted: 7 June 2019; Published: 12 June 2019

Abstract: Tetrel bonds, the purportedly non-covalent interaction between a molecule that contains an atom of group 14 and an anion or (more generally) an atom or molecule with lone electron pairs, are under intense scrutiny. In this work, we perform an interacting quantum atoms (IQA) analysis of several simple complexes formed between an electrophilic fragment (A) (CH_3F, CH_4, CO_2, CS_2, SiO_2, SiH_3F, SiH_4, GeH_3F, GeO_2, and GeH_4) and an electron-pair-rich system (B) (NCH, NCO^-, OCN^-, F^-, Br^-, CN^-, CO, CS, Kr, NC^-, NH_3, OC, OH_2, SH^-, and N_3^-) at the aug-cc-pvtz coupled cluster singles and doubles (CCSD) level of calculation. The binding energy (E_{bind}^{AB}) is separated into intrafragment and inter-fragment components, and the latter in turn split into classical and covalent contributions. It is shown that the three terms are important in determining E_{bind}^{AB}, with absolute values that increase in passing from electrophilic fragments containing C, Ge, and Si. The degree of covalency between A and B is measured through the real space bond order known as the delocalization index (δ^{AB}). Finally, a good linear correlation is found between δ^{AB} and E_{xc}^{AB}, the exchange correlation (xc) or covalent contribution to E_{bind}^{AB}.

Keywords: energy partition; interacting quantum atoms; quantum theory of atoms in molecules; delocalization index; covalent interaction; self-energy

1. Introduction

There has been a growing interest in the last years in the field of non-covalent interactions (NCI) [1], which have been shown to be critical for the correct description of the structure and properties of different molecules and materials [2,3], including nanomaterials [4,5], molecular solids [6,7], surfaces [8,9], and biological systems [10–12]. From the many types of interactions that are usually classified as non-covalent, hydrogen bonding A−H···D [13,14], where A is a group more electronegative that H and D is an entity able to act as an electron donor, is undoubtedly the best-known by all chemists. Besides hydrogen and halogen bonding [15,16] (possibly the best-known type of NCI after the former), other purported NCIs involving atoms of groups 14, 15, and 16 (and even rare gas atoms [17,18]) have recently received the names of tetrel, pnictogen [19,20], and chalcogen bonding, respectively, although some of these complexes were identified and characterized by different experimental techniques long before they were given these names [19]. In all of them, the 14, 15, or 16 group element, acting as an electron acceptor or electrophilic site, seeks the nucleophilic part of another system, for instance an atomic or molecular anion (F^-, Br^-,..., CN^-, NC^-, N_3^-,...), a π−electron pair of a Lewis base, or a non-bonding electron pair of an arbitrary molecule. As far as tetrel bonds are concerned, and to name just a few works, Bürgi et al. pioneered the study and description of nucleophilic additions to carbonyl C-atoms or $n \to \pi^*$ interactions [21–23], recognized by several authors as important to biology [24–26], and Thomas et al. found experimental evidence for carbon bonding (an interaction where a carbon atom acts as an electrophilic site toward a variety of nucleophiles) in the solid state from X-ray charge density analysis [27]. Southern and Bryce presented results for NMR parameters of a series of model compounds in which a tetrel bond between a methyl C atom and the N or O atom of several functional groups is found [28]. Scilabra has shown that contacts between a Ge or Sn atom with different lone-pair-possessing atoms in crystal structures are quite common [29], and Mitzel and Losehand found crystal structures of $Si(ONMe_2)_4$ with a short-distance Si–N bond [30]. A particularly relevant mention is also the work of Bauzá, Frontera, and Mooibroek, who pioneered tetrel interactions and wrote several interesting reviews on the topic [31–35].

From the theoretical side, Scheiner [36] made a comparison of halide receptors based on H, halogen, chalcogen, pnicogen, and tetrel bonds by means of molecular electrostatic potential (MEP) maps and natural bond orbital (NBO) analyses [37,38] at the density functional theory (DFT) M06-2X//aug-cc-pDVZ level of calculation, and Alkorta et al. performed MP2//aug-cc-pVTZ energetic studies, calculating harmonic vibrational frequencies, EOM-CCSD spin-spin coupling constants, and NBO analyses in complexes of CO_2 with azoles [39], azines [40], and carbenes as electron pair donors to CO_2 [40]. Carbon bonding in the X–C$\cdots$Y (X=O/F, Y=O/S/F/Cl/Br/N/P) and X–C$\cdots\pi$ (X=F,Cl,Br,CN) systems were studied by Mani and Arunan [41,42], started with MP2/6-311+G(3df,2p) and MP2//aug-cc-pvtz calculations to optimize the geometry, and followed this up with CCSD(T) calculations to estimate the interaction energy and NBO, quantum theory of atoms in molecules (QTAIM) [43,44], and MEP analyses. Interestingly, the formation of the type of tetrel interaction called carbon bonding had been previously proposed by Grabowski as a preliminary step necessary in SN_2 reactions [45].

Mixed theoretical/experimental studies have also been carried out by Sethio, Oliveira, and Kraka, aimed at a quantitative assessment of tetrel bonding utilizing vibrational spectroscopy [46]. Finally, several other theoretical papers have been published in recent years to determine the influence that a substitution of the ligands have on the tetrel bond strength [47–53].

As far as we know, in the already extensive literature existing today regarding theoretical studies of tetrel bonding systems, there is no publication in which a detailed energy partition analysis of these compounds has been performed. We will carry out this study in this work. Specifically, we will use the interacting quantum atoms (IQA) method [54–57] to analyze about thirty complexes formed between an electrophilic fragment of the set CH_3F, CH_4, CO_2, CS_2, SiO_2, SiH_3F, SiH_4, GeH_3F, GeO_2, and GeH_4 and an electron-pair-rich system of the set NCH, NCO^-, OCN^-, F^-, Br^-, CN^-, CO, CS, Kr, NC^-, NH_3, OC, OH_2, SH^-, and N_3^-. IQA is a real space orbital invariant energy partition method inspired by the QTAIM that exactly recovers the total energy of a molecule by splitting its total energy in terms of intra-atomic and interatomic components. It is general in the sense that any type of wavefunction may in principle be analyzed with it. All that is required is to have at our disposal the one- and (diagonal) two-particle density matrices. Hartree–Fock, complete active space (CAS), full-CI, CCSD and EOM-CCSD wavefunctions have been analyzed to date using IQA. DFT calculations can also be used within the IQA partition, at least formally, taking the Kohn–Sham determinant of the system as the approximate wavefunction and performing a physically sound scaling of the interactions [58,59].

The degree of detail with which IQA allows us to scrutinize the energetic interactions is really high. However, in this work we will not use all of these potentialities. This means that we will only split the different energy contributions at the fragment, and not the atomic level. We will thus worry neither about analyzing how the net energy of a given fragment is distributed among its atoms nor on how the atoms of this fragment interact with each other. Discussions related to the geometry of the fragments will not be considered either. Once the geometry of the molecules has been optimized (as discussed in the next section), all subsequent energetic analyses will refer to these geometries.

The rest of the article has been divided as follows. The theoretical methods used in our analyses are briefly discussed in Section 2. Some computational details related to the above methods are given in Section 3. The results and their discussion are presented in Section 4. Finally, the more relevant conclusions of this work are given in Section 5.

2. Theoretical Methods

In this section, we describe very briefly the interacting quantum atoms (IQA) [54–57] approach that has been used to obtain all the data and energetic quantities that will be discussed in Section 4. We also give some relevant details of the coupled cluster method up to single and double excitations (CCSD) [60] that we have employed to derive the wavefunctions that are fed into the IQA method, and comment, also very briefly, on some points regarding the computation, within the IQA scheme, of the binding energy of a supermolecule AB from its IQA energetic quantities and those of the isolated fragments A and B.

The interacting quantum atoms (IQA) method is an energy partition scheme that is based on the exhaustive partition of the real space occupied by a molecule according to the quantum theory of atoms in molecules (QTAIM) [43]. IQA exactly recovers the total energy of a molecule and can be applied, in principle, to any level of theory as soon as the one-particle, $\rho_1(\mathbf{r}_1, \mathbf{r}_1')$, and (diagonal) two-particle, $\rho_2(\mathbf{r}_1, \mathbf{r}_2)$, density matrices are available. The total electronic Born–Oppenheimer energy of a molecule reads as [61]

$$E = \int \hat{h} \rho_1(\mathbf{r}_1, \mathbf{r}_1') d\mathbf{r}_1 + \frac{1}{2} \iint \frac{\rho_2(\mathbf{r}_1, \mathbf{r}_2)}{r_{12}} d\mathbf{r}_1 d\mathbf{r}_2 + E_{\text{nuc}}, \quad (1)$$

where $\hat{h}$ is the monoelectronic operator that includes the kinetic energy and nuclear attraction terms, $\hat{h}_i = \hat{t}_i - \sum_A Z_A/r_{iA}$, and $E_{\text{nuc}} = \sum_{A>B} Z_A Z_B R_{AB}^{-1}$ is the total nuclear repulsion energy. If the physical space R^3 is partitioned according to QTAIM [43,44], $R^3 = \cup_A \Omega_A$, where Ω_A represents the atomic basin of atom A, it is clear that the monoelectronic energy in Equation (1) can be split into as many contributions as the total number of atoms of the molecule (say n), and the bielectronic energy into n^2 terms. Doing so, we obtain the IQA energy partition

$$E = \sum_A \int_{\Omega_A} \hat{h} \rho_1(\mathbf{r}_1, \mathbf{r}_1') d\mathbf{r}_1 + \frac{1}{2} \sum_{A,B} \int_{\Omega_A} \int_{\Omega_B} \frac{\rho_2(\mathbf{r}_1, \mathbf{r}_2)}{r_{12}} d\mathbf{r}_1 d\mathbf{r}_2 + E_{\text{nuc}}.$$

Grouping together intra- ($A = B$) and inter-atomic ($A \neq B$) terms,

$$E = \sum_A E_{\text{net}}^A + \sum_{A>B} E_{\text{int}}^{AB} \quad (2)$$

$$= \sum_A T^A + V_{\text{ne}}^{AA} + V_{\text{ee}}^{AA} + \sum_{A>B} V_{\text{nn}}^{AB} + V_{\text{ne}}^{AB} + V_{\text{ne}}^{BA} + V_{\text{ee}}^{AB}, \quad (3)$$

where E_{net}^A is the net or self-energy of atom A, which collects all the energy terms involving exclusively the nucleus and electrons within this atom, and E_{int}^{AB} is the total interatomic energy between atoms A and B. In Equation (3), V_{nn}^{AB}, V_{ne}^{AB}, and V_{ee}^{AB} when $A \neq B$ represent the nucleus–nucleus, nucleus–electron, and electron–electron interactions associated to the pair of atoms A, B, and V_{ne}^{AA} is the interaction of the electrons inside Ω_A with the nucleus of this atomic basin. V_{ee}^{AA} is the total electron repulsion of the electrons inside Ω_A among themselves. When A or B or both represent groups of atoms or molecules instead of single atoms, Equations (2) and (3) remain almost valid, with minor modifications that involve only the self-energy term E_{net}^A [62]. By splitting $\rho_2(\mathbf{r}_1, \mathbf{r}_2)$ into its classical and exchange-correlation components, E_{int}^{AB} may be written as $E_{\text{int}}^{AB} = V_{\text{cl}}^{AB} + V_{\text{xc}}^{AB}$, where V_{cl}^{AB} is the electrostatic interaction between all particles (nuclei plus electrons) inside A with all particles inside B, and V_{xc}^{AB} represents the purely quantum-mechanical or covalent interaction. In the Hartree–Fock (HF) approximation, only the Fermi correlation is taken into account, which leads to $V_{\text{xc}}^{AB} = V_{\text{x}}^{AB}$ in the IQA method, which contains only exchange interactions. However, for many correlated methods it is still possible to write formally $V_{\text{xc}}^{AB} = V_{\text{x}}^{AB} + V_{\text{corr}}^{AB}$, provided that the pure-exchange two-particle density matrix, $\rho_2^x(\mathbf{r}_1, \mathbf{r}_2)$, is taken as the Dirac–Fock expression $\rho_2^x(\mathbf{r}_1, \mathbf{r}_2) = -\rho_1(\mathbf{r}_1, \mathbf{r}_2) \times \rho_1(\mathbf{r}_2, \mathbf{r}_1)$, and $\rho_2^{\text{corr}}(\mathbf{r}_1, \mathbf{r}_2)$ is simply defined as $\rho_2^{\text{corr}}(\mathbf{r}_1, \mathbf{r}_2) = \rho_2^{\text{xc}}(\mathbf{r}_1, \mathbf{r}_2) - \rho_2^x(\mathbf{r}_1, \mathbf{r}_2)$. Among the several post-HF levels of theory including dynamical correlation energy contributions (absolutely necessary to address the study of the systems considered in this work), we have chosen the coupled cluster method including only single and double excitations [60]. Other approaches, such as the second-order Møller–Plesset perturbation theory (MP2) [63], which overestimates the dispersion energy interactions [64], have not been considered. In the CCSD method, the reference wavefunction of the system is the HF determinant, and the total energy E is written as the sum of the energy of this reference wavefunction plus the correlation energy of the molecule, $E = E_{HF} + E_{corr}$. The latter can be expressed in terms of the CCSD amplitudes t_{ij}^{ab} as

$$E_{corr} = \sum_{iajb} t_{ij}^{ab} (ia|jb), \qquad (4)$$

where $(ia|jb)$ are two electron integrals in the molecular orbital basis (MO), the Mulliken convention has been used, and i, j and a, b refer to occupied and virtual orbitals, respectively. As the total energy E itself, E_{corr} can be partitioned à la IQA, leading to intra-atomic, $E_{corr}^{net,A}$, and interatomic, $E_{corr}^{int,AB}$ terms. Other details of the IQA implementation within the CCSD method are described elsewhere [65].

The binding energy between two atoms, fragments or molecules A and B is defined by

$$E_{bind}^{AB} = E^{AB} - E^{A} - E^{B}, \qquad (5)$$

where E^{AB}, E^{A}, and E^{B} are the total energies of AB, A, and B, respectively. If these three total energies are separated into their HF and correlation components, E_{bind}^{AB} results:

$$\begin{aligned} E_{bind}^{AB} &= (E_{HF}^{AB} - E_{HF}^{A} - E_{HF}^{B}) + (E_{corr}^{AB} - E_{corr}^{A} - E_{corr}^{B}) & (6) \\ &= E_{HF,bind}^{AB} + E_{corr,bind}^{AB}. & (7) \end{aligned}$$

Assuming that the geometries of A and B are the same as in the supermolecule AB, E_{bind}^{AB} in the IQA method is given by

$$\begin{aligned} E_{bind}^{AB} &= E_{def}^{A} + E_{def}^{B} + V_{cl}^{AB} + (V_{x}^{AB} + E_{corr}^{AB}) & (8) \\ &= E_{def}^{A} + E_{def}^{B} + V_{cl}^{AB} + V_{xc}^{AB}, & (9) \end{aligned}$$

where each deformation energy $E_{def}^{R} = E_{net}^{R} - E^{R}$ (R = A, B) represents the energy change suffered by R when it passes from being isolated to interacting with the other fragment(s). In case the geometry of R has changed in going from the isolated state to the supermolecule, the so-called preparation energy, E_{prep}^{R}, defined as $E_{prep}^{R} = E^{R}(\text{supermolecule geometry}) - E^{R}(\text{isolated geometry})$, must be added to E_{bind}. On the other hand, it is customary in some energy partition methods, such as the energy decomposition analysis (EDA) method [66–68], to associate the term of Pauli exchange-repulsion (xr) with the increase of energy that takes place as a consequence of the antisymmetrization and normalization of the direct product of fragments' wavefunctions. Here, however, we will reserve this name to the sum of E_{def}^{A}, E_{def}^{B}, and V_{x}^{AB}, i.e.,

$$E_{xr}^{AB} = E_{def}^{A} + E_{def}^{B} + V_{x}^{AB}. \qquad (10)$$

Clearly, the origin of E_{xr}^{AB} in IQA is strictly different from that in EDA. However, in spite of this, the IQA E_{xr}^{AB} energy corresponds, in many ways, to other conventional exchange-repulsion terms [57,69]. For instance, as in other schemes, this energy turns out to be usually (but not necessarily) positive (see below). For that reason, we have decided to keep the name of exchange-repulsion for the energetic term defined in Equation (10). After using this definition in Equation (8), E_{bind} takes the form

$$E_{bind}^{AB} = V_{cl}^{AB} + E_{xr}^{AB} + E_{corr}^{AB}. \qquad (11)$$

In all the calculations presented in the following section, the basis set superposition error (BSSE), inherent to the calculation E_{bind}, has been corrected in the IQA scheme by using the Boys and Bernardi counterpoise method [70] to compute the total energies of the isolated fragments, E^R.

3. Computational Details

The calculations of this work have been done as follows. In a first step, the geometries of all the studied systems were optimized at the density functional theory (DFT) level with the WB97X-D functional [71] and the aug-cc-pvtz basis set [72] using the gamess package [73]. Then, single-point CCSD calculations at the optimized geometries were carried out with a locally modified copy of the PySCF code [74] using the same basis set. Core orbitals were frozen, and for truncating the virtual space, the frozen natural orbital approximation (FNO) with a cutoff in the natural occupations of 10^{-4} was used [75]. All the interaction energies include the BSSE correction [70]. The CCSD amplitudes t_{ij}^{ab} and the one- and two-particle density matrices were also obtained with PySCF [74].

The IQA energy partitioning was performed with our in-house program promolden [76]. The necessarily numerical IQA integrations were done using β-spheres for all the atoms, with radii between 0.1 and 0.3 bohr. Restricted angular Lebedev quadratures with 3074 points and 451-point Gauss–Chebyshev mapped radial grids were used inside the β-spheres, with L expansions cut at $l = 8$. Outside the β-spheres, extended 5810-point Lebedev, 551 mapped radial point Gauss–Legendre quadratures, and L expansions up to $l = 10$ were selected.

4. Results and Discussion

A graphical rendering of the optimized complexes studied in this work appears in Figure 1. The full set of atomic Cartesian coordinates is collected in the supplementary information. Since we have not carried out a systematic exploration of all possible local energy minima, there is no guarantee that the geometry depicted in the figure corresponds to the global minimum. For nine of the 31 complexes, both fragments are connected by a solid line ($SiH_4 \cdots F^-$, $GeH_4 \cdots F^-$, $SiH_3F \cdots N_3^-$, $GeH_3F \cdots N_3^-$, $SiO_2 \cdots NCH$, $SiO_2 \cdots CO$, $SiO_2 \cdots CS$, $SiO_2 \cdots Br^-$, $GeO_2 \cdots Br^-$). Under the rendering conditions that we have used, this implies that the two linked atoms are separated by a distance less than the sum of their covalent radii plus 0.025 Å. In the remaining 22 systems, there is no connection line between any pair of atoms $a \in A$ and $b \in B$. As we will see, the nine connected complexes are those with a delocalization index δ^{AB} (a measure of the covalent bond order in real space, see Table 2) greater than 0.5, similar to that of a typical polar covalent bond. It seems that, other factors aside, a clear correlation exists between δ^{AB} and the distance between the connected atoms of both fragments.

Given the numerical character of all the IQA integrated quantities [77,78], we want to check, first of all, the reliability and consistency of our results. We collect in Table 1 the binding energy of the different $A \cdots B$ systems computed directly as the total CCSD energy of the dimer, E^{AB}, minus the sum of the CCSD energies of both monomers, E^A and E^B ($E_{\text{bind}}(\text{CCSD})$, Equation (5)), the IQA binding energy obtained from Equation (8) ($E_{\text{bind}}(\text{IQA})$), its difference (diff), and the total IQA integrated charges, Q_A, Q_B, and $Q = Q_A + Q_B$. Almost systematically, the exact value of Q is very well reproduced by our IQA integrations. In all of the systems except $GeH_3F \cdots N_3^-$, $CH_3F \cdots N_3^-$ and $SiO_2 \cdots Br^-$, the error is less than or equal to 0.001e. As binding energies are regarded, the absolute error is lower than 0.5 Kcal/mol in 20 of the 30 systems and greater than this number in the remaining cases. Although it cannot be inferred from the numbers in the table, it can be said that almost 100% the error associated with the computation of $E_{\text{bind}}(\text{IQA})$ is due to its HF contribution, $E_{\text{HF,bind}}(\text{IQA})$ (see Equation (7)), since the IQA integrations of E_{corr}^{AB}, E_{corr}^A, and E_{corr}^B (see Equation (6)) reproduce their CCSD-analogous quantities extraordinarily well. Be that as it may, the fact that the $CO_2 \cdots Kr$ system in the IQA partition, contrarily to the CCSD calculation, is predicted to be unstable with respect to the isolated fragments, should not be taken too seriously given that the difference between $E_{\text{bind}}(\text{IQA})$ and $E_{\text{bind}}(\text{CCSD})$ in this system (0.52 Kcal/mol) is comparable to the average error of the numerical IQA integrations. Nonetheless, we believe that the present results are overall quite satisfactory, although we do not want to deny that the weakest point of the IQA energy partitioning method lies possibly in the existing difficulties of further reducing the errors associated with the numerical integrations of the method.

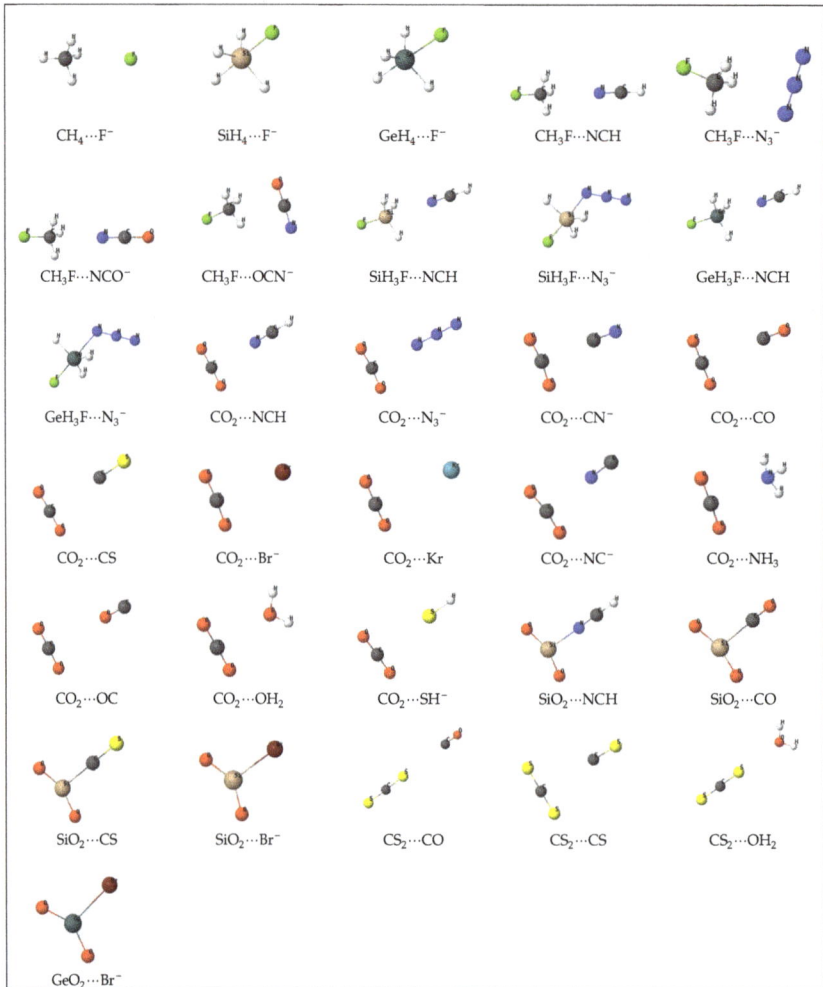

Figure 1. AB complexes studied in this work.

The electronic and total (electronic plus nuclear) charge of every atom (or fragment) of a molecular system, obtained by integrating the electronic or total density inside the corresponding atomic (or fragment) basin, is one of the main outcomes of the QTAIM methodology. It can be seen in Table 1 that the electronic charge transferred from fragment B to fragment A is relatively small in all of the studied systems for which the isolated fragment B is neutral. Only in $CO_2 \cdots NH_3$, $SiO_2 \cdots CO$, and $SiO_2 \cdots NCH$ is this transfer greater than 0.01e, whereas in the cases of $CO_2 \cdots OC$ and $CS_2 \cdots CS$, both A and B remain almost neutral after the complex is formed from the isolated fragments. Conversely, when the isolated fragment B is negatively charged, its ability to transfer electrons to the acceptor fragment increases notably. The most representative examples of this behavior are the systems $GeH_3F \cdots N_3^-$, $GeO_2 \cdots Br^-$, $GeH_4 \cdots F^-$, and $SiO_2 \cdots Br^-$. Given the generally greater polarizability of the valence electrons in anions as compared to neutral molecules, this is not a surprising result. Regarding the azide anion N_3^-, we observe in Table 1 how its ability to transfer electrons to the acceptor fragment MH$_3$F (M=C,Si,Ge) increases on descending in a group. The gain of electrons by the SiO$_2$ fragment in the three of the four systems of Table 1 in which this fragment appears is also greater than that corresponding to the CO$_2$ molecule in the equivalent complexes. Only in CO$_2$···CS is the CO$_2$ molecule more negatively charged than the SiO$_2$ molecule in SiO$_2$···CS.

Table 1. Columns 2–7 collect the binding energy obtained from Equations (5), E_{bind}(CCSD) ≡ (CCSD), and (8), E_{bind}(IQA) ≡ (IQA), its difference (diff = E_{bind}(IQA) − E_{bind}(CCSD)), and the total interacting quantum atom (IQA) integrated charges, Q_A, Q_B, and $Q = Q_A + Q_B$ Columns 8–14 show the different contributions to E_{bind}(IQA), according to Equations (8)–(11). Energy units in (Kcal/mol); CCSD—coupled cluster singles and doubles).

System (A···B)	(CCSD)	(IQA)	diff	Q_A	Q_B	Q	E^A_{def}	E^B_{def}	V^{AB}_{cl}	V^{AB}_x	V^{AB}_{xc}	E^{AB}_{int}	E^{AB}_{xr}
CH$_4$···F$^-$	−3.42	−3.41	0.01	−0.0112	−0.9891	−1.0003	8.42	3.77	−6.27	−8.46	−9.33	−15.60	3.73
SiH$_4$···F$^-$	−60.17	−59.87	0.31	−0.0426	−0.9579	−1.0005	123.02	87.24	−183.63	−82.43	−86.50	−270.13	127.83
GeH$_4$···F$^-$	−40.17	−38.89	1.28	−0.0830	−0.9170	−1.0001	79.50	46.57	−92.28	−69.62	−72.69	−164.97	56.46
CH$_3$F···NCH	−1.52	−1.36	0.16	−0.0033	0.0032	−0.0000	3.37	3.47	−1.94	−5.53	−6.25	−8.19	1.31
CH$_3$F···N$_3^-$	−9.21	−8.78	0.44	−0.0200	−0.9789	−0.9989	11.18	5.05	−8.47	−14.46	−16.54	−25.01	1.77
CH$_3$F···NCO$^-$	−9.89	−9.69	0.21	−0.0188	−0.9809	−0.9997	10.11	6.32	−11.73	−12.96	−14.39	−26.12	3.47
CH$_3$F···OCN$^-$	−8.37	−8.31	0.06	−0.0147	−0.9854	−1.0001	10.24	5.51	−9.28	−12.97	−14.78	−24.06	2.78
SiH$_3$F···NCH	−3.06	−3.02	0.04	−0.0052	0.0043	−0.0008	8.44	10.29	−5.06	−14.99	−16.68	−21.75	3.74
SiH$_3$F···N$_3^-$	−40.91	−41.45	−0.54	−0.0678	−0.9282	−0.9960	73.85	70.33	−103.27	−77.13	−82.36	−185.64	67.05
GeH$_3$F···NCH	−3.65	−2.99	0.66	−0.0031	0.0041	0.0010	9.20	9.77	−5.92	−14.31	−16.04	−21.96	4.66
GeH$_3$F···N$_3^-$	−36.72	−36.31	0.40	−0.1110	−0.8878	−0.9988	46.32	39.64	−53.01	−65.10	−69.26	−122.27	20.86
CO$_2$···NCH	−1.60	−2.08	−0.48	−0.0056	0.0059	0.0003	3.55	4.13	−2.28	−6.58	−7.47	−9.76	1.09
CO$_2$···N$_3^-$	−6.79	−6.46	0.34	−0.0236	−0.9757	−0.9994	12.43	8.07	−9.87	−15.31	−17.09	−26.96	5.20
CO$_2$···CN$^-$	−8.25	−8.64	−0.39	−0.0564	−0.9428	−0.9992	14.77	11.78	−10.12	−23.26	−25.07	−35.19	3.29
CO$_2$···CO	−0.65	−1.14	−0.49	−0.0058	0.0058	0.0000	2.71	2.88	−0.76	−5.31	−5.97	−6.73	0.28
CO$_2$···CS	−1.29	−1.83	−0.54	−0.0105	0.0096	−0.0008	3.84	4.19	−1.68	−7.27	−8.19	−9.87	0.77
CO$_2$···Br$^-$	−5.60	−4.99	0.62	−0.0401	−0.9600	−1.0001	9.08	6.79	−4.97	−14.01	−15.89	−20.86	1.87
CO$_2$···Kr	−0.48	+0.03	0.52	−0.0035	0.0028	−0.0007	0.58	2.33	−0.08	−2.29	−2.79	−2.88	0.61
CO$_2$···NC$^-$	−8.34	−8.89	−0.55	−0.0333	−0.9662	−0.9996	14.29	10.83	−12.70	−19.43	−21.31	−34.01	5.69
CO$_2$···NH$_3$	−2.09	−2.41	−0.31	−0.0158	0.0155	−0.0003	6.63	7.39	−4.12	−11.06	−12.32	−16.43	2.97
CO$_2$···OC	−0.44	−0.91	−0.47	0.0001	−0.0000	0.0001	1.11	1.40	−0.57	−2.39	−2.85	−3.42	0.12
CO$_2$···OH$_2$	−2.24	−2.52	−0.28	−0.0051	0.0052	0.0000	5.42	5.87	−3.86	−8.86	−9.94	−13.81	2.43
CO$_2$···SH$^-$	−3.99	−4.32	−0.33	−0.0244	−0.9758	−1.0003	6.64	3.96	−3.67	−9.68	−11.25	−14.93	0.93
SiO$_2$···NCH	−22.43	−21.00	1.42	−0.0174	0.0185	0.0010	57.82	77.23	−89.17	−62.94	−66.89	−156.06	72.12
SiO$_2$···CO	−9.61	−8.69	0.92	−0.0112	0.0114	0.0002	49.33	69.34	−57.46	−66.28	−69.91	−127.37	52.39
SiO$_2$···CS	−30.17	−28.78	1.39	−0.0082	0.0086	0.0003	75.50	92.86	−107.86	−85.39	−89.29	−197.15	82.98
SiO$_2$···Br$^-$	−78.06	−76.19	1.87	−0.1552	−0.8433	−0.9985	60.08	84.17	−134.94	−79.98	−85.51	−220.45	64.27
CS$_2$···CO	−0.77	−0.86	−0.09	−0.0008	0.0010	0.0002	0.96	1.64	−0.09	−2.90	−3.36	−3.45	−0.31
CS$_2$···CS	−0.95	−0.46	0.49	−0.0007	0.0006	−0.0001	3.95	3.57	−0.40	−6.37	−7.57	−7.98	1.14
CS$_2$···OH$_2$	−1.48	−1.97	−0.40	0.0012	−0.0003	0.0009	2.44	3.17	−1.88	−4.99	−5.70	−7.58	0.61
GeO$_2$···Br$^-$	−65.37	−65.29	0.08	−0.2923	−0.7068	−0.9991	23.17	48.63	−44.92	−88.70	−92.17	−137.09	−16.90

We will analyze now the different energetic contributions to the binding energy of the studied complexes. A first point to remark is that the electron relaxation that takes place within the A and B fragments when they pass from the isolated state to their final position in the supermolecule leads systematically to positive values of the deformation energies. This behavior is general whenever the net energies of A and B in the isolated state (E^A, E^B) and in the supermolecule (E^A_{net}, E^B_{net}) are computed with the same electronic structure method and there is no charge transfer from A to B or from B to A. In the present calculations, we have seen that this transfer is actually very small (except in the very few cases cited above where the isolated fragment B is an anion and, even in these cases, we have seen that the B→A electron transfer is not too large). Hence, the electronic reorganization that takes place when the supermolecule is formed from the isolated fragments is always accompanied by an increase in the deformation energy contribution to the binding energy.

There is no general rule to uncover which of the two deformation energies, E^A_{def} or E^B_{def}, is the dominant of the two in each system. Actually, there seems to be a tendency for both fragments to have deformation energies of a similar magnitude. For a given acceptor fragment A, its deformation energy obviously depends on the companion donor fragment B. With the exception of CO$_2$···SH$^-$, $E^{CO_2}_{def}$ in the CH$_3$F···B and CO$_2$···B supermolecules is much greater when the isolated fragment B is negative than when it is neutral. This result is not at all surprising, for it seems reasonable to think that the ability of B to alter the electronic distribution of A is greater in the first case than in the second.

Another point that is worth noting is that the ability of a given donor fragment B to alter the electron distribution of A (and consequently, to increase its deformation energy) increases in the order C > Ge > Si, where M=(C, Si, Ge) is the atom of group 14 included in fragment A. This can be easily seen in Table 1 by analyzing the deformation energy of A in the series $CH_3F\cdots N_3^- \rightarrow SiH_3F\cdots N_3^- \rightarrow GeH_3F\cdots N_3^-$, $CH_4\cdots F^- \rightarrow SiH_4\cdots F^- \rightarrow GeH_4\cdots F^-$, and $CO_2\cdots Br^- \rightarrow SiO_2\cdots Br^- \rightarrow GeO_2\cdots Br^-$. An exception of this rule is the series $CH_3F\cdots NCH \rightarrow SiH_3F\cdots NCH \rightarrow GeH_3F\cdots NCH$, in which the deformation energy of the SiH_3F fragment is marginally smaller than that of the GeH_3F fragment. The deformation energy of the donor fragment B follows the same order.

The classical interaction between the fragments A and B, V_{cl}^{AB}, is always stabilizing. The range of values of this energetic interaction goes from almost negligible in some systems (e.g., −0.1 Kcal/mol in $CO_2\cdots Kr$ and $CS_2\cdots CO$) up to a few tens of Kcal/mol in other cases. As expected, given that the point charge interaction is generally the most important contribution to V_{cl}^{AB} and that it usually dominates over all the higher-order multipolar interactions, this classical interaction tends to be more negative when both fragments are significantly charged. There are, however, several cases in which this statement is not fulfilled at all. For instance, V_{cl}^{AB} in the $SiO_2\cdots CS$ system takes a value as large as −108 Kcal/mol, despite the fact that the absolute value of the charges of the SiO_2 and CS fragments are smaller than 0.01e. In other complexes, such as $CO_2\cdots CN^-$ and $CO_2\cdots Br^-$, the situation is the opposite. In these two cases, both fragments have a non-negligible charge, and the classical interaction between them is, however, relatively small. These facts suggest that in many of the studied systems, V_{cl}^{AB} has important multipolar contributions and that nothing conclusive can be said about the magnitude of this energetic component looking exclusively at the values of the net charges of both fragments. The second intergroup contribution to E_{bind}^{AB} and E_{int}^{AB} is the exchange-correlation interaction energy, V_{xc}^{AB}. Its exchange contribution, V_x^{AB}, also appears in Table 1, and the difference between both quantities gives the intergroup correlation binding energy, E_{corr}^{AB}. The comparison between V_{xc}^{AB} and V_x^{AB} indicates that E_{corr}^{AB} is, in general, rather small and, of course, much less important than either of them. This does not mean that the intergroup correlation energy in some of the systems is not comparable to the value of the binding energy itself: E_{bind}^{AB} comes from the sum of several quantities, some of them possibly quite large, but the final result can be very small and comparable to one or more of the quantities that have been added.

Regarding the values of V_{xc}^{AB} (or V_x^{AB}), we must note that, similarly to V_{cl}^{AB}, the exchange-correlation energy is always a stabilizing contribution to the binding energy of the complex. In fact, the absolute values of V_{xc}^{AB} are greater than their corresponding classical interactions in 25 of the 31 studied complexes. Five of the 6 exceptions are easy to understand as they correspond to complexes in which both fragments have relatively high charges. Only $SiO_2\cdots CS$ challenges this explanation. In any case, both V_{xc}^{AB} and V_{cl}^{AB} are in general important in determining the final value of V_{int}^{AB}. Since the exchange-correlation interaction energy, V_{xc}^{AB}, is associated with covalency while V_{cl}^{AB} describes ionicity, both types of interactions (covalent- and ionic-like energies) are necessary for a proper and accurate description of the complexes analyzed in the present work.

The comparison between the classical and exchange-correlation energies of Table 1 for equivalent complexes in which the central atom of the electrophilic fragment is M=C, Si, or Ge is very illuminating. For instance, for the nine AB complexes formed with A=(CH_4, SiH_4, GeH_4) and B=(F^-, N_3^-, NCH), both V_{cl}^{AB} and V_{xc}^{AB} increase in the order Si > Ge > C when B=F^- or N_3^-, while both quantities are much smaller and rather similar for the C, Si, and Ge cases when B=NCH. The explanation for this behavior is relatively simple: The M−X distance, R_{M-X}, where X is the atom of the donor fragment that is closer to M, decreases noticeably in the order C > Ge > Si when B=F^- (3.04, 2.00, and 1.76 Å, respectively) or B=N_3^- (3.03, 2.23, 2.04 Å), while R_{M-X} is larger and not so different in the three cases when B=NCH (3.22, 2.98, and 2.96 Å for M=C, Ge, and Si, respectively). Thus, the value of R_{M-X} determines, to a large extent, the magnitude of the classical and exchange-correlation interaction energies. (The distances between all the inequivalent atomic pairs are collected in the supplementary information.) Actually, the relative magnitudes of the deformation energies E_{def}^A and E_{def}^B for these nine complexes can also be explained based almost exclusively on the value of R_{M-X}. In turn, R_{M-X} correlates quite well with the total charge of M, +0.14 (C), +3.12 (Si), and +2.10 (Ge) when B=F^- and +0.69 (C), +3.17 (Si), and +2.22 (Ge) when B=N_3^-.

It has been recently shown that there is a theoretical link between the conventional concept of bond order and the energetics of chemical interactions [79,80]. Expanding V_{xc}^{AB} as a multipolar series, the zero-th order term in the expansion (that dominates V_{xc}^{AB}) is nothing but a distance-scaled bond order,

$$V_{xc}^{AB} \simeq -\frac{\delta^{AB}}{2R_{AB}}, \qquad (12)$$

where $\delta^{AB} = -2\int_{\Omega_A}\int_{\Omega_B} \rho_2^{xc}(\mathbf{r}_1,\mathbf{r}_2)d\mathbf{r}_1 d\mathbf{r}_2$ is the delocalization index between the atoms A and B [81], a measure in real space of the bond order between both atoms. To explore to what extent the above equation is satisfied when A and B are fragments instead of single atoms, we have computed the δ^{AB} values for the studied complexes (Table 2) and plotted V_{xc}^{AB} versus δ^{AB} in Figure 2.

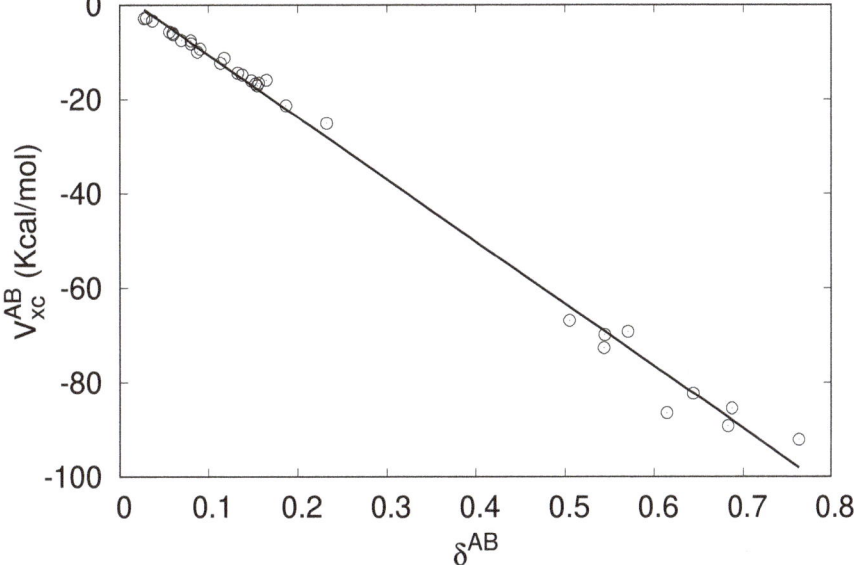

Figure 2. V_{xc}^{AB} versus δ^{AB} values for the complexes of Table 1.

Although Equation (12) is approximate even when A and B are atoms and the average distance between fragments (say R_{AB}) can be different in each of the studied complexes, a linear correlation exists between V_{xc}^{AB}, the exchange-correlation interfragment energy, and δ^{AB}, the covalent bond order between these fragments [80]. The more significant deviations from the trend in the lower part of Figure 2 (high δ^{AB} values) is due to two reasons. The first one is that Equation (12) is only approximate. As the fragments are formed with heavier and/or more polarizable atoms, higher multipolar contributions to V_{xc}^{AB} become more important, making Equation (12) increasingly inaccurate. Secondly, in these cases, the multipolar expansion itself, regardless of the maximum order to which it is carried out, is no longer valid for short-range energy components (not only in V_{xc}^{AB} but also in V_{cl}^{AB}) are essential due penetration energy contributions. Of course, the analysis of the degree of compliance of Equation (12) in the present context can be refined. For instance, labeling $a_1, a_2, \ldots$ and $b_1, b_2, \ldots$ the atoms of A and B, respectively, V_{xc}^{AB} and δ^{AB} are exactly given by $V_{xc}^{AB} = \sum_{i\in A}\sum_{j\in B} V_{xc}^{a_i b_j}$ and $\delta^{AB} = \sum_{i\in A}\sum_{j\in B} \delta^{a_i b_j}$, and the exchange-correlation energy between atoms a_i and b_j can be approximated as $-\delta^{a_i b_j}/(2R_{a_i b_j})$. Hence,

$$V_{\text{xc}}^{AB} \simeq -\sum_{i \in A} \sum_{j \in B} \frac{\delta^{a_i b_j}}{2R_{a_i b_j}}. \qquad (13)$$

This equation is a computationally cheap form to approximately evaluate V_{xc}^{AB}, since the calculation of each $\delta^{a_i b_j}$ requires only a three-dimensional integration, whereas V_{xc}^{AB} needs a six-dimensional numerical quadrature, much more complicated in all aspects.

The CCSD delocalization indexes in Table 2 range from very small values, highlighting that the interaction between fragments A and B is basically non-covalent, up to values well above 0.5, which are typical of some prototype polar-covalent single bonds. These latter values occur specifically in complexes in which the electron acceptor fragment contains the Si or Ge atom (except $GeH_3F\cdots NCH$ and $SiH_3F\cdots NCH$). These results show that the assertion that tetrel bonds are just another category of non-covalent interactions is not correct, at least if this affirmation is solely based on the value of the bond order between the two fragments involved. On the other hand, it is apparent from Figure 2 that there is a gap in the center of the V_{xc}^{AB} versus δ^{AB} trend. It is possible that the reason for this gap is not the representativeness of the sample, although a wider exploration of complexes with the same electrophilic fragments as the ones used here but with many other electron-pair rich systems would be necessary to confirm this.

Table 2. CCSD and density functional theory (DFT) delocalization indexes, δ^{AB}.

A⋯B	CCSD	DFT	A⋯B	CCSD	DFT
$CH_4\cdots F^-$	0.0910	0.1281	$CO_2\cdots Br^-$	0.1649	0.2030
$SiH_4\cdots F^-$	0.6142	0.7611	$CO_2\cdots Kr$	0.0299	0.0357
$GeH_4\cdots F^-$	0.5431	0.6878	$CO_2\cdots NC^-$	0.1871	0.2364
$CH_3F\cdots NCH$	0.0602	0.0752	$CO_2\cdots NH_3$	0.1137	0.1405
$CH_3F\cdots N_3^-$	0.1564	0.1877	$CO_2\cdots OC$	0.0274	0.0338
$CH_3F\cdots NCO^-$	0.1328	0.1672	$CO_2\cdots OH_2$	0.0878	0.1086
$CH_3F\cdots OCN^-$	0.1378	0.1700	$CO_2\cdots SH^-$	0.1178	0.1425
$SiH_3F\cdots NCH$	0.1526	0.1837	$SiO_2\cdots NCH$	0.5048	0.6375
$SiH_3F\cdots N_3^-$	0.6431	0.7535	$SiO_2\cdots CO$	0.5439	0.7054
$GeH_3F\cdots NCH$	0.1479	0.1770	$SiO_2\cdots CS$	0.6823	0.9346
$GeH_3F\cdots N_3^-$	0.5702	0.6671	$SiO_2\cdots Br^-$	0.6868	0.8568
$CO_2\cdots NCH$	0.0694	0.0860	$CS_2\cdots CO$	0.0368	0.0431
$CO_2\cdots N_3^-$	0.1546	0.1919	$CS_2\cdots CS$	0.0803	0.0930
$CO_2\cdots CN^-$	0.2320	0.2910	$CS_2\cdots OH_2$	0.0560	0.0677
$CO_2\cdots CO$	0.0596	0.0726	$GeO_2\cdots Br^-$	0.7631	0.9514
$CO_2\cdots CS$	0.0806	0.0980			

Since the exchange-correlation density can not be rigorously defined in DFT, the concept of delocalization index does not have a solid physical basis in that context. Nonetheless, the DFT δ^{AB}s can be formally calculated from the Kohn–Sham determinant of the system. Their values also appear in Table 2. In all cases, $\delta^{AB}(\text{DFT}) > \delta^{AB}(\text{CCSD})$; i.e., DFT tends to exacerbate the bond order between the fragments A and B. Thus, the assertion of the above paragraph relative to the classification of tetrel bonds as covalent or non-covalent interactions becomes reinforced when DFT is used to obtain the delocalization indexes.

Adding the V_{cl}^{AB} and V_{xc}^{AB} energies, we obtain E_{int}^{AB}, the total interaction energy between A and B. Taking into account our previous comments regarding the relative (and comparable) values V_{cl}^{AB} and V_{xc}^{AB}, it is clear that E_{int}^{AB} is more stabilizing than each of its two contributions individually. Were it not for the damping and destabilizing effect caused by the deformation energies, some of the fragments of the investigated complexes would be strongly binded. However, since $E_{\text{bind}}^{AB} = E_{\text{def}}^{A} + E_{\text{def}}^{B} + E_{\text{int}}^{AB}$, the final values of E_{bind}^{AB} (second or third column in Table 1) are, with some exceptions, relatively small.

The sum of the deformation energies of the fragments plus the exchange interaction energy (E_{xr}^{AB}, Equation (10)) plays, in the IQA method [57,69], a role very similar to the sum of the Pauli repulsion energy, ΔE_{Pauli}, plus the orbital relaxation term, ΔE_{orb}, in the energy decomposition analysis (EDA) method [66–68]. Actually, when the fragments interact but overlap very weakly, E_{cl}^{AB} tends to the classical electrostatic EDA term, V_{elstat}^{AB}, and E_{xr}^{AB} must converge to ΔE_{Pauli}. The values of E_{xr}^{AB} in Table 1 are positive in all cases except in the $CS_2\cdots CO$ complex, where it is marginally negative (-0.31 Kcal/mol, not very significant due to the inherent inaccuracy of the IQA numerical integration) and in the $GeO_2\cdots Br^-$ system (-16.9 Kcal/mol). The negative and not negligible value of E_{xr}^{AB} in this last case highlights that the hypothesis of weak overlap between both fragments, necessary for $E_{\text{xr}}^{AB} \simeq \Delta E_{\text{Pauli}}$, is very far from being satisfied in $GeO_2\cdots Br^-$. The high δ^{AB} value in Table 2, fairly similar to that of a typical simple covalent bond (and the largest of all calculated delocalization indexes), further reinforces this claim. There is a very clear separation between the complexes containing Si or Ge in the acceptor fragment and those in which this fragment is CH_4, CH_3F, CO_2, or CS_2. When the element of group 14 is C, E_{xr}^{AB} is never greater than 6.0 Kcal/mol, whereas E_{xr}^{AB} in the complexes with SiO_2 is several tens of Kcal/mol and as large as 127.83 Kcal/mol in the $SiH_4\cdots F^-$ the complex. Although we have already commented that when E_{xr}^{AB} is very large and positive, E_{cl}^{AB} also happens to be large and negative, the compensation is not perfect, and consequently, the values of E_{bind}^{AB} for the complexes containing Si or Ge are, in general, the greater. In the case of $GeO_2\cdots Br^-$, both E_{xr}^{AB} and V_{cl}^{AB} are negative, and this makes the value of the binding energy for it almost the most stabilizing of all the systems analyzed, with the exception of $SiO_2\cdots Br^-$. Negative exchange-repulsion terms can only be interpreted as being due to strong covalency.

Among the Si- and Ge-containing complexes, $SiH_3F\cdots NCH$ and $GeH_3F\cdots NCH$ present some peculiarities. Their inter-fragment Pauli exchange-repulsion energies are very small (3.74 and 4.66 Kcal/mol, respectively), just like their classical (-5.06 and -5.92 Kcal/mol), exchange-correlation (-16.68 and -16.04 Kcal/mol), and deformation energy (8.44, 10.29, and 9.20, 9.77 Kcal/mol) contributions. In fact, taking a look at the C-containing complexes in the acceptor fragment, we observe that when the electron donor group is NCH, all of these energy components tend to be lower than in the case of other acceptor groups. Two examples of this are the classical interaction energy, V_{cl}^{AB}, in the $CH_3F\cdots NCH$ and $CO_2\cdots NCH$ complexes, with -1.94 and -2.28 Kcal/mol, respectively. These numbers should not lead us to believe that the interactions between two individual atoms, one of each fragment, are also small. For instance, the $C\cdots N$ and $C\cdots C$ classical energies in $CO_2\cdots NCH$ are -366.04 and 284.32 Kcal/mol, respectively, and the $O\cdots N$ and $O\cdots C$ energies are about 179.06 and -139.57 Kcal/mol, respectively. When the full $C\cdots NCH$ V_{cl}^{AB} interaction is computed, its value becomes -39.00 Kcal/mol, an order of magnitude lower than the figures commented above. If all the interactions between the atoms of the electron donor fragment and those of the acceptor fragment are added together, the quantity -1.94 Kcal/mol that appears in Table 1 is obtained. This type of analysis can be done with the classical components of the interaction of any of the systems in the Table and the conclusions would be the same: individual atom-atom energies can be, in general, quite large. However, due to the almost electroneutrality of the fragments in many cases, they tend to cancel out in the final picture. As we have recently stressed, the meaning of Coulombic terms in the computation of intermolecular or interfragment energies is simple, but a considerable effort is still necessary before it is fully understood. As a final note, we want to emphasize that while measure of the intrinsic bond strength between two molecules, atoms, or fragments can be obtained from the plain E_{int}^{AB} values [82], the calculation of the total binding energy unavoidably requires that the deformation energies be added to E_{int}^{AB}.

5. Conclusions

The interacting quantum atoms (IQA) methodology has been used to carry out a detailed energy partition of about thirty tetrel bonds formed between different electron-acceptor fragments (A) containing a C, Si, or Ge atom, and several neutral and anionic electron-donor fragments (B). The geometries of all the complexes were fully optimized at the DFT level, and all subsequent IQA analyses were performed at the CCSD level.

Almost every energetic quantity contributing to the total binding energy between A and B, E_{bind}^{AB}, is separated in the IQA method into intra-atomic and interatomic components. Adding together all the one- and two-center terms belonging to a given fragment, one obtains its net- or self-energy, E_{net}^R (R = A, B). When the total energies of the isolated fragments are computed at the same computational level as the complexes and subtracted from E_{net}^R, the fragment deformation energies E_{def}^R appear. Their computed values are systematically positive, and the greater or smaller value of each E_{def}^R gives a measure of the degree of electronic reorganization suffered by the fragment upon complex formation. Due to their positive values, the deformation energies are destabilizing contributions to E_{bind}^{AB}. Complexes containing a C atom in the acceptor fragment are those with the smaller deformation energies, and those that contain a Si atom have greater E_{def}^R values than their analogues with germanium.

A detailed analysis of all IQA energy contributions leads us to conclude that, overall, the IQA energy quantities obtained for the complexes in which the charge-acceptor fragment (A) contains a C atom are smaller than when the atom of group 14 is Si, which, in general but with some exceptions, are usually greater than when the complex contains Ge instead of Si. In agreement with several authors, there are plenty of examples of tetrel interactions that can hardly be classified as non-covalent interactions. In some extreme cases, like in the $GeO_2 \cdots Br^-$ system, all real space indicators point toward a standard strong polar-covalent interaction. This situation is similar to that found in other recently defined bonds, where a full window of interaction energies going from very weak to considerably strong links have been found.

The IQA energy partition method used in this work is fully framed in the context of quantum chemical topology. Among its possible advantages over other existing schemes, its orbital invariance is possibly the most important of all. The IQA method can be applied independently of the electronic structure method used to construct the wave function that describes the molecular system under study. Accurate electronic structure methods, such as full interaction configuration (full-CI), multireference singles and doubles interaction configuration (MR-CISD), or the CCSD method used in this work, can be applied as easily as a mean field scheme, such as the Hartree–Fock method. Actually, in order for IQA to be used, it only requires the knowledge of the one-particle and (diagonal) two-particle density matrices, although molecular descriptions at the DFT level are also possible in the IQA context. Finally, although IQA has, to date, been applied almost exclusively in the ground electronic state, we have also recently started to use it in excited electronic states [83,84].

Author Contributions: A.C.C., E.F. and A.M.P. worked equally on the conceptualization, methodology, and formal analysis; E.F. wrote the manuscript, and A.C.C. and A.M.P. edited and reviewed it; J.L.C-S. did the bibliography analysis and performed the calculations.

Funding: We acknowledge financial support form Spanish MINECO/FEDER, grant PGC2018-095953-B-I00 and Principado de Asturias Government grant FC-GRUPIN- IDI/2018/000117.

Conflicts of Interest: The authors declare no conflict of interest.

Abbreviations

The following abbreviations are used in this manuscript:

QTAIM	Quantum theory of atoms in molecules
IQA	Interacting quantum atoms
NBO	Natural bond orbital
MP2	Second-order Møller–Plesset
MEP	Molecular electrostatic potential
EDA	Energy decomposition analysis
DFT	Density functional theory
CCSD	Singles and doubles coupled cluster
HF	Hartree–Fock
FNO	Frozen natural orbital
BSSE	Basis set superposition error
EOM	Equation of motion
full-CI	Full interaction configuration
MR-CISD	Multireference singles and doubles interaction configuration

References

1. Atwood, J.L. *Encyclopedia of Supramolecular Chemistry*; M. Dekker: New York, NY, USA, 2004.
2. DiStasio, R.A.; Gobre, V.V.; Tkatchenko, A. Many-body van der Waals interactions in molecules and condensed matter. *J. Phys. Condens. Matter* **2014**, *26*, 213202. [CrossRef] [PubMed]
3. Hermann, J.; DiStasio, R.A.; Tkatchenko, A. First-Principles Models for van der Waals Interactions in Molecules and Materials: Concepts, Theory, and Applications. *Chem. Rev.* **2017**, *117*, 4714–4758. [CrossRef] [PubMed]
4. Rance, G.A.; Marsh, D.H.; Bourne, S.J.; Reade, T.J.; Khlobystov, A.N. van der Waals Interactions between Nanotubes and Nanoparticles for Controlled Assembly of Composite Nanostructures. *ACS Nano* **2010**, *4*, 4920–4928. [CrossRef] [PubMed]
5. Reimers, J.R.; Ford, M.J.; Marcuccio, S.M.; Ulstrup, J.; Hush, N.S. Competition of van der Waals and chemical forces on gold-sulfur surfaces and nanoparticles. *Nat. Rev. Chem.* **2017**, *1*. [CrossRef]
6. Kronik, L.; Tkatchenko, A. Understanding Molecular Crystals with Dispersion-Inclusive Density Functional Theory: Pairwise Corrections and Beyond. *Acc. Chem. Res.* **2014**, *47*, 3208–3216. [CrossRef] [PubMed]
7. LeBlanc, L.M.; Weatherby, J.A.; Otero-de-la Roza, A.; Johnson, E.R. Non-Covalent Interactions in Molecular Crystals: Exploring the Accuracy of the Exchange-Hole Dipole Moment Model with Local Orbitals. *J. Chem. Theory Comput.* **2018**, *14*, 5715–5724. [CrossRef] [PubMed]
8. Maurer, R.J.; Ruiz, V.G.; Tkatchenko, A. Many-body dispersion effects in the binding of adsorbates on metal surfaces. *J. Chem. Phys.* **2015**, *143*, 102808. [CrossRef]
9. Ruiz, V.G.; Liu, W.; Tkatchenko, A. Density-functional theory with screened van der Waals interactions applied to atomic and molecular adsorbates on close-packed and non-close-packed surfaces. *Phys. Rev. B* **2016**, *93*, 035118. [CrossRef]
10. Otero-de-la Roza, A.; Cao, B.H.; Price, I.K.; Hein, J.E.; Johnson, E.R. Predicting the Relative Solubilities of Racemic and Enantiopure Crystals by Density-Functional Theory. *Angew. Chem. Int. Ed.* **2014**, *53*, 7879–7882. [CrossRef]
11. Mohebifar, M.; Johnson, E.R.; Rowley, C.N. Evaluating Force-Field London Dispersion Coefficients Using the Exchange-Hole Dipole Moment Model. *J. Chem. Theory Comput.* **2017**, *13*, 6146–6157. [CrossRef]
12. He, X.; Fusti-Molnar, L.; Cui, G.; Merz, K.M. Importance of dispersion and electron correlation in ab initio protein folding. *J. Phys. Chem. B* **2009**, *113*, 5290–5300. [CrossRef] [PubMed]
13. Grabowski, S.J. What Is the Covalency of Hydrogen Bonding? *Chem. Rev.* **2011**, *111*, 2597–2625. [CrossRef] [PubMed]
14. Arunan, E.; Desiraju, G.R.; Klein, R.A.; Sadlej, J.; Scheiner, S.; Alkorta, I.; Clary, D.C.; Crabtree, R.H.; Dannenberg, J.J.; Hobza, P.; et al. Definition of the hydrogen bond (IUPAC Recommendations 2011). *Pure Appl. Chem.* **2011**, *83*, 1637–1641. [CrossRef]
15. Cavallo, G.; Metrangolo, P.; Milani, R.; Pilati, T.; Priimagi, A.; Resnati, G.; Terraneo, G. The halogen bond. *Chem. Rev.* **2016**, *116*, 2478–2601. [CrossRef] [PubMed]
16. Neaton, J.B. A direct look at halogen bonds. *Science* **2017**, *358*, 167–168. [CrossRef] [PubMed]
17. Luyckx, R.; Coulon, P.; Lekkerkerker, H.N. Dispersion forces between noble gas atoms. *J. Chem. Phys.* **1978**, *69*, 2424–2427. [CrossRef]
18. Legon, A.C.; Sharapa, D.; Clark, T. Dispersion and polar flattening: noble gas–halogen complexes. *J. Mol. Model.* **2018**, *24*. [CrossRef] [PubMed]
19. Legon, A.C. Tetrel, pnictogen and chalcogen bonds identified in the gas phase before they had names: A systematic look at non-covalent interactions. *Phys. Chem. Chem. Phys.* **2017**, *19*, 14884–14896. [CrossRef]
20. Brammer, L. Halogen bonding, chalcogen bonding, pnictogen bonding, tetrel bonding: Origins, current status and discussion. *Faraday Discuss.* **2017**, *203*, 485–507. [CrossRef]
21. Bürgi, H.B. Chemical reaction coordinates from crystal structure data. I. *Inorg. Chem.* **1973**, *12*, 2321–2325. [CrossRef]
22. Bürgi, H.B.; Dunitz, J.D.; Schefter, E. Geometrical Reaction Coordinates. II. Nucleophilic Addition to a Carbonyl Group. *J. Am. Chem. Soc.* **1973**, *95*, 5065–5067. [CrossRef]
23. Bürgi, H.B.; Dunitz, J.D.; Lehn, J.M.; Wipff, G. Stereochemistry of reaction paths at carbonyl centres. *Tetrahedron* **1974**, *30*, 1563–1572. [CrossRef]

24. Maccallum, P.H.; Poet, R.; Milner-White, E.J. Coulombic attractions between partially chargedmain-chain atoms stabilise the right-handed twist found in most β-strands. *J. Mol. Biol.* **1995**, *248*, 374–384. [CrossRef]
25. Bartlett, G.J.; Choudhary, A.; Raines, R.T.; Woolfson, D.N. $n \rightarrow \pi^*$ interactions in proteins. *J. Mol. Biol.* **2010**, *6*, 615–620. [CrossRef] [PubMed]
26. Harder, M.; Kuhn, B.; Diederich, F. Efficient Stacking on Protein Amide Fragments. *ChemMedChem* **2013**, *8*, 397–404. [CrossRef] [PubMed]
27. Thomas, S.P.; Pavan, M.S.; Row, T.N.G. Experimental evidence for 'carbon bonding' in the solid state from charge density analysis. *Chem. Commun.* **2014**, *50*, 49–51. [CrossRef]
28. Southern, S.A.; Bryce, D.L. NMR Investigations of Noncovalent Carbon Tetrel Bonds. Computational Assessment and Initial Experimental Observation. *J. Phys. Chem. A* **2015**, *119*, 11891–11899. [CrossRef]
29. Scilabra, P.; Kumar, V.; Ursini, M.; Resnati, G. Close contacts involving germanium and tin in crystal structures: experimental evidence of tetrel bonds. *J. Mol. Model.* **2018**, *24*, 37. [CrossRef]
30. Mitzel, N.W.; Losehand, U. β-Donor Bonds in Compounds Containing SiON Fragments. *Angew. Chem. Int. Ed.* **1997**, *36*, 2807–2809. [CrossRef]
31. Bauzá, A.; Mooibroek, T.J.; Frontera, A. Tetrel-Bonding Interaction: Rediscovered Supramolecular Force? *Angew. Chem. Int. Ed.* **2013**, *52*, 12317–12321. [CrossRef]
32. Bauzá, A.; Mooibroek, T.J.; Frontera, A. Influence of ring size on the strength of carbon bonding complexes between anions and perfluorocycloalkanes. *Phys. Chem. Chem. Phys.* **2014**, *16*, 19192–19197. [CrossRef] [PubMed]
33. Bauzá, A.; Frontera, A.; Mooibroek, T.J. 1,1,2,2-Tetracyanocyclopropane (TCCP) as supramolecular synthon. *Phys. Chem. Chem. Phys.* **2016**, *18*, 1693–1698. [CrossRef] [PubMed]
34. Bauzá, A.; Mooibroek, T.J.; Frontera, A. The Bright Future of Unconventional σ/π-Hole Interactions. *ChemPhysChem* **2015**, *16*, 2496–2517. [CrossRef] [PubMed]
35. Bauzá, A.; Mooibroek, T.J.; Frontera, A. Tetrel Bonding Interactions. *Chem. Rec.* **2016**, *16*, 473–487. [CrossRef] [PubMed]
36. Scheiner, S. Comparison of halide receptors based on H, halogen, chalcogen, pnicogen, and tetrel bonds. *Faraday Discuss.* **2017**, *203*, 213–226. [CrossRef]
37. Weinhold, F.; Landis, C.R. Natural bond orbitals and extensions of localized bonding concepts. *Chem. Educ. Res. Pract.* **2001**, *2*, 91–104. [CrossRef]
38. Glendening, E.D.; Landis, C.R.; Weinhold, F. Natural bond orbital methods. *Wiley Interdiscip. Rev. Comput. Mol. Sci.* **2012**, *2*, 1–42. [CrossRef]
39. Bene, J.D.; Elguero, J.; Alkorta, I. Complexes of CO_2 with the Azoles: Tetrel Bonds, Hydrogen Bonds and Other Secondary Interactions. *Molecules* **2018**, *23*, 906. [CrossRef]
40. Bene, J.E.D.; Alkorta, I.; Elguero, J. Carbenes as Electron-Pair Donors To CO2 for C···C Tetrel Bonds and C–C Covalent Bonds. *J. Phys. Chem. A* **2017**, *121*, 4039–4047. [CrossRef]
41. Mani, D.; Arunan, E. The X-C...Y (X = O/F, Y = O/S/F/Cl/Br/N/P) *carbon bond* and hydrophobic interactions. *Phys. Chem. Chem. Phys.* **2013**, *15*, 14377–14383. [CrossRef]
42. Mani, D.; Arunan, E. The $X - C ... \pi$ (X=F,Cl,Br,CN) Carbon Bond. *J. Phys. Chem. A* **2014**, *118*, 10081–10089. [CrossRef] [PubMed]
43. Bader, R.F.W. *Atoms in Molecules: A Quantum Theory*; Oxford University Press: Oxford, UK, 1990.
44. Bader, R.F.W. A quantum theory of molecular structure and its applications. *Chem. Rev.* **1991**, *91*, 893–928. [CrossRef]
45. Grabowski, S.J. Tetrel bond—σ-hole bond as a preliminary stage of the SN_2 reaction. *Phys. Chem. Chem. Phys.* **2014**, *16*, 1824–1834. [CrossRef] [PubMed]
46. Sethio, D.; Oliveira, V.; Kraka, E. Quantitative Assessment of Tetrel Bonding Utilizing Vibrational Spectroscopy. *Molecules* **2018**, *23*, 2763. [CrossRef] [PubMed]
47. Xu, H.; Cheng, J.; Yang, X.; Liu, Z.; Bo, X.; Li, Q. Interplay between the σ-tetrel bond and σ-halogen bond in PhSiF3···4-iodopyridine···N-base. *RSC Adv.* **2017**, *7*, 21713–21720. [CrossRef]
48. Esrafili, M.D.; Kiani, H.; Mohammadian-Sabet, F. Tuning of carbon bonds by substituent effects: An ab initio study. *Mol. Phys.* **2016**, *114*, 3658–3668. [CrossRef]
49. Esrafili, M.; Mousavian, P. Strong Tetrel Bonds: Theoretical Aspects and Experimental Evidence. *Molecules* **2018**, *23*, 2642. [CrossRef] [PubMed]

50. Liu, M.; Li, Q.; Li, W.; Cheng, J. Tetrel bonds between PySiX3 and some nitrogenated bases: Hybridization, substitution, and cooperativity. *J. Mol. Graph. Model.* **2016**, *65*, 35–42. [CrossRef]
51. Liu, M.; Li, Q.; Cheng, J.; Li, W.; Li, H.B. Tetrel bond of pseudohalide anions with XH3F (X = C, Si, Ge, and Sn) and its role in SN2 reaction. *J. Chem. Phys.* **2016**, *145*, 224310. [CrossRef] [PubMed]
52. Li, Q.; Guo, X.; Yang, X.; Li, W.; Cheng, J.; Li, H.B. A σ-hole interaction with radical species as electron donors: does single-electron tetrel bonding exist? *Phys. Chem. Chem. Phys.* **2014**, *16*, 11617–11625. [CrossRef]
53. Mitzel, N.W.; Losehand, U. β-Donor Interactions of Exceptional Strength in N,N-Dimethylhydroxylaminochlorosilane, ClH2SiONMe2. *J. Am. Chem. Soc.* **1998**, *120*, 7320–7327. [CrossRef]
54. Blanco, M.A.; Pendás, A.M.; Francisco, E. Interacting quantum atoms: A correlated energy decomposition scheme based on the quantum theory of atoms in molecules. *J. Chem. Theory Comput.* **2005**, *1*, 1096–1109. [CrossRef] [PubMed]
55. Francisco, E.; Pendás, A.M.; Blanco, M.A. A molecular energy decomposition scheme for atoms in molecules. *J. Chem. Theory Comput.* **2006**, *2*, 90–102. [CrossRef] [PubMed]
56. Pendás, A.M.; Francisco, E.; Blanco, M.A.; Gatti, C. Bond paths as privileged exchange channels. *Chemistry* **2007**, *13*, 9362–9371. [CrossRef] [PubMed]
57. Pendás, A.M.; Blanco, M.A.; Francisco, E. Steric repulsions, rotation barriers, and stereoelectronic effects: A real space perspective. *J. Comput. Chem.* **2009**, *30*, 98–109. [CrossRef]
58. Francisco, E.; Casals-Sainz, J.L.; Rocha-Rinza, T.; Martín Pendás, A. Partitioning the DFT exchange-correlation energy in line with the interacting quantum atoms approach. *Theor. Chem. Acc.* **2016**, *135*. [CrossRef]
59. Keith, T.A. *AIMAll (Version 12.06.03)*; TK Gristmill Software: Overland Park, KS, USA, 2012.
60. Shavitt, I.; Bartlett, R.J. *Many-Body Methods in Chemistry and Physics. MBPT and Coupled-Cluster Theory*, 1st ed.; Cambridge Molecular Science; Cambridge University Press: New York, NY, USA, 2009.
61. McWeeny, R.; Sutcliffe, B.T. *Methods of Molecular Quantum Mechanics*; Theoretical Chemistry; A Series of Monographs; Academic Press Inc.: Cambridge, MA, USA, 1969.
62. Martín Pendás, A.; Blanco, M.A.; Francisco, E. Chemical Fragments in Real Space: Definitions, Properties, and Energetic Decompositions. *J. Comput. Chem.* **2007**, *28*, 161–184. [CrossRef]
63. Møller, C.; Plesset, M.S. Note on an Approximation Treatment for Many-Electron Systems. *Phys. Rev.* **1934**, *46*, 618. [CrossRef]
64. Tkatchenko, A.; DiStasio, R.A.; Head-Gordon, M.; Scheffler, M. Dispersion-corrected Møller–Plesset second-order perturbation theory. *J. Chem. Phys.* **2009**, *131*, 094106. [CrossRef]
65. Holguín-Gallego, F.J.; Chávez-Calvillo, R.; García-Revilla, M.; Francisco, E.; Pendás, Á.M.; Rocha-Rinza, T. Electron correlation in the interacting quantum atoms partition via coupled-cluster lagrangian densities. *J. Comput. Chem.* **2016**, *37*, 1753–1765. [CrossRef]
66. Ziegler, T.; Rauk, A. On the calculation of bonding energies by the Hartree Fock Slater method. *Theor. Chim. Acta* **2006**, *46*, 1–10. [CrossRef]
67. Morokuma, K.; Kitaura, K. Energy Decomposition Analysis of Molecular Interactions. *Chem. Appl. Atom. Mol. Electrostat. Potentials* **2013**, 215–242. [CrossRef]
68. Su, P.; Li, H. Energy decomposition analysis of covalent bonds and intermolecular interactions. *J. Chem. Phys.* **2009**, *131*, 014102. [CrossRef] [PubMed]
69. Martín Pendás, A.; Blanco, M.A.; Francisco, E. The nature of the hydrogen bond: a synthesis from the interacting quantum atoms picture. *J. Chem. Phys.* **2006**, *125*, 184112. [CrossRef] [PubMed]
70. Boys, S.F.; Bernardi, F. The calculation of small molecular interactions by the differences of separate total energies. Some procedures with reduced errors. *Mol. Phys.* **1970**, *19*, 553–566. [CrossRef]
71. Chai, J.D.; Head-Gordon, M. Long-range corrected hybrid density functionals with damped atom–atom dispersion corrections. *Phys. Chem. Chem. Phys.* **2008**, *10*, 6615–6620. [CrossRef] [PubMed]
72. Dunning, T.H. Gaussian basis sets for use in correlated molecular calculations. I. The atoms boron through neon and hydrogen. *J. Chem. Phys.* **1989**, *90*, 1007. [CrossRef]
73. Schmidt, M.W.; Baldridge, K.K.; Boatz, J.A.; Elbert, S.T.; Gordon, M.S.; Jensen, J.H.; Koseki, S.; Matsunaga, N.; Nguyen, K.A.; Su, S.; et al. General atomic and molecular electronic structure system. *J. Comput. Chem.* **1993**, *14*, 1347–1363. [CrossRef]

74. Sun, Q.; Berkelbach, T.C.; Blunt, N.S.; Booth, G.H.; Guo, S.; Li, Z.; Liu, J.; McClain, J.D.; Sayfutyarova, E.R.; Sharma, S.; et al. PySCF: The Python-based simulations of chemistry framework. *Wiley Interdiscip. Rev. Comput. Mol. Sci.* **2018**, *8*, e1340. [CrossRef]
75. Taube, A.G.; Bartlett, R.J. Frozen natural orbital coupled-cluster theory: Forces and application to decomposition of nitroethane. *J. Chem. Phys.* **2008**, *128*, 164101. [CrossRef]
76. Martín Pendás, A.; Francisco, E. Promolden. A QTAIM/IQA Code. Unpublished work.
77. Martín Pendás, A.; Blanco, M.A.; Francisco, E. Two-electron integrations in the quantum theory of atoms in molecules. *J. Chem. Phys.* **2004**, *120*, 4581. [CrossRef] [PubMed]
78. Martín Pendás, A.; Francisco, E.; Blanco, M.A. Two-electron integrations in the Quantum Theory of Atoms in Molecules with correlated wave functions. *J. Comput. Chem.* **2005**, *26*, 344. [CrossRef] [PubMed]
79. Rafat, M.; Popelier, P.L.A. *The Quantum Theory of Atoms in Molecules. From Solid State to DNA and Drug Design*; WIley-VCH: Hoboken, NJ, USA, 2007; p. 121.
80. Francisco, E.; Menéndez Crespo, D.; Costales, A.; Martín Pendás, A. A Multipolar Approach to the Interatomic Covalent Interaction Energy. *J. Comput. Chem.* **2017**. [CrossRef] [PubMed]
81. Bader, R.F.W.; Stephens, M.E. Spatial localization of the electronic pair and number distributions in molecules. *J. Am. Chem. Soc.* **1975**, *97*, 7391–7399. [CrossRef]
82. Menéndez-Crespo, D.; Costales, A.; Francisco, E.; Martín Pendás, Á. Real-Space In Situ Bond Energies: Toward A Consistent Energetic Definition of Bond Strength. *Chem. Eur. J.* **2018**, *24*, 9101–9112. [CrossRef] [PubMed]
83. Sainz, J.L.; Jara-Cortés, J.; Hernández-Trujillo, J.; Guevara-Vela, J.; Francisco, E.; Martín Pendás, A. Chemical Bonding in Excited States: Electron Localization, Delocalization and Statistics in Real Space. *ChemRxiv* **2019**. [CrossRef]
84. Fernández-Alarcón, A.; Casals-Sainz, J.; Guevara-Vela, J.M.; Costales, A.; Francisco, E.; Martín Pendás, A.; Rocha-Rinza, T. Partition of electronic excitation energies: The IQA/EOM-CCSD method. *Phys. Chem. Chem. Phys.* **2019**. [CrossRef]

© 2019 by the authors. Licensee MDPI, Basel, Switzerland. This article is an open access article distributed under the terms and conditions of the Creative Commons Attribution (CC BY) license (http://creativecommons.org/licenses/by/4.0/).

Article

Calculation of $V_{S,max}$ and Its Use as a Descriptor for the Theoretical Calculation of pKa Values for Carboxylic Acids

Guillermo Caballero-García, Gustavo Mondragón-Solórzano, Raúl Torres-Cadena, Marco Díaz-García, Jacinto Sandoval-Lira and Joaquín Barroso-Flores *

Centro Conjunto de Investigación en Química Sustentable UAEM-UNAM, Unidad San Cayetano, Carretera Toluca—Atlacomulco km 14.5, Personal de la UNAM, Toluca 50200, Mexico; gcaballerog91@gmail.com (G.C.-G.); gustavms93@gmail.com (G.M.-S.); torres.cadena.raul@gmail.com (R.T.-C.); diaz.marco95@outlook.com (M.D.-G.); jsandovalira@gmail.com (J.S.-L.)
* Correspondence: jbarroso@unam.mx; Tel.: +52-722-276-6610 (ext. 7754)

Academic Editor: Steve Scheiner
Received: 16 October 2018; Accepted: 25 December 2018; Published: 26 December 2018

Abstract: The theoretical calculation of pKa values for Brønsted acids is a challenging task that involves sophisticated and time-consuming methods. Therefore, heuristic approaches are efficient and appealing methodologies to approximate these values. Herein, we used the maximum surface electrostatic potential ($V_{S,max}$) on the acidic hydrogen atoms of carboxylic acids to describe the H-bond interaction with water (the same descriptor that is used to characterize σ-bonded complexes) and correlate the results with experimental pKa values to obtain a predictive model for other carboxylic acids. We benchmarked six different methods, all including an implicit solvation model (water): Five density functionals and the Møller–Plesset second order perturbation theory in combination with six different basis sets for a total of thirty-six levels of theory. The ωB97X-D/cc-pVDZ level of theory stood out as the best one for consistently reproducing the reported pKa values, with a predictive power of 98% correlation in a test set of ten other carboxylic acids.

Keywords: pKa; hydrogen bond; maximum surface potential; σ–hole

---

1. Introduction

The hydrogen bond is a strong, directional, non-covalent interaction responsible for a large number of chemical phenomena spanning from chemistry to biochemistry [1–3], which has become a paradigm amongst the toolbox of chemical concepts [4,5]. Hydrogen bonding is also a major driving force of chemical reactivity. For instance, the deprotonation of Brønsted acids in aqueous media, where the reaction constant, K_a, and its associated logarithmic quantity, pKa = −log$K$a, are an intrinsic characteristic of each acid. This process occurs through the abstraction of the acidic hydrogen atom using a water molecule (Equation (1)). It is commonly regarded that the deprotonation reaction is mainly promoted by electrostatic interactions, given the partial positive charge on the acid hydrogen. Characterization of a protic acid through the pKa values is of practical importance and usefulness in various steps of the chemical design rationale, and therefore, it is an important quantity.

$$HA + H_2O \rightarrow H_3O^+ + A^- \tag{1}$$

The accurate prediction of pKa values for carboxylic acids by means of computational methods covers a wide range of potential applications from chemical design and biochemistry research to drug development [6–8]. However, calculating the equilibrium constant for the deprotonation reaction of a Brønsted acid implies the calculation of the Gibbs Free Energy change, (−ΔG), which in turn entails the calculation of extremely accurate solvation energies for all species involved (Scheme 1). Calculation of accurate solvation free energies remains challenging, since it requires the use of sophisticated and computationally intensive methods, such as G3MP2 or CBS-QB3 [9], mostly due to a poor description of the solute–solvent interactions [10]. This method is highly sensitive to even slight deviations, since an error of only 1.36 kcal/mol—Barely above chemical accuracy—Leads to a unit error in the pK_a, making it impractical for large molecules.

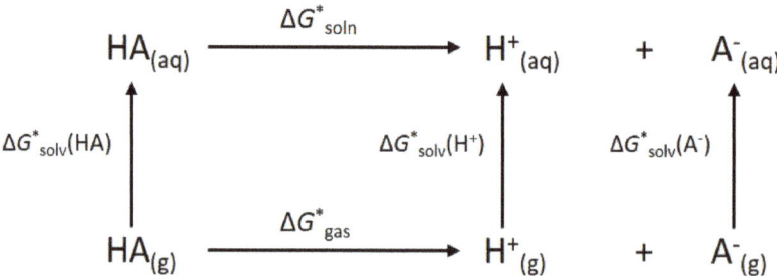

Scheme 1. Thermodynamic cycle for a protic acid deprotonation.

Since the interaction of water molecules with the acidic hydrogen atom is not isotropic, some parallels between hydrogen and σ–hole bonded systems arise. The formation of these directional interactions implies the presence of an electrostatic potential maximum located on the opposite side of the O–H σ–bond, which can be quantified by the maximum surface electrostatic potential, $V_{S,max}$. By assuming that the deprotonation of a carboxylic acid begins with the formation of a hydrogen bond with a water molecule, RCOOH···H$_2$O, we propose the use of $V_{S,max}$ as a suitable descriptor for the strength of this interaction, which in turn correlates with the corresponding pK_a values, in a similar fashion to how a σ-hole-based interaction is quantified in halogen or tetrel bonds. Previously, the nucleophilicities and electrophilicities of Lewis acids and bases, respectively, have been derived from the interpolation of the mutual dissociation energies [11].

Previous efforts for deriving suitable pK_a descriptors from ab initio or DFT descriptors have been successfully published, in some cases mixing implicit and explicit solvation models [12]. Electrostatic properties, such as the total molecular electrostatic potential (MEP) on the acidic hydrogen, combined with the sum of the valence Natural Atomic Orbital (NAO) energies on the acidic atom and the leaving proton for amino acids and nucleotides exhibits a correlation coefficient R^2 = 0.91 [13] with the experimental pK_a values. Monard and Jensen used various kinds of atomic charges of the conjugated phenolates, alkoxides, or thiolates, with the best correlations being observed for the atomic electrostatic charges from a Natural Population Analysis (NPA) calculated at the B3LYP/3-21G (R^2 = 0.995) and M06-2X/6-311G (R^2 = 0.986) levels of theory for alcohols and thiols, in implicit solvent, respectively. Other efforts include correlations on the excited states of photoacids [14] using Time Dependent DFT at the ωB97X-D/6-31G(d) level of theory for a family of hydroxyl-substituted aromatic compounds. QSPR models have yielded, for instance, a three parameters model which uses the MEP maxima, the number of carboxylic acid and amine groups for phenols, at the HF/6-31G(d,p) and B3LYP/6-31G(d,p) levels of theory (R^2 = 0.96) [15]. It involves a four parameters linear equation comprising the highest normal mode vibrational frequency, the partial positive and negative charges divided by the total surface area and a reactivity index, defined in terms of a population analysis on the frontier orbital HOMO (R^2 ca. 0.95 sic.) [16] for N-Base ligands at the semi empirical AM1 level of theory, as well as a Principal Components Analysis (PCA) for organic and inorganic acids (RMSE = 0.0195) [17]. Moreover, genetic algorithms (GA) and neural networks (NN) have employed frontier orbital energies for a chemical space of sixty commercial drugs [18] (GA, R^2 = 0.703; NN, R^2 = 0.929). Thus far, the only major commercial program capable of including the effects of molecular conformations on the estimation of pKa values is 'Jaguar pKa' [19,20]. For more thorough reviews on the development of pKa descriptors, please refer to References [21–23].

Non-covalent interactions like tetrel [24], pnicogen [25], chalcogen [26,27], carbon [28], and halogen [29–31] bonds offer some resemblances to H-bonded systems, both in structural and reactivity terms. All these forms of bonding correspond to directional, intermolecular non-covalent interactions of an electrostatic nature involving elements in groups 14 through 17, respectively. These atoms behave as electrophiles through the interaction with either n or π electrons from Lewis bases [32,33]. The formation of these non-covalent interactions stems from a similar origin, via the presence of σ–holes [34–36], a localized region of positive electrostatic potential on the surface of the bridge atom (prominently present in atoms of group 17), and opposite to the internuclear axis of one of the covalent σ bonds, hence the σ–holes. A stretch of this label has been applied to hydrogen bonding, despite the absence of p electrons on hydrogen atoms and the high polarizability of the hydrogen bonds [37–39].

Energetically, the strength of these interactions increases as the bridging atom increases its atomic number, the electronegativity of the atom bonded opposite to the non-covalent bond, and the number of electron-withdrawing groups bonded to the bridging atom. Tetrel bonds, for instance, are stabilizing interactions in nature [40,41] that form cooperative networks [26,42–46], a feature that is used as a powerful tool for the design of crystal structures [47–49]. The stabilization arising from these interactions ranges from 1 kcal/mol to 50 kcal/mol [50]. Therefore, the formation and strength of these interactions closely depends on the polarization of the electron density surrounding the bridging atom. In the particular case of tetrel bonds, these factors have been extensively investigated by Scheiner, who has further assessed the electronic [50,51] and steric [52] contributions.

Several computational studies on the nature of tetrel bonds have been published so far, from their strict quantum treatment [53] to their charge transfer dynamics in the attoseconds regime [54], and the tunneling bond-breaking processes promoted by σ-holes [55].

Thus, the importance of the study of non-covalent interactions has large implications for crystal engineering [56], biochemistry, and the understanding of chemical reactivity [57–59]. In our research group, we have reported the chemical reduction of a trichloromethyl group into a methyl group via the attack of σ–holes on chlorine atoms by thiophenolate anions, a reaction mechanism which is extensible to other trichloromethyl compounds [60].

Herein, we presented a benchmark of linear models which correlate the $V_{S,max}$ calculations with various DFT methods, and used MP2 as a reference (see methodology section), to the pKa values of carboxylic acids. Physically, the obtained value of $V_{S,max}$ on the acidic hydrogen atom reflects the attractive interaction between it and a water molecule, and thus in turn can be used to describe the deprotonation process in electrostatic terms.

2. Results

Thirty (30) different carboxylic acids with reported pKa values were selected from Lange's Handbook of Chemistry [61], and they were optimized and the surface electrostatic potential calculated (see methods section for full details). The structures of the acids are shown in Figure 1. The levels of theory used were obtained from the combination of the following functionals: ωB97X-D (A) [62], B3LYP (B) [63], LC-ωPBE (C) [64], M06-2X (D) [65] and PBE0 (E) [66], as well as the Møller-Plesset second-order perturbation theory, MP2 (F), and the following basis sets, 6-31+G(d,p) (1), 6-311++G(d,p) (2), cc-pVDZ (3), cc-pVTZ (4), aug-cc-pVTZ (5), and Def2-TZVP (6).

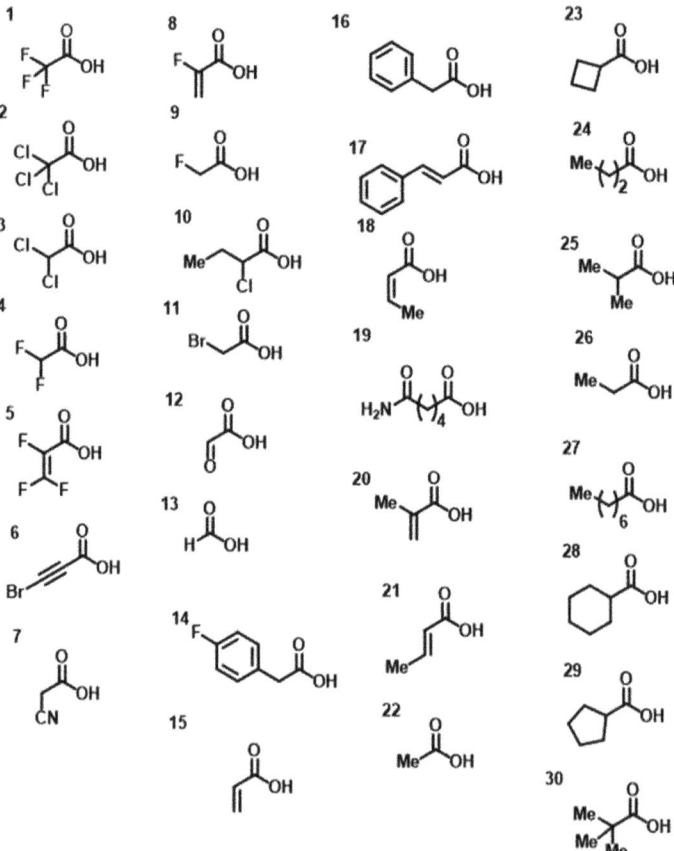

Figure 1. Thirty carboxylic acids comprising the chemical space under study.

In total, thirty-six levels of theory were used to calculate the electronic structure of the thirty carboxylic acids, which comprised the chemical space under study for a total of 1080 different wave functions, upon which the maximum surface potential, $V_{S,max}$, was calculated and plotted against the experimental pKa_{exp} value. Our model was based on simple linear regressions to obtain the best fittings. The $V_{S,max}$ on each acidic hydrogen atom was used for the correlations, as an example, Figure 2 depicts the location of $V_{S,max}$ on the acid hydrogen atom for compound 14. This value was calculated on the isodensity surface $\varrho = 0.001$ a.u., and it was used as a descriptor for the magnitude of the attractive interaction RCOOH$\cdots$H$_2$O.

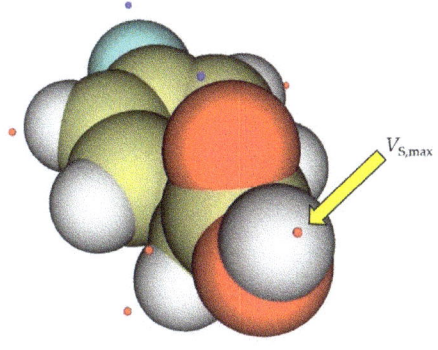

Figure 2. Maximum surface electrostatic potential, $V_{S,max}$, over the acidic hydrogen atom shown for compound 14 taking an isodensity value of 0.001 a.u. (isosurface not shown). Red dots represent positive $V_{S,max}$ values and the blue dots represent negative $V_{S,max}$ values.

All the correlation coefficients, slopes, and intercepts for all thirty-six levels of theory are collected in Table 1.

Table 1. Linear regression parameters obtained for the pKa vs $V_{S,max}$ plots. Intercept units in kcal/mol.

Level of Theory	Slope	Intercept	R^2	Level of Theory	Slope	Intercept	R^2
A1	−0.1954	16.1237	0.9626	D1	−0.1987	16.3352	0.9598
A2	−0.1975	15.8213	0.9645	D2	−0.1947	15.5073	0.9598
A3	−0.2185	16.1879	0.9680	D3	−0.2201	16.2212	0.9653
A4	−0.2113	16.3958	0.9627	D4	−0.2082	16.2411	0.9577
A5	−0.2063	16.3542	0.9594	D5	−0.2027	16.0780	0.9535
A6	−0.2131	16.3967	0.9589	D6	−0.2095	15.9993	0.9534
B1	−0.1902	15.1863	0.9494	E1	−0.1953	15.8840	0.9553
B2	−0.1909	14.8019	0.9515	E2	−0.1953	15.3858	0.9511
B3	−0.2191	15.5109	0.9570	E3	−0.2167	15.8799	0.9616
B4	−0.2072	15.4614	0.9521	E4	−0.2118	16.1281	0.9536
B5	−0.1983	15.1505	0.9457	E5	−0.2038	15.8232	0.9490
B6	−0.2030	15.1449	0.9277	E6	−0.2125	15.9739	0.9485
C1	−0.2001	16.4372	0.9654	F1	−0.1996	15.8814	0.9613
C2	−0.1996	15.9700	0.9647	F2	−0.2123	15.9191	0.9625
C3	−0.2272	16.6499	0.9682	F3	−0.2285	16.3778	0.9702
C4	−0.2166	16.7945	0.9633	F4	−0.2198	16.3264	0.9661
C5	−0.2085	16..4776	0.9597	F5	−0.2094	16.0399	0.9550
C6	−0.2162	16.4972	0.9579	F6	−0.2187	16.1635	0.9616

The obtained linear model is shown in Figure 3 for method (A) only, the plots with the rest of the methods (B)–(F) are presented in the Supporting Information section (Figures S3, S5, S7, S9 and S11).

A physical interpretation of the trends observed in Figure 3 can be rationalized in terms of the polarization of the O–H bond in the carboxylic acid motif. When the electron density of this bond was more polarized towards the oxygen atom, then the hydrogen atom possessed a more positive electrostatic potential, at the same time it was more labile and readily available for water to abstract it, thus having a lower pKa.

To further analyze the obtained models, a comparison between the experimental and calculated pKa values was made by calculating the $\Delta pKa = pKa_{exp} − pKa_{cal}$. Figure 4 shows these plots for the results obtained with the functional (A), where the corresponding ΔpKa plots for the other levels of theory are collected in the Supporting Information section (Figures S4, S6, S8, S10 and S12).

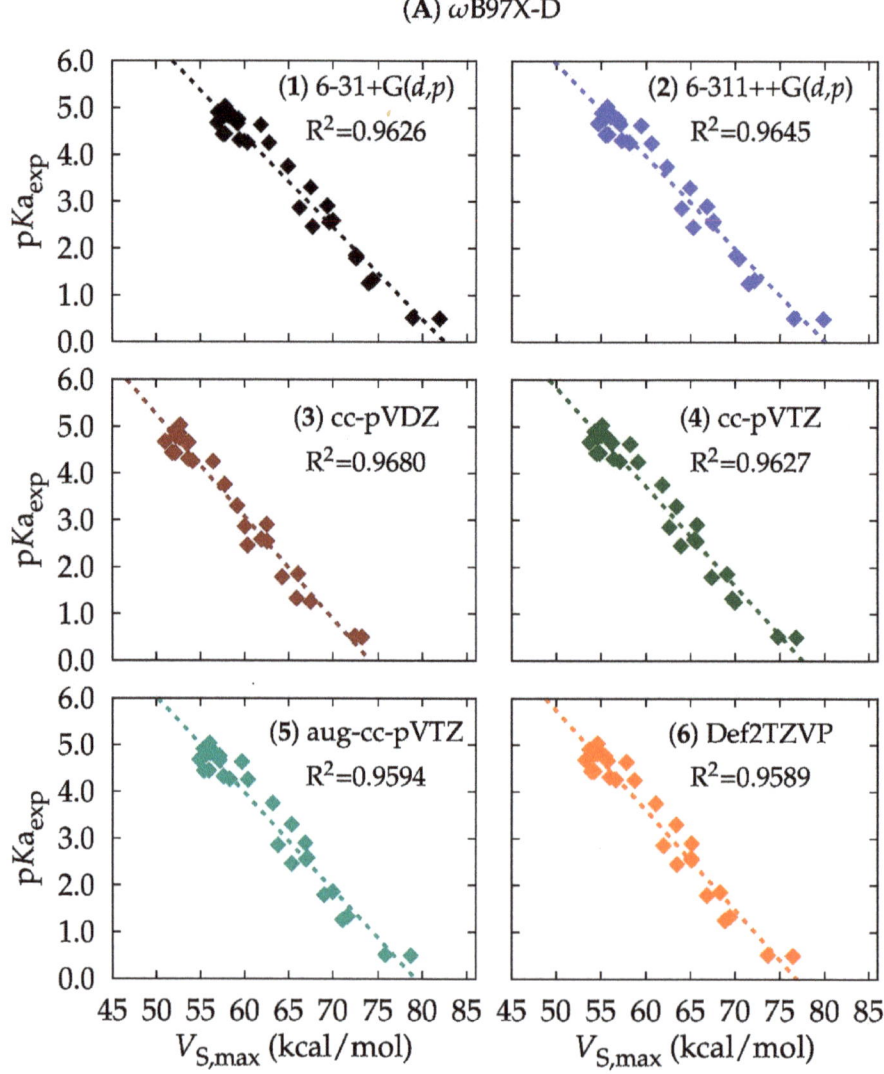

Figure 3. Linear correlations between pKa$_{exp}$ against $V_{S,max}$ for DFT method (A), with the six basis sets (1) through (6).

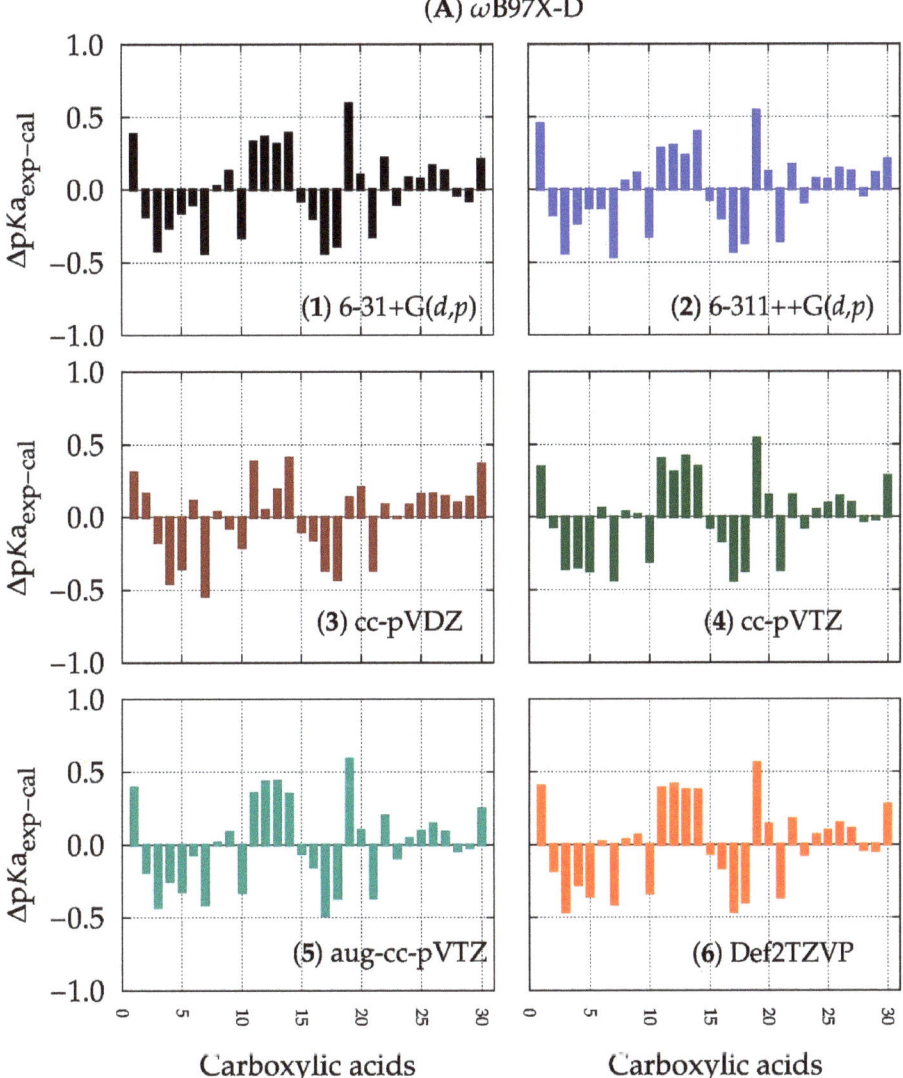

Figure 4. ΔpKa = pKa$_{exp}$ − pKa$_{cal}$ for DFT method (**A**) with the six basis sets (1) through (6).

The set of models obtained for functionals (C) and (A) had the highest correlation values across the basis sets employed (see the discussion in Section 3 for further results analysis). Particularly, the A3 level of theory (ωB97X-D/cc-pVDZ) exhibited simultaneously, a high correlation ($R^2 = 0.9680$) and the lowest ΔpKa values. Table S8 shows the pKa intervals for all levels of theory and it can be observed that all (C) models have a ΔpKa interval above 1.0 units, whereas all (A) models have ΔpKa intervals below 1.0 unit, which means an accuracy of ±0.5 pKa units. Considering these results and the calculation parameters supplied (isodensity and grid values), we proposed the following equation:

$$pKa = -0.2185\, V_{S,max} + 16.1879 \qquad (2)$$

To assess the predictive capabilities of our model, given by Equation (2), we built a test set with ten carboxylic acids (Figure 5), with pKa values that lay within the range covered by the original training set. The experimental pKa values were reported in Reference [61] and were reproduced in Table 2, together with the calculated values for the test set and the differences, which lay in the range of $\Delta pKa = \pm 0.3$ units. Figure 6 shows the remarkable correlation between the experimental and calculated values with a correlation coefficient $R^2 = 0.9801$.

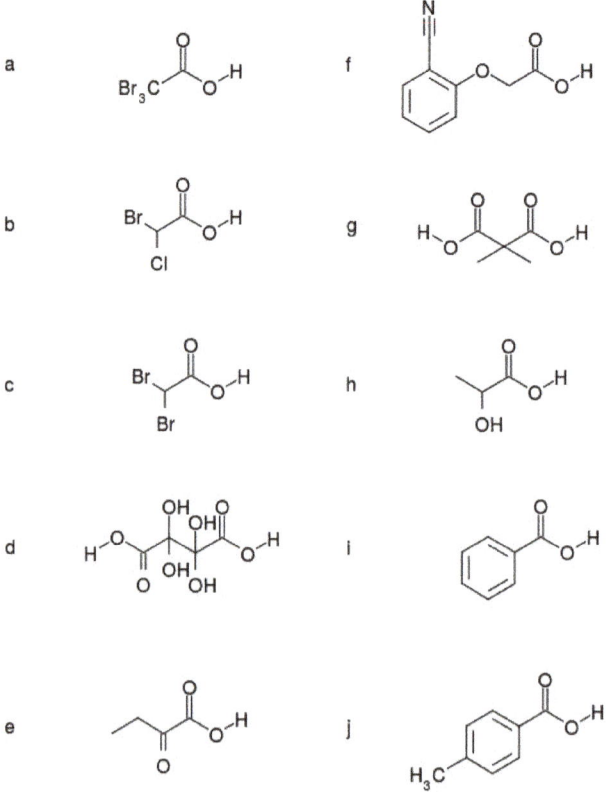

Figure 5. Carboxylic acids used as a test set for Equation (2).

Table 2. $V_{S,max}$ calculated with the A3 model. Experimental and calculated pKa values for compounds a–j and the differences.

	$V_{S,max}$	pKa_{exp} *	pKa_{cal} **	$\Delta pKa_{exp - cal}$
a	70.3733	0.7200	0.8097	−0.0897
b	66.8745	1.3900	1.5743	−0.1843
c	66.2754	1.4700	1.7052	−0.2352
d	65.6112	1.9000	1.8503	0.0497
e	62.0236	2.3600	2.6343	−0.2743
f	61.9719	2.9500	2.6456	0.3044
g	59.0914	3.1600	3.2750	−0.1150
h	54.9561	3.9100	4.1787	−0.2687
i	54.6370	4.2200	4.2484	0.0284
j	52.8925	4.3600	3.6296	−0.2696

* See Reference [61]. ** Values obtained with Equation (2).

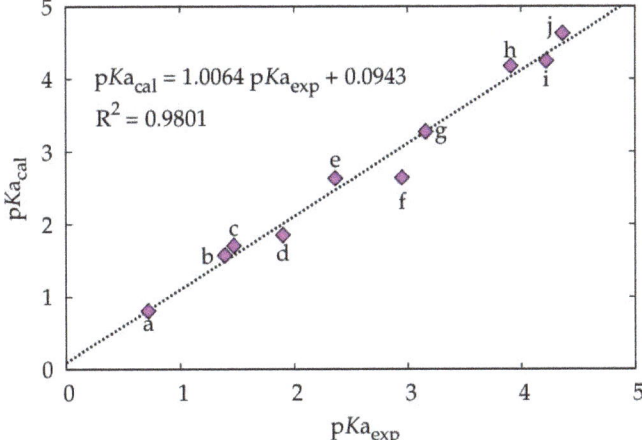

Figure 6. Correlation between the experimental and calculated pKa values.

3. Discussion

3.1. Computational Method: DFT or Ab Initio?

As a comparison standard, the Møller-Plesset second-order perturbation theory, MP2 (F), was included in the study, not only to assess its accuracy, but to compare the DFT and at least one wave function method as well. From all the tested levels of theory, the highest R^2 correlation coefficients (Table 1) between $V_{S,max}$ and pKa_{exp} values were obtained consistently with the ab initio MP2 method. Nevertheless, the DFT functional ωB97X-D functional (A), yielded comparably similar results at a fraction of the computational cost. The lowest correlation coefficients were obtained with the B3LYP functional (B), which, despite being one of the most popular ones to model organic molecules, could be describing the surface electrostatic potential inadequately. A similar performance to that of B3LYP was observed for the PBE0 functional (E), which in turn, was slightly improved when long range corrections were included in the case of LC-ωPBE (C). The latter functional was thought to yield much better results due to this long-range correlation term; however, that was not the case.

The M06-2X functional (D) also showed to be properly describing the surface electrostatic potentials, as shown in the high correlation coefficients. This was plausibly because of the dispersion terms included in its formulation. The ωB97X-D functional included an empirical dispersion term which was added a posteriori to correct the energy, but not the electron density [67].

Although the M06-2X functional is widely used and regarded as probably the best functional to model organic reactions [65], in this case, it yielded a larger discrepancy in the ΔpKa plots than the plots obtained with ωB97X-D (Figure 4 and Figure S8).

The fact that the MP2/cc-pVDZ and the ωB97X-D/cc-pVDZ yielded comparable results showed that for the case of modeling surface electrostatic potentials, a computationally expensive method may not always be preferred, as very similar or even better results can be obtained with a less demanding approach in just a fraction of the time.

3.2. Basis Set: Is Larger Better?

Most of the reported benchmarks to model organic molecules deal with the selection of the proper DFT functional methods [68–70]. However, little attention is paid to the basis set, or more precisely, to the proper functional/basis set combination (i.e., the level of theory). For further details, refer to Figures S13–S18, where the obtained models are organized by basis set.

Four out of the six methods yielded the strongest $V_{S,max}$-pKa correlations when using the relatively medium size cc-pVDZ basis set. Surprisingly, the M06-2X functional presented the largest ΔpKa deviations when combined with the largest basis set aug-cc-pVTZ (Figures S7 and S8).

In the case of the MP2 calculations (Figure S12), increasing the so-called quality of the basis set may not be beneficial in all cases. When comparing the split-valence Pople's basis sets, practically the same correlation was found with the double-ζ set and the corresponding triple-ζ quality one, 0.9613 versus 0.9625, respectively. On the other hand, the Dunning–Huzinaga basis showed a decrease in correlation when increasing the set size from cc-pVDZ to cc-pVTZ, 0.9702 and 0.9661, respectively. However, the ΔpKa deviations were practically consistent among the MP2 levels of theory.

In terms of the difference between the experimental and correlated pKa values, the A3 level of theory yielded the smallest ΔpKa deviations, with most of the differences kept under 0.5 pKa units, showing that, for this case, a larger basis set size may not always be better.

3.3. A Final Remark

In the thirty-six levels of theory tested in this study, the calculation of the $V_{S,max}$ of three compounds (6, 7, and 12) required an average of conformers, where the angle (D1 = O=C-O-H) was either 0.0° or 180.0°. The conformation D1 = 180.0° was stable due to strong delocalization effects from nearby Π bonds to the σ^*_{O-H} orbital in the acidic hydrogen atom or intramolecular hydrogen bonding with Lewis basic motifs (Figure 7). For such kind of compounds, further improvements are required in the methodology for our linear models.

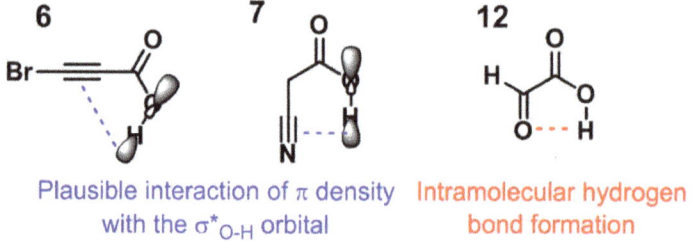

Figure 7. Intramolecular interactions for compounds 6, 7, and 12, conformers at D1 = 180.0°.

So far, the applicability domain of these regressions is limited by the pKa data used to construct the models (0.5 < pKa < 5.0). Caution must be taken when using the linear models presented herein for molecules outside this range.

4. Materials and Methods

Geometry optimizations and wave function printouts for the 30 carboxylic acids were performed using the Gaussian 09 rev. E01 suite of programs as in Reference [71], at each of the different levels of theory (see text). All calculations included the Conductor-like Polarizable Continuum Model (CPCM) implicit solvation model (water) as described in References [72,73]. The radii for cavity construction was the UFF default which takes the radii from the UFF (Universal Force Field) scaled by 1.1 with explicit spheres for hydrogen atoms. Frequency analyses were performed at the end of each geometry optimization at the same level of theory to verify that the found geometries corresponded to the energy minima. The ultrafine integration grid was used in all the calculations.

The maximum surface potential ($V_{S,max}$) calculations were performed on the wave function files with the 'MultiWFN' program, version 3.3.8 as in Reference [74], using an isodensity value of 0.001 a.u. All the computed values were collected in the Supporting Information (Tables S1–S6).

5. Conclusions

$V_{S,max}$ is a scalar quantity that characterizes a σ-hole, and according to our calculations, it has also proven to be a suitable descriptor to be correlated with the pKa value of carboxylic acids, yielding differences in pKa of high accuracy. ΔpKa = ±0.30 when the ωB97X-D/cc-pVDZ level of theory was used to calculate the associated electron density upon which the $V_{S,max}$ value was obtained. By means of straightforward DFT calculations with a simple implicit solvation model (CPCM), the value of the $V_{S,max}$ could be calculated and Equation (2) obtained herein, could be used to estimate the pKa values without the need for a full thermodynamic cycle calculation; thus, avoiding long computations of solvation free energies and other costly quantities which require high accuracy methods.

The ωB97X-D/cc-pVDZ level of theory (A3) yielded the lowest ΔpKa values, standing as the best choice for estimating the pKa of any given acid through the calculation of the $V_{S,max}$. Hence, we highly recommend this level of theory for geometry optimization and wave function file print. Care must be taken as the pKa value sought after should be between 0.5 and 5.0 pH units, for this is the applicability domain of our resulting equations, given the chemical space covered herein.

Further testing is needed for these regression models to become universal. However, it is important to stress that $V_{S,max}$ has turned out to be a powerful descriptor for predicting the pKa values of carboxylic acids as it is reflected by low, yet distinguishable differences across all methods studied herein. The presence of intramolecular non-covalent interactions, for example, hydrogen bonding, as well as highly electron-delocalizing groups within the chemical space, are key features to consider in the inclusion of an average of the $V_{S,max}$ for the most stable conformers. Our proposed descriptor is also dependent of the isodensity value for the definition of the surface upon which it is calculated, and it is highly recommended to keep the value suggested by Bader et al. [75] of ϱ = 0.001 a.u. However, by taking these considerations into account as part of the parametrical requirements of Equation (2), then extremely accurate pKa results are obtained in a straightforward fashion.

Supplementary Materials: The following are available online, Tables S1–S6: Calculated $V_{S,max}$ values for carboxylic H atoms to the different levels of theory studied, Table S7: Reported pKa values for carboxylic acids studied. Figures S1, S3, S5, S7, S9 and S11: Correlation of pKa$_{exp}$ vs $V_{S,max}$. Figures S2, S4, S6, S8, S10 and S12: Difference between the experimental and calculated pKa values (ΔpKa$_{exp-cal}$).

Author Contributions: Conceptualization, J.B.-F. and G.C.-G.; methodology, J.B.-F.; G.C.-G.; and J.S.-L.; validation, J.B.-F. and J.S.-L.; formal analysis, J.B.-F.; G.C.-G.; G.M.-S.; and J.S.-L.; data curation, G.C.-G.; G.M.-S.; R.T.-C.; and M.D.-G.; writing—original draft preparation, G.C.-G and G.M.-S.; writing—review and editing, J.B.-F.; G.C.-G.; and J.S.-L.; visualization, G.C.-G.; J.S.-L.; and G.M.-S.; supervision, J.B.-F.; project administration, J.B.-F.

Funding: This research received no external funding.

Acknowledgments: We thank DGTIC-UNAM for granting us access to their supercomputing facilities known as 'Miztli'. We also thank Citlalit Martínez for keeping our local computational facilities running properly. J.S.L. thanks DGAPA-UNAM for the postdoctoral scholarship. G.M.S. thanks CONACyT for the funding awarded (awardee number 771371).

Conflicts of Interest: The authors declare no conflict of interest.

References

1. Grabowski, S.J. What Is the Covalency of Hydrogen Bonding? *Chem. Rev.* **2011**, *111*, 2597–2625. [CrossRef] [PubMed]
2. Zhao, G.J.; Han, K.L. Hydrogen bonding in the electronic excited state. *Acc. Chem. Res.* **2012**, *45*, 404–413. [CrossRef] [PubMed]
3. Mejía, S.; Hernández-Pérez, J.M.; Sandoval-Lira, J.; Sartillo-Piscil, F. Looking inside the intramolecular C-H···O hydrogen bond in lactams derived from α-methylbenzylamine. *Molecules* **2017**, *22*, 361. [CrossRef] [PubMed]
4. Yourdkhani, S.; Jabłoński, M. Revealing the physical nature and the strength of charge-inverted hydrogen bonds by SAPT(DFT), MP2, SCS-MP2, MP2C, and CCSD(T) methods. *J. Comput. Chem.* **2017**, *38*, 773–780. [CrossRef] [PubMed]

5. Jabłoński, M. Binding of X–H to the lone-pair vacancy: Charge-inverted hydrogen bond. *Chem. Phys. Lett.* **2009**, *477*, 374–376. [CrossRef]
6. Cruciani, G.; Milletti, F.; Storchi, L.; Sforna, G.; Goracci, L. In silico pKa prediction and ADME profiling. *Chem. Biodivers.* **2009**, *6*, 1812–1821. [CrossRef]
7. Krieger, E.; Dunbrack, R.; Hooft, R.; Krieger, B. Computational Drug Discovery and Design. In *Methods in Molecular Biology*; Baron, R., Ed.; Springer: New York, NY, USA, 2012; Volume 819, ISBN 978-1-61779-464-3.
8. Kim, M.O.; McCammon, J.A. Computation of pH-dependent binding free energies. *Biopolymers* **2016**, *105*, 43–49. [CrossRef]
9. Liptak, M.D.; Gross, K.C.; Seybold, P.G.; Feldgus, S.; Shields, G.C. Absolute pKa Determinations for Substituted Phenols. *J. Am. Chem. Soc.* **2002**, *124*, 6421–6427. [CrossRef]
10. Ho, J.; Coote, M.L. First-principles prediction of acidities in the gas and solution phase. *Wiley Interdiscip. Rev. Comput. Mol. Sci.* **2011**, *1*, 649–660. [CrossRef]
11. Alkorta, I.; Legon, A. Nucleophilicities of Lewis Bases B and Electrophilicities of Lewis Acids A Determined from the Dissociation Energies of Complexes B···A Involving Hydrogen Bonds, Tetrel Bonds, Pnictogen Bonds, Chalcogen Bonds and Halogen Bonds. *Molecules* **2017**, *22*, 1786. [CrossRef]
12. Thapa, B.; Schlegel, H.B. Density Functional Theory Calculation of pKa's of Thiols in Aqueous Solution Using Explicit Water Molecules and the Polarizable Continuum Model. *J. Phys. Chem. A* **2016**, *120*, 5726–5735. [CrossRef] [PubMed]
13. Zhao, D.; Rong, C.; Yin, D.; Liu, S. Molecular Acidity of Building Blocks of Biological Systems: A Density Functional Reactivity Theory Study. *J. Theor. Comput. Chem.* **2013**, *12*, 1350034. [CrossRef]
14. Wang, Y.F.; Cheng, Y.C. Molecular electrostatic potential on the proton-donating atom as a theoretical descriptor of excited state acidity. *Phys. Chem. Chem. Phys.* **2018**, *20*, 4351–4359. [CrossRef] [PubMed]
15. Virant, M.; Drvarič Talian, S.; Podlipnik, Č.; Hribar-Lee, B. Modelling the Correlation Between Molecular Electrostatic Potential and pKa on Sets of Carboxylic Acids, Phenols and Anilines. *Acta Chim. Slov.* **2017**, 560–563. [CrossRef]
16. Palaz, S.; Türkkan, B.; Eroğlu, E. A QSPR Study for the Prediction of the pKa of N-Base Ligands and Formation Constant Kc of Bis(2,2'-bipyridine)Platinum(II)-N-Base Adducts Using Quantum Mechanically Derived Descriptors. *ISRN Phys. Chem.* **2012**, *2012*, 1–11. [CrossRef]
17. Veyseh, S.; Hamzehali, H.; Niazi, A.; Ghasemi, J.B. Application of multivariate image analysis in QSPR study of pKa of various acids by principal components-least squares support vector machine. *J. Chil. Chem. Soc.* **2015**, *60*, 2985–2987. [CrossRef]
18. Noorizadeh, H.; Farmany, A.; Noorizadeh, M. pKa modelling and prediction of drug molecules through GA-KPLS and L-M ANN. *Drug Test. Anal.* **2013**, *5*, 103–109. [CrossRef] [PubMed]
19. Bochevarov, A.D.; Watson, M.A.; Greenwood, J.R.; Philipp, D.M. Multiconformation, Density Functional Theory-Based pKa Prediction in Application to Large, Flexible Organic Molecules with Diverse Functional Groups. *J. Chem. Theory Comput.* **2016**, *12*, 6001–6019. [CrossRef]
20. Yu, H.S.; Watson, M.A.; Bochevarov, A.D. Weighted Averaging Scheme and Local Atomic Descriptor for pKa Prediction Based on Density Functional Theory. *J. Chem. Inf. Model.* **2018**, *58*, 271–286. [CrossRef]
21. Seybold, P.G.; Shields, G.C. Computational estimation of pKa values. *Wiley Interdiscip. Rev. Comput. Mol. Sci.* **2015**, *5*, 290–297. [CrossRef]
22. Niu, Y.; Lee, J.K. pKa Prediction. In *Applied Theoretical Organic Chemistry*; Tantillo, D.J., Ed.; World Scientific: London, UK, 2018; pp. 6540–6544. ISBN 978-1-78634-408-3.
23. Matta, C.F. Modeling biophysical and biological properties from the characteristics of the molecular electron density, electron localization and delocalization matrices, and the electrostatic potential. *J. Comput. Chem.* **2014**, *35*, 1165–1198. [CrossRef] [PubMed]
24. Bauzá, A.; Mooibroek, T.J.; Frontera, A. Tetrel-Bonding Interaction: Rediscovered Supramolecular Force? *Angew. Chemie Int. Ed.* **2013**, *52*, 12317–12321. [CrossRef] [PubMed]
25. Gholipour, A. Mutual interplay between pnicogen–π and tetrel bond in $PF_3 \perp X$–Pyr... SiH_3CN complexes: NMR, SAPT, AIM, NBO, and MEP analysis. *Struct. Chem.* **2018**, *29*, 1255–1263. [CrossRef]
26. Guo, X.; Liu, Y.W.; Li, Q.Z.; Li, W.Z.; Cheng, J.B. Competition and cooperativity between tetrel bond and chalcogen bond in complexes involving F_2CX (X = Se and Te). *Chem. Phys. Lett.* **2015**, *620*, 7–12. [CrossRef]
27. Liu, M.; Li, Q.; Li, W.; Cheng, J.; McDowell, S.A.C. Comparison of hydrogen, halogen, and tetrel bonds in the complexes of HArF with YH_3X (X = halogen, Y = C and Si). *RSC Adv.* **2016**, *6*, 19136–19143. [CrossRef]

28. Mani, D.; Arunan, E. The X-C···Y (X = O/F, Y = O/S/F/Cl/Br/N/P) "carbon bond" and hydrophobic interactions. *Phys. Chem. Chem. Phys.* **2013**, *15*, 14377–14383. [CrossRef]
29. Cavallo, G.; Metrangolo, P.; Milani, R.; Pilati, T.; Priimagi, A.; Resnati, G.; Terraneo, G. The halogen bond. *Chem. Rev.* **2016**, *116*, 2478–2601. [CrossRef]
30. Auffinger, P.; Hays, F.; Westhof, E.; Ho, P.S. Halogen bonds in biological molecules. *Proc. Natl. Acad. Sci. USA.* **2004**, *101*, 16789–16794. [CrossRef]
31. Politzer, P.; Murray, J.S.; Clark, T. Halogen bonding and other σ-hole interactions: A perspective. *Phys. Chem. Chem. Phys.* **2013**, *15*, 11178–11189. [CrossRef]
32. Legon, A.C. Tetrel, pnictogen and chalcogen bonds identified in the gas phase before they had names: A systematic look at non-covalent interactions. *Phys. Chem. Chem. Phys.* **2017**, *19*, 14884–14896. [CrossRef]
33. Edwards, A.J.; Mackenzie, C.F.; Spackman, P.R.; Jayatilaka, D.; Spackman, M.A. Intermolecular interactions in molecular crystals: What's in a name? *Faraday Discuss.* **2017**, *203*, 93–112. [CrossRef] [PubMed]
34. Liu, M.; Li, Q.; Scheiner, S. Comparison of tetrel bonds in neutral and protonated complexes of pyridine TF_3 and furan TF_3 (T = C, Si, and Ge) with NH_3. *Phys. Chem. Chem. Phys.* **2017**, *19*, 5550–5559. [CrossRef] [PubMed]
35. Zierkiewicz, W.; Michalczyk, M.; Scheiner, S. Comparison between tetrel bonded complexes stabilized by σ and π hole interactions. *Molecules* **2018**, *23*, 1416. [CrossRef] [PubMed]
36. Kolář, M.H.; Hobza, P. Computer Modeling of Halogen Bonds and Other σ-Hole Interactions. *Chem. Rev.* **2016**, *116*, 5155–5187. [CrossRef] [PubMed]
37. Shields, Z.P.; Murray, J.S.; Politzer, P. Directional tendencies of halogen and hydrogen bonds. *Int. J. Quantum Chem.* **2010**, *110*, 2823–2832. [CrossRef]
38. Scheiner, S. Assembly of Effective Halide Receptors from Components. Comparing Hydrogen, Halogen, and Tetrel Bonds. *J. Phys. Chem. A* **2017**, *121*, 3606–3615. [CrossRef]
39. Tang, Q.; Li, Q. Interplay between tetrel bonding and hydrogen bonding interactions in complexes involving F_2XO (X = C and Si) and HCN. *Comput. Theor. Chem.* **2014**, *1050*, 51–57. [CrossRef]
40. Del Bene, J.E.; Alkorta, I.; Elguero, J. Exploring the $(H_2C=PH_2)^+$:N-Base Potential Surfaces: Complexes Stabilized by Pnicogen, Hydrogen, and Tetrel Bonds. *J. Phys. Chem. A* **2015**, *119*, 11701–11710. [CrossRef]
41. Esrafili, M.D.; Vakili, M.; Javaheri, M.; Sobhi, H.R. Tuning of tetrel bonds interactions by substitution and cooperative effects in $XH_3Si···NCH···HM$ (X = H, F, Cl, Br; M = Li, Na, BeH and MgH) complexes. *Mol. Phys.* **2016**, *114*, 1974–1982. [CrossRef]
42. Marín-Luna, M.; Alkorta, I.; Elguero, J. Cooperativity in Tetrel Bonds. *J. Phys. Chem. A* **2016**, *120*, 648–656. [CrossRef]
43. Solimannejad, M.; Orojloo, M.; Amani, S. Effect of cooperativity in lithium bonding on the strength of halogen bonding and tetrel bonding: $(LiCN)n···ClYF_3$ and $(LiCN)n···YF_3Cl$ (Y= C, Si and n = 1–5) complexes as a working model. *J. Mol. Model.* **2015**, *21*. [CrossRef] [PubMed]
44. Mahmoudi, G.; Bauzá, A.; Frontera, A.; Garczarek, P.; Stilinović, V.; Kirillov, A.M.; Kennedy, A.; Ruiz-Pérez, C. Metal-organic and supramolecular lead(II) networks assembled from isomeric nicotinoylhydrazone blocks: The effects of ligand geometry and counter-ion on topology and supramolecular assembly. *CrystEngComm* **2016**, *18*, 5375–5385. [CrossRef]
45. Wei, Y.; Cheng, J.; Li, W.; Li, Q. Regulation of coin metal substituents and cooperativity on the strength and nature of tetrel bonds. *RSC Adv.* **2017**, *7*, 46321–46328. [CrossRef]
46. Esrafili, M.D.; Mohammadian-Sabet, F. Cooperativity of tetrel bonds tuned by infstituent effects. *Mol. Phys.* **2016**, *114*, 1528–1538. [CrossRef]
47. George, J.; Dronskowski, R. Tetrel bonds in infinite molecular chains by electronic structure theory and their role for crystal stabilization. *J. Phys. Chem. A* **2017**, *121*, 1381–1387. [CrossRef] [PubMed]
48. Mahmoudi, G.; Dey, L.; Chowdhury, H.; Bauzá, A.; Ghosh, B.K.; Kirillov, A.M.; Seth, S.K.; Gurbanov, A.V.; Frontera, A. Synthesis and crystal structures of three new lead(II) isonicotinoylhydrazone derivatives: Anion controlled nuclearity and dimensionality. *Inorganica Chim. Acta* **2017**, *461*, 192–205. [CrossRef]
49. Mahmoudi, G.; Gurbanov, A.V.; Rodríguez-Hermida, S.; Carballo, R.; Amini, M.; Bacchi, A.; Mitoraj, M.P.; Sagan, F.; Kukulka, M.; Safin, D.A. Ligand-Driven Coordination Sphere-Induced Engineering of Hybrid Materials Constructed from $PbCl_2$ and Bis-Pyridyl Organic Linkers for Single-Component Light-Emitting Phosphors. *Inorg. Chem.* **2017**, *56*, 9698–9709. [CrossRef] [PubMed]

50. Scheiner, S. Systematic Elucidation of Factors That Influence the Strength of Tetrel Bonds. *J. Phys. Chem. A* **2017**, *121*, 5561–5568. [CrossRef]
51. Dong, W.; Li, Q.; Scheiner, S. Comparative strengths of tetrel, pnicogen, chalcogen, and halogen bonds and contributing factors. *Molecules* **2018**, *23*, 1681. [CrossRef]
52. Scheiner, S. Steric Crowding in Tetrel Bonds. *J. Phys. Chem. A* **2018**, *122*, 2550–2562. [CrossRef]
53. Laconsay, C.J.; Galbraith, J.M. A valence bond theory treatment of tetrel bonding interactions. *Comput. Theor. Chem.* **2017**, *1116*, 202–206. [CrossRef]
54. Chandra, S.; Bhattacharya, A. Attochemistry of Ionized Halogen, Chalcogen, Pnicogen, and Tetrel Noncovalent Bonded Clusters. *J. Phys. Chem. A* **2016**, *120*, 10057–10071. [CrossRef] [PubMed]
55. Kozuch, S.; Nandi, A.; Sucher, A. Ping-Pong Tunneling Reactions: Can Fluoride Jump at Absolute Zero? *Chem.—A Eur. J.* **2018**. [CrossRef]
56. Servati Gargari, M.; Stilinovic, V.; Bauzá, A.; Frontera, A.; McArdle, P.; Van Derveer, D.; Ng, S.W.; Mahmoudi, G. Design of Lead(II) Metal-Organic Frameworks Based on Covalent and Tetrel Bonding. *Chem.—A Eur. J.* **2015**, *21*, 17951–17958. [CrossRef]
57. Grabowski, S.J. Tetrel bond-σ-hole bond as a preliminary stage of the S_N2 reaction. *Phys. Chem. Chem. Phys.* **2014**, *16*, 1824–1834. [CrossRef]
58. Szabó, I.; Császár, A.G.; Czakó, G. Dynamics of the $F^- + CH_3Cl \rightarrow Cl^- + CH_3F$ S_N2 reaction on a chemically accurate potential energy surface. *Chem. Sci.* **2013**, *4*, 4362. [CrossRef]
59. Stei, M.; Carrascosa, E.; Kainz, M.A.; Kelkar, A.H.; Meyer, J.; Szabó, I.; Czakó, G.; Wester, R. Influence of the leaving group on the dynamics of a gas-phase S_N2 reaction. *Nat. Chem.* **2015**, *8*, 1–6. [CrossRef]
60. Caballero-García, G.; Romero-Ortega, M.; Barroso-Flores, J. Reactivity of electrophilic chlorine atoms due to σ-holes: A mechanistic assessment of the chemical reduction of a trichloromethyl group by sulfur nucleophiles. *Phys. Chem. Chem. Phys.* **2016**, *18*, 27300–27307. [CrossRef]
61. Lange, N.A.; Speight, J.G. *Lange's Handbook of Chemistry*; McGraw-Hill: New York, NY, USA, 2005; ISBN 0071432205.
62. Chai, J.D.; Head-Gordon, M. Long-range corrected hybrid density functionals with damped atom–atom dispersion corrections. *Phys. Chem. Chem. Phys.* **2008**, *10*, 6615. [CrossRef]
63. Stephens, P.J.; Devlin, F.J.; Chabalowski, C.F.; Frisch, M.J. Ab Initio Calculation of Vibrational Absorption and Circular Dichroism Spectra Using Density Functional Force Fields. *J. Phys. Chem.* **1994**, *98*, 11623–11627. [CrossRef]
64. Vydrov, O.A.; Scuseria, G.E.; Perdew, J.P. Tests of functionals for systems with fractional electron number. *J. Chem. Phys.* **2007**, *126*, 154109. [CrossRef] [PubMed]
65. Zhao, Y.; Truhlar, D.G. The M06 suite of density functionals for main group thermochemistry, thermochemical kinetics, noncovalent interactions, excited states, and transition elements: Two new functionals and systematic testing of four M06-class functionals and 12 other function. *Theor. Chem. Acc.* **2008**, *120*, 215–241. [CrossRef]
66. Perdew, J.P.; Burke, K.; Ernzerhof, M. Generalized Gradient Approximation Made Simple. *Phys. Rev. Lett.* **1997**, *78*, 1396. [CrossRef]
67. Risthaus, T.; Grimme, S. Benchmarking of London dispersion-accounting density functional theory methods on very large molecular complexes. *J. Chem. Theory Comput.* **2013**, *9*, 1580–1591. [CrossRef] [PubMed]
68. Thanthiriwatte, K.S.; Hohenstein, E.G.; Burns, L.A.; Sherrill, C.D. Assessment of the Performance of DFT and DFT-D Methods for Describing Distance Dependence of Hydrogen-Bonded Interactions. *J. Chem. Theory Comput.* **2011**, *7*, 88–96. [CrossRef] [PubMed]
69. Wolters, L.P.; Schyman, P.; Pavan, M.J.; Jorgensen, W.L.; Bickelhaupt, F.M.; Kozuch, S. The many faces of halogen bonding: A review of theoretical models and methods. *Wiley Interdiscip. Rev. Comput. Mol. Sci.* **2014**, *4*, 523–540. [CrossRef]
70. Jensen, F. Method Calibration or Data Fitting? *J. Chem. Theory Comput.* **2018**, *14*, 4651–4661. [CrossRef]
71. Frisch, M.J.; Trucks, G.W.; Schlegel, H.B.; Scuseria, G.E.; Robb, M.A.; Cheeseman, J.R.; Scalmani, G.; Barone, V.; Mennucci, B.; Petersson, G.A.; et al. *Gaussian 09*; Revision D. 01; Gaussian Inc.: Wallingford, CT, USA, 2009.
72. Cossi, M.; Rega, N.; Scalmani, G.; Barone, V. Energies, structures, and electronic properties of molecules in solution with the C-PCM solvation model. *J. Comput. Chem.* **2003**, *24*, 669–681. [CrossRef] [PubMed]
73. Barone, V.; Cossi, M. Quantum Calculation of Molecular Energies and Energy Gradients in Solution by a Conductor Solvent Model. *J. Phys. Chem. A* **1998**, *102*, 1995–2001. [CrossRef]

74. Lu, T.; Chen, F. Multiwfn: A multifunctional wavefunction analyzer. *J. Comput. Chem.* **2012**, *33*, 580–592. [CrossRef]
75. Bader, R.F.W.; Carroll, M.T.; Cheeseman, J.R.; Chang, C. Properties of atoms in molecules: Atomic volumes. *J. Am. Chem. Soc.* **1987**, *109*, 7968–7979. [CrossRef]

Sample Availability: Not available.

 © 2018 by the authors. Licensee MDPI, Basel, Switzerland. This article is an open access article distributed under the terms and conditions of the Creative Commons Attribution (CC BY) license (http://creativecommons.org/licenses/by/4.0/).

Article

Quantitative Assessment of Tetrel Bonding Utilizing Vibrational Spectroscopy

Daniel Sethio †, Vytor Oliveira † and Elfi Kraka *

Computational and Theoretical Chemistry Group, Department of Chemistry, Southern Methodist University, 3215 Daniel Avenue, Dallas, TX 75275-0314, USA; sethio.daniel@gmail.com (D.S.); vytor3@gmail.com (V.O.)
* Correspondence: ekraka@gmail.com; Tel.: +1-214-768-1609
† These authors contributed equally to this work.

Academic Editor: Steve Scheiner
Received: 2 October 2018; Accepted: 20 October 2018; Published: 25 October 2018

Abstract: A set of 35 representative neutral and charged tetrel complexes was investigated with the objective of finding the factors that influence the strength of tetrel bonding involving single bonded C, Si, and Ge donors and double bonded C or Si donors. For the first time, we introduced an intrinsic bond strength measure for tetrel bonding, derived from calculated vibrational spectroscopy data obtained at the CCSD(T)/aug-cc-pVTZ level of theory and used this measure to rationalize and order the tetrel bonds. Our study revealed that the strength of tetrel bonds is affected by several factors, such as the magnitude of the σ-hole in the tetrel atom, the negative electrostatic potential at the lone pair of the tetrel-acceptor, the positive charge at the peripheral hydrogen of the tetrel-donor, the exchange-repulsion between the lone pair orbitals of the peripheral atoms of the tetrel-donor and the heteroatom of the tetrel-acceptor, and the stabilization brought about by electron delocalization. Thus, focusing on just one or two of these factors, in particular, the σ-hole description can only lead to an incomplete picture. Tetrel bonding covers a range of -1.4 to -26 kcal/mol, which can be strengthened by substituting the peripheral ligands with electron-withdrawing substituents and by positively charged tetrel-donors or negatively charged tetrel-acceptors.

Keywords: noncovalent interactions; weak interactions; tetrel bonding; intrinsic bond strength; local stretching force constant; CCSD(T)

1. Introduction

Noncovalent interactions (NCIs) have received increasing attention in the last two decades [1–3] due to their technological and fundamental importance in physics, chemistry, and biology [4–6]. Despite of the fact that NCIs are weak compared to covalent bonds (about an order of magnitude smaller), the importance of NCIs absolutely cannot be neglected [7–9]. They are ubiquitous and play a significant role in determining the properties of matter from small molecules to supramolecular systems like DNA and proteins [7,10]. They stabilize molecular structures [11,12], construct supramolecular materials [13], lower the activation energy of chemical reactions [14], and regulate the properties of crystal materials [15]. A series of different types of NCIs has been reported, namely, hydrogen bonds [16–22], aerogen bonds (group 18) [23–26], halogen bonds (Group 17) [27–32], chalcogen bonds (Group 16) [33–37], pnicogen bonds (Group 15) [38–42], tetrel bonds (Group 14) [43–49], and triel bonds (Group 13) [50,51].

Recently, tetrel bonding has found many applications due to its unique properties, such as strength, directionality, and origin of attraction [52]. Tetrel bonds play an important role in crystal engineering and supramolecular chemistry as a new potential molecular linker [44,53] and in dynamical processes such as protein folding and ligand–acceptor interactions [54–56]. Tetrel bonds also play an important role in the preliminary stages of SN2 reactions [57] and hydrophobic interactions [58,59].

The formation of tetrel bonds can be understood as an interaction between an electron-deficient tetrel atom of a Lewis acid (tetrel donor, T-donor) and an electron-rich of a Lewis base (tetrel acceptor, T-acceptor) (see Figure 1) [53]. The Lewis base (T-acceptor) can be any electron-rich entity possesing a lone pair [60–63], a π-system [55,64], an anion [44,65], etc. To explain the formation of a tetrel bond via σ-hole interactions, Politzer, Murray, and Clark suggested an interaction between a region of positive electrostatic potential as a result of diminished electron density on the tetrel atom (T-donor) and a region of negative electrostatic potential on an electron-rich atom (T-acceptor) [3,58,66–68]. The diminished electron density on the tetrel atom occurs as a result of electrons being mostly localized in the bonding region, which leaves a deficiency of electronic density in the outer lobe of the *p*-type valence orbital along the extension of the covalent bonding on the tetrel atom [69].

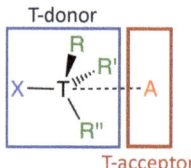

Figure 1. Schematic representation of tetrel complexes between the electron-deficient tetrel atom of a Lewis acid (tetrel donor, T-donor, T = C, Si, Ge) and the electron-rich tetrel atom of a Lewis base (tetrel acceptor, T-acceptor, A = FH, OH$_2$, NH$_3$, Cl$^-$).

A series of experimental studies was conducted to identify and characterize tetrel bonding. The first convincing evidence of tetrel bonding was reported by Jönson and co-workers in 1975, where they observed that the carbon atoms of the carbon dioxide dimer can attractively interact with the lone pair of the oxygen of water [70] which was confirmed nine years later via microwave spectroscopic analysis by Klemperer and co-workers [71]. Recently, Guru-Row and co-workers provided experimental evidence of tetrel bonding based on an X-ray charge density analysis [43]. They revealed the existence of a bond path connecting the oxygen atom with the −CH$_3$ carbon atom in R$_3$N$^+$−CH$_3$···OH complexes [43]. Mitzel and co-workers discussed Si···N tetrel bonding in the crystalline Si(ONMe$_2$)$_4$ [72,73]. Evidence of tetrel bonding has also been observed by NMR spectroscopy [74]. The chemical shifts, quadrupolar couplings, and *J*-coupling are sensitive to the presence of tetrel bonding. For example, the *J*-coupling constant for (secondary) tetrel bonds has a magnitude of about 3 Hz [75].

The strength of tetrel bonding can be enhanced by cooperative effects [76–78] in conjunction with hydrogen bonding [79,80], halogen bonding [80], chalcogen bonding [81,82], lithium bonding [83], or with other tetrel bonding [84]. Cooperative effects in tetrel bonding [85–87] play an important role in crystal materials, chemical reactions, and biological systems [78,88,89]. Thus, the understanding of the strength and the nature of tetrel bonding is the key to understanding its properties. The molecular electrostatic potential (V$_s$) and its maximum value (V$_{s,max}$) are commonly used measures to quantify the strength of the σ-hole interaction [58,90,91]. A limited correlation between the interaction energies and the value of the V$_{s,max}$ has been reported by several authors [27,28,61,66,92]. However, very recently, Scheiner and co-workers pointed out that the maximum magnitude of the molecular electrostatic potential is not an ideal bond strength indicator [93,94]. Therefore, there is an urgent need for a qualified intrinsic bond strength descriptor, which we address in the present work.

One of the most common measures for quantifying the chemical bond strength is the bond dissociation energy (BDE) or the bond dissociation enthalpy (BDH). It has been shown that the BDE or BDH has the limitation of describing the intrinsic strength of a bond [95–98] because it includes the geometry relaxation of the fragments as well as the reorganization of the electron density. The intrinsic bond strength based on the local mode force constants k^a measures the bond strength with only infinitesimal changes in the electronic structure of the molecule, thus excluding misleading additional contributions from the relaxation of the fragments. Many examples show that a chemical bond may have a large value of k^a but a low BDE, vice versa [40,96,97].

Vibrational spectroscopy is an important tool that has been used to identify and characterize small-to-medium-sized molecules [99–101]. However, normal vibrational modes are of limited use as bond strength measure due to mode–mode coupling. A major breakthrough was achieved by the work of Konkoli and Cremer where the use of vibrational spectroscopy as an intrinsic bond strength measure via local vibrational modes was refined [102,103]. The intrinsic strength of chemical bonds is probed using the associated local stretching force constants k^a [104–106]. The local stretching force constants k^a have been successfully used to determine the intrinsic bond strength of covalent bonds such as CC bonds [105,107–109], NN bonds [110], NF bonds [98], CO bonds [111], and CX (X = F, Cl, Br, I) bonds [112–115] as well as weak chemical interactions such as hydrogen bonding [18–22], halogen bonding [30,116–118], pnicogen bonding [40–42], chalcogen bonding [96,97], and recently, BH···π interactions [119,120]. In this study, we investigate the strength and the nature of the tetrel bonds for a representative set of 35 complexes (see Figure 2) and also compare tetrel bonding with halogen and chalcogen bonding.

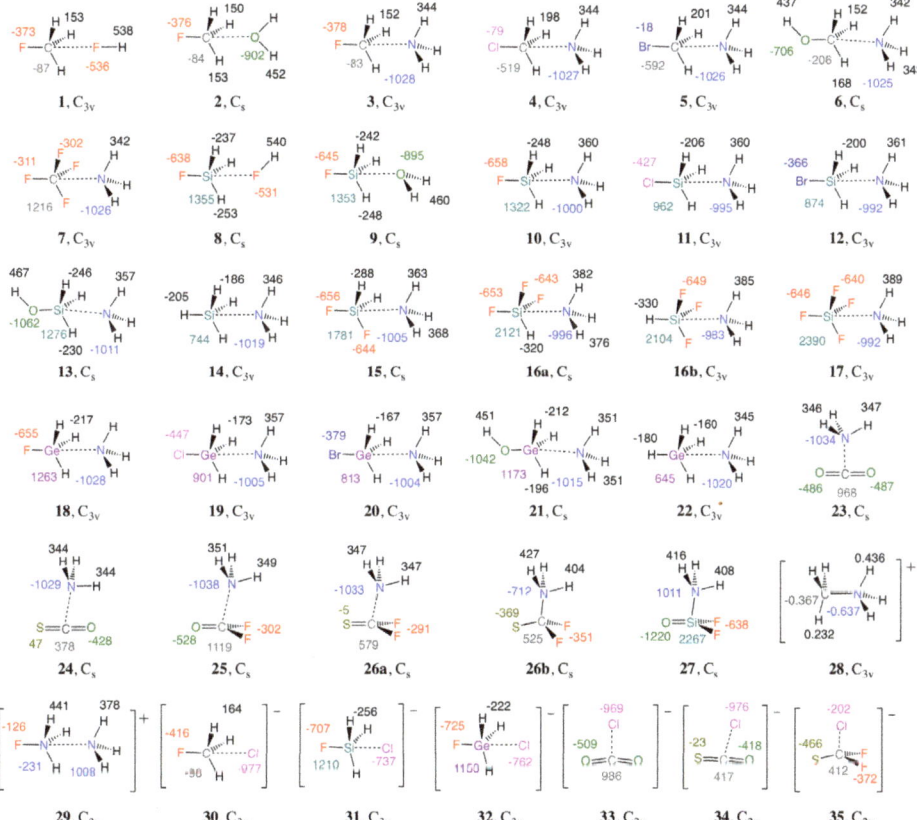

Figure 2. Schematic representation of complexes **1-35** with atomic charges (in me) from the natural population analysis calculated at the CCSD(T)/aug-cc-pVTZ level of theory. Colors are used to correlate charges to specific atoms.

The main objectives of the present work are (i) to quantify the impact of changing the tetrel atom, its substituents, and the tetrel acceptor on the tetrel bond strength; (ii) to better understand the interplay between various electronic effects such as electrostatics, covalent contributions to tetrel bonding, exchange-repulsion between the tetrel acceptor and the peripheral ligands (R, R', and R'') of the tetrel-donor, etc.; (iii) to compare the strength of tetrel bonds with halogen (XBs), chalcogen (ChB), and pnicogen bonds (PnB); and (iv) to develop an effective strategy to tune the strength of the tetrel bond.

2. Computational Methods

To evaluate the key factors that influence the strength of the tetrel bonds, geometry optimizations and normal vibrational modes of complexes **1–35** (see Figure 2), monomers **36–60** (see Table 2, Figure 3, and Supporting Information Figure S1), and reference molecules **R1–R2** were calculated using the coupled cluster theory with singles, doubles, and perturbative triples (CCSD(T)) [121,122] in combination with the Dunning's aug-cc-pVTZ basis set [123–125] which contains diffuse basis functions for describing the charge distribution of hetero-atoms, anions, and also, the dispersion interactions in tetrel bonds. All CCSD(T) calculations were carried out using a convergence criterion of 10^{-7} Hartree bohr^{-1} for geometry and a threshold of 10^{-9} for self-consistent field and CC-amplitudes.

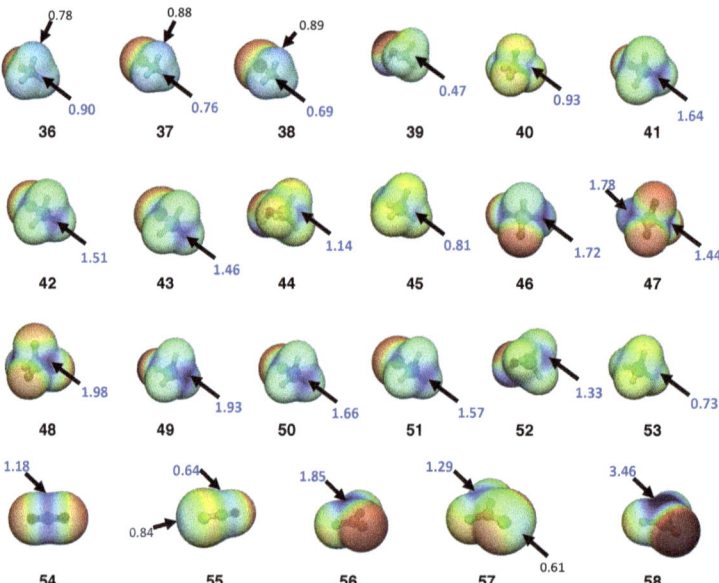

Figure 3. Molecular electrostatic potential of neutral tetrel-donors mapped onto the 0.001 a.u electron density surface. Blue and red correspond, respectively, to the positive and negative potential. The extreme values are ±1.9 eV. The $V_{S_{max}}$ at the tetrel σ or π-hole are given in bold blue, while the $V_{S_{max}}$ at the H (**36, 37, 38**) and at the chalcogen atoms (**55, 57**) are shown in black. Calculated at the CCSD(T)/aug-cc-pVTZ level of theory.

Normal vibrational modes were converted into local vibrational modes using the Konkoli–Cremer method [102–104,107]. The electronic and mass coupling between normal vibrational modes were eliminated using the mass-decoupled analogue of the Wilson equation [107,126]. The resulting local vibrational modes, which were free from any mode-mode coupling, were associated with a given internal coordinate q_n (bond length, bond angle, dihedral angle, etc.), which could be connected to normal vibrational modes in an one-to-one relationship via the Adiabatic Connection Scheme (ACS) [104,108]. The local force constant k^a, obtained from the corresponding local vibrational mode, was used to measure the intrinsic bond strength of the tetrel bonds.

For simplification, the local force constant k^a was converted to the bond strength order (BSO n) using a power relationship of the generalized Badger rule [127]:

$$n = a(k^a)^b. \qquad (1)$$

The constants a = 0.418 and b = 0.564 were obtained by two references of well-defined bond order, namely the FF bond in F_2 (n = 1.0) and the three-center-four-electron bond in F_3^- (n = 0.5), as previously done in a study of halogen bonds [30] and other noncovalent interactions [117].

The binding energy was separated into two contributions:

$$\Delta E = E_{int} + E_{def}. \qquad (2)$$

E_{int} is the interaction energy for the frozen geometry of the monomers, and the deformation energy (E_{def}) is the energetic difference between the monomers' frozen geometry and their minimum energy geometry. The counterpoise correction (CP) [128] is usually used to eliminate the basis set superposition error (BSSE) present in E_{int}. However, BSSE often compensates for the error caused by an incomplete basis set; consequently, uncorrected E_{int} values can be closer to the complete basis set limit (CBS) [117,129]. To test if this was the case, we compared CP-corrected and uncorrected CCSD(T)/aug-cc-pVTZ results to the ones obtained using the more saturated pentuple zeta basis set aug-cc-pV5Z [123] (see Supporting Information Table S1). The latter results were obtained by employing the domain-based local pair natural orbital (DLPNO) [130,131] approximation to CCSD(T) using a tight convergence criteria to ensure that the errors caused by the DLPNO approximation were negligible. It turned out that the uncorrected CCSD(T)/aug-cc-pVTZ values of E_{int} were, on average, closer to the CP-corrected and uncorrected DLPNO-CCSD(T)/aug-cc-pV5Z results than the CP-corrected CCSD(T)/aug-cc-pVTZ values (see Supporting Information Tables S1 and S2). Therefore, in the next sections, only ΔE values without counterpoise correction are discussed (see Table 1).

Table 1. Summary of energetics, electron density, energy density, geometric, bond strength order, and vibrational spectroscopy data for complexes 1–35 *.

#	Complex (symm.)	ΔE	ΔE_{cp}	E_{def}	r TA	r XT	CT	ρ_b TA	H_b TA	k^a TA	n TA	k^a XT	n XT
\multicolumn{14}{c}{Neutral tetrel bonds involving C donors}
1	FCH$_3$···FH (C$_{3v}$)	−1.50	−1.29	0.01	2.972	1.392	2	0.034	0.012	0.045	0.073	5.018	1.038
2	FCH$_3$···OH$_2$ (C$_s$)	−2.10	−1.87	0.02	3.035	1.394	3	0.041	0.012	0.055	0.081	4.956	1.030
3	FCH$_3$···NH$_3$ (C$_{3v}$)	−2.25	−2.05	0.02	3.218	1.395	5	0.040	0.009	0.049	0.076	4.912	1.025
4	ClCH$_3$···NH$_3$ (C$_{3v}$)	−2.08	−1.88	0.02	3.289	1.798	6	0.037	0.008	0.043	0.071	2.943	0.768
5	BrCH$_3$···NH$_3$ (C$_{3v}$)	−2.00	−1.80	0.02	3.304	1.953	6	0.037	0.008	0.041	0.069	2.515	0.703
6	(HO)CH$_3$···NH$_3$ (C$_s$)	−1.38	−1.21	0.01	3.362	1.429	3	0.031	0.008	0.032	0.060	4.652	0.994
7	CF$_4$···NH$_3$ (C$_{3v}$)	−1.62	−1.24	0.06	3.426	1.328	1	0.030 ^{a}	0.007 ^{a}	0.044	0.072	5.926	1.140
\multicolumn{14}{c}{Neutral tetrel bonds involving Si donors}
8	FSiH$_3$···FH (C$_s$)	−2.28	−1.85	0.06	2.964	1.617	9	0.055	0.005	0.062	0.087	4.970	1.032
9	FSiH$_3$···OH$_2$ (C$_s$)	−4.20	−3.61	0.35	2.774	1.623	25	0.092	0.002	0.088	0.106	4.762	1.007
10	FSiH$_3$···NH$_3$ (C$_{3v}$)	−6.80	−5.94	2.11	2.523	1.637	81	0.179	−0.033	0.103	0.116	4.209	0.940
11	ClSiH$_3$···NH$_3$ (C$_{3v}$)	−6.13	−5.41	2.02	2.580	2.117	84	0.165	−0.024	0.073	0.095	1.941	0.607
12	BrSiH$_3$···NH$_3$ (C$_{3v}$)	−6.11	−5.35	2.23	2.566	2.290	90	0.170	−0.027	0.066	0.090	1.505	0.526
13	(HO)SiH$_3$···NH$_3$ (C$_s$)	−4.13	−3.61	0.68	2.825	1.680	42	0.108	−0.003	0.070	0.093	4.065	0.921
14	SiH$_4$···NH$_3$ (C$_{3v}$)	−2.27	−1.97	0.15	3.202	1.490	18	0.060	0.004	0.049	0.076	2.793	0.746
15	SiF$_2$H$_2$···NH$_3$ (C$_s$)	6.99	−5.73	4.74	2.400	1.613	95	0.225	−0.066	0.083	0.103	4.573	0.985
16a	SiF$_3$H···NH$_3$ (C$_s$)	−7.66	−5.77	11.77	2.205	1.617	139	0.320	−0.126	0.249	0.191	4.698	1.000
16b	HSiF$_3$···NH$_3$ (C$_{3v}$)	−6.30	−4.14	21.22	2.104	1.474	172	0.390	−0.149	0.493	0.280	2.974	0.772
17	SiF$_4$···NH$_3$ (C$_{3v}$)	−11.40	−8.86	21.15	2.072	1.609	176	0.419	−0.164	0.678	0.335	5.046	1.041

Table 1. Cont.

#	Complex (symm.)	ΔE	ΔE$_{cp}$	E$_{def}$	r TA	r XT	CT	ρ_b TA	H$_b$ TA	k^a TA	n TA	k^a XT	n XT
	Neutral tetrel bonds involving Ge donors												
18	FGeH$_3$···NH$_3$ (C$_{3v}$)	−7.77	−7.18	1.40	2.624	1.816	44	0.169	−0.008	0.149	0.143	4.125	0.929
19	ClGeH$_3$···NH$_3$ (C$_{3v}$)	−6.22	−5.75	1.07	2.755	2.216	64	0.134	−0.001	0.103	0.116	1.921	0.604
20	BrGeH$_3$···NH$_3$ (C$_{3v}$)	−6.01	−5.53	1.07	2.776	2.375	66	0.132	0.000	0.097	0.112	1.591	0.543
21	(HO)GeH$_3$···NH$_3$ (C$_s$)	−4.58	−4.18	0.50	2.910	1.818	39	0.101	0.004	0.089	0.107	3.49	0.845
22	GeH$_4$···NH$_3$ (C$_{3v}$)	−1.99	−1.79	0.09	3.323	1.550	15	0.052	0.005	0.047	0.074	2.580	0.713
	Neutral tetrel bonds involving double bonded C or Si donors												
23	CO$_2$···NH$_3$ (C$_s$)	−3.09	−2.84	0.11	2.922	1.167	5	0.107	0.002	0.079	0.100	15.183	1.938
24	SCO···NH$_3$ (C$_s$)	−1.97	−1.69	0.02	3.209	1.573	3	0.046	0.009	0.047	0.074	7.081	1.260
25	CF$_2$O···NH$_3$ (C$_s$)	−5.55	−4.82	0.27	2.687	1.178	12	0.113	0.005	0.122	0.127	14.393	1.880
26a	CF$_2$S···NH$_3$ (C$_s$)	−3.91	−3.23	0.11	2.897	1.607	9	0.078	0.008	0.086	0.105	6.397	1.190
26b	CF$_2$S···NH$_3$ (C$_s$)	1.45	4.28	24.13	1.587	1.701	545	1.388	−1.339	1.414	0.508	3.828	0.891
27	SiF$_2$O···NH$_3$ (C$_s$)	−44.14	−42.16	7.96	1.917	1.529	229	0.569	−0.224	1.838	0.589	8.803	1.425
	Charge-assisted interactions												
28	CH$_3$$^+$···NH$_3$ (C$_{3v}$)	−110.25 [b]	−109.01	24.95	1.511	1.087	329	1.517	−1.952	3.766	0.882	5.458	1.088
29	FNH$_3$$^+$···NH$_3$ (C$_{3v}$)	−23.14	−22.77	0.43	2.619	1.374	35	0.142	0.012	0.364	0.236	5.226	1.062
30	FCH$_3$···Cl$^−$ (C$_{3v}$)	−9.77	−9.34	0.39	3.179	1.419	23	0.064	0.010	0.128	0.131	4.155	0.933
31	FSiH$_3$···Cl$^−$ (C$_{3v}$)	−20.73	−19.49	12.03	2.504	1.703	263	0.277	−0.115	0.370	0.238	2.793	0.746
32	FGeH$_3$···Cl$^−$ (C$_{3v}$)	−26.10	−25.09	10.71	2.566	1.892	238	0.290	−0.069	0.455	0.268	2.451	0.693
33	CO$_2$···Cl$^−$ (C$_s$)	−7.45	−6.99	1.44	2.920	1.170	31	0.107	0.002	0.109	0.120	14.879	1.916
34	SCO···Cl$^−$ (C$_s$)	−5.36	−4.96	0.52	3.143	1.581	24	0.073	0.006	0.079	0.100	6.568	1.208
35	CF$_2$S···Cl$^−$ (C$_s$)	−16.81	−13.83	32.63	1.898	1.725	798	1.031	−0.593	1.100	0.441	3.414	0.835

* Binding energies (ΔE), counterpoise corrected binding energies ΔE$_{cp}$ and monomers' deformation energies upon complexation (E$_{def}$) in kcal/mol. XT bond distance r(XT) and tetrel bond distance r(TA) in Å. Density at the TA critical point ρ_b in e/Å^3, energy density at the TA critical point H$_b$ in Hartree/Å^3. Natural population analysis (NPA) charge transfer in mili-electrons (me). TA and XT local stretching force constant (k^a) in mdyn/Å and bond strength order (BSO) n values. Computed at the CCSD(T)/aug-cc-pVTZ level of theory.
[a] Calculated at a cage critical point (see Ref. [147]). [b] Covalent bond, see text.

CCSD(T) calculations were performed using the CFOUR program [132,133], whereas DLPNO-CCSD(T) calculations were done in ORCA 4.0 [134]. Analytical vibrational frequencies were used to verify that each equilibrium geometry obtained by CCSD(T) corresponded to a geometry minimum. The charge distribution was calculated with the natural population analysis (NPA) within the Natural Bond Orbital (NBO) scheme [135,136] using the NBO6 program [137,138]. The electron density $\rho(r)$ and the energy density H(r) at the T···A (T: tetrel atom and A: tetrel-acceptor) electron density critical point (r) were calculated using the AIMALL program [139]. The molecular electrostatic potentials of T-donors and T-acceptors mapped onto the 0.001 e/bohr^3 electron density surface were calculated using the Multiwfn3.5 [140] program. Noncovalent interaction (NCI) plots were calculated using the NCIplot program [141]. NBO charges, as well as, H(r), $\rho(r)$ and V(r) were derived from CCSD(T) response densities obtained from CFOUR calculations with the help of MOLBO and Molden2AIM scripts [142]. Local mode force constants and frequencies were calculated using the COLOGNE18 program package [143].

3. Results and Discussion

Table 1 summarizes the complex binding energy ΔE, the counterpoise corrected binding energy ΔE$_{cp}$, the monomers' deformation energy E$_{def}$, the distance r(TA) between the tetrel-donor atom (T = C, Si or Ge) and an heteroatom (A = F, O, N or Cl$^−$) of the tetrel-acceptor (T-acceptor), also called the tetrel bond (TB) distance, the distance r(XT) between the T donor atom and the donor group or atom (X = H, F, Cl, Br, OH, =O or =S), the intermonomer charge transfer (CT) obtained from the natural population analysis (NPA), the electron density ρ_b and the energy density H$_b$ at the density critical point associated with TB, the local stretching force constant of TA (k^a(TA)) and XT (k^a(XT)) and the BSO n of the TA and XT. The calculated NBO atomic charges are given in Figure 2. T donor properties such as the maximum electrostatic potential at the σ-hole region of the tetrel atom (V$_{s_{max}}$), the total dipole moment, and the isotropic polarizability are listed in Table 2. The BSO n values of all TB are given as functions of their local stretching force constant $k^a(T\cdots A)$ in Figure 4. Similar to previous studies [30,34,40,116–118,144], we determined the covalent character of the TB by utilizing the energy density H_b at the density critical point of the TB (Figure 5); electrostatic interactions were characterized by having positive H_b values, whereas, according to the Cremer–Kraka criterion, [145,146], covalent interactions have negative H_b values, indicating that the accumulated electron density at the interactions region stabilizes the complex. Although the relationship between BSO n and H_b is scattered, the TB strength tends to increase with the increasing covalent character of the interaction, especially among neutral complexes.

Table 2. Geometry, vibrational spectroscopy data, and values of the electrostatic potential for the monomers *.

#	Monomers	$V_{s_{max}}(X)$	r(XT)	k^a(XT)	n(XT)	Dipole	α_{iso}
36	F–CH$_3$	0.90	1.389	5.107	1.048	1.88	2.5
37	Cl–CH$_3$	0.76	1.792	3.068	0.786	1.92	4.3
38	Br–CH$_3$	0.69	1.948	2.616	0.718	1.86	5.4
39	HO–CH$_3$	0.47	1.426	4.749	1.006		3.1
40	F–CF$_3$	0.93	1.321	6.204	1.170	0.00	2.8
41	F–SiH$_3$	1.64	1.613	5.120	1.049	1.38	4.1
42	Cl–SiH$_3$	1.51	2.072	2.799	0.746	1.41	6.2
43	Br–SiH$_3$	1.46	2.238	2.321	0.672	1.38	7.4
44	HO–SiH$_3$	1.14	1.664	4.517	0.978		4.9
45	H–SiH$_3$	0.81	1.483	2.903	0.762	0.00	4.6
46	F–SiH$_2$F	1.72	1.597	5.497	1.092		3.5
47a	F–SiF$_2$H	1.78	1.583	5.884	1.135		3.8
47b	H–SiF$_3$	1.44	1.458	3.273	0.815	1.43	3.8
48	F–SiF$_3$	1.98	1.571	6.281	1.178	0.00	3.3
49	F–GeH$_3$	1.93	1.793	4.951	1.030	2.25	4.7
50	Cl–GeH$_3$	1.66	2.175	2.491	0.699	2.04	6.9
51	Br–GeH$_3$	1.57	2.330	2.091	0.633	1.93	8.1
52	HO–GeH$_3$	1.33	1.802	3.872	0.896		5.5
53	H–GeH$_3$	0.73	1.542	2.693	0.730	0.00	5.2
54	O=CO	1.18	1.167	15.613	1.969	0.00	2.6
55	S=CO	0.64	1.575	7.227	1.275	0.68	5.2
56	O=CF$_2$	1.85	1.177	14.680	1.902	1.00	2.8
57	S=CF$_2$	1.29	1.603	6.626	1.214	0.16	5.2
58	O=SiF$_2$	3.46	1.517	9.243	1.465	2.31	4.0
59	CH$_3^+$	10.01				0.00	1.3
60	F–NH$_3^+$	8.58	1.368	5.642	1.109	4.78	1.7

* Maximum electrostatic potential at the σ-hole of X ($V_{s_{max}}(X)$) in eV. XT bond distance r(XT) in Å, XT local stretching force $k^a(XT)$ in mdyn/Å, XT bond strength order n(XT). Dipole moment in Debye and static isotropic polarizability in Å^3. All values were calculated with CCSD(T)/aug-cc-pVTZ.

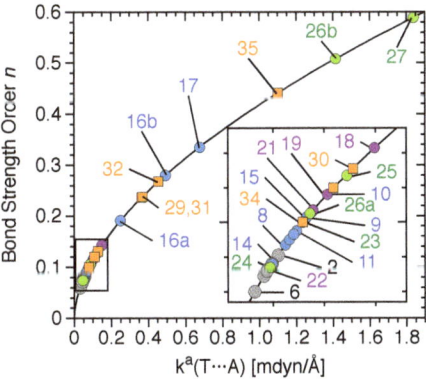

Figure 4. Power relationship between the relative bond strength order (BSO) n and the local stretching force constants k^a of the TA interaction of complexes **1–35**. C donors are gray, Si donors are blue, Ge donors are purple, double bonded donors are green, and charge-assisted TBs are orange. Complex **28** is not shown. Calculated at the CCSD(T)/aug-cc-pVTZ level of theory.

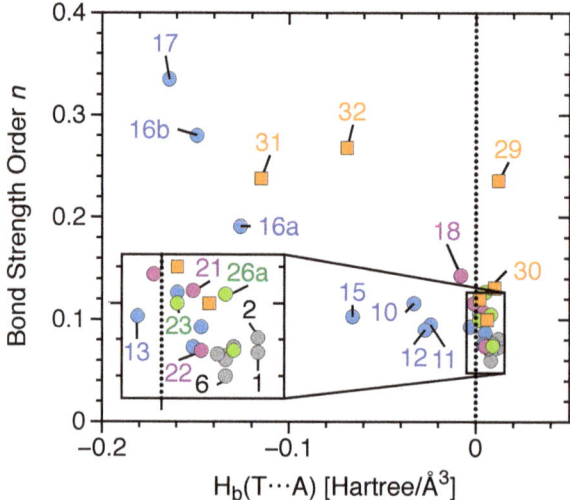

Figure 5. Comparison of the relative bond strength order (BSO) n and the energy density at the bond critical point H_b of the tetrel bond of complexes. C tetrel bonds are gray, Si tetrel bonds are blue, Ge tetrel bonds are purple, double bonded tetrel bonds are green, and anionic tetrel bonds are orange. Complexes **26b–28** and **35** are not shown. Calculated at the CCSD(T)/aug-cc-pVTZ level of theory.

3.1. Tetrel Bonds (TB) in Neutral Complexes

TBs involving C donors: Tetrel bonds (TBs) involving a neutral sp^3 hybridized carbon as a tetrel donor and neutral tetrel acceptors (complexes **1–7**) have weak interactions (BSO $n \leq 0.081$; $\Delta E \geq -2.25$ kcal/mol). The energy density at the TB density critical point is destabilizing ($H_b \geq 0.007$ Hartree/Å^3), indicating that these TBs are electrostatic in nature. TB with C donors that have peripheral H ligands (**1–6**) are not only stabilized by a lone pair of the T acceptor lp(A)-σ-hole electrostatic attraction but also by the electrostatic attraction between the positive charge at the hydrogen atoms (Hs) and the negative charge at lp(A) (see the NPA atomic charges in Figure 2 and the monomers' electrostatic potential in Table 2 and Figure 3). The presence of the later interactions were also verified by noncovalent interaction plots (NCIs plots), which showed a weak (attractive) electrostatic interaction (see Supporting Information Figure S2). Complex **7** (CF$_4$···NH$_3$) shows that an attractive interaction can be formed even in the absence of positively charged Hs on the T-donor. However, in **7**, there is no electron density path connecting the N of the T acceptor to the C of the T donor indicating the formation of a very weak dispersive interaction, as pointed by Grabowski [147]. The calculated spectra of **7** clearly shows an intermonomer stretching vibration of A$_1$ symmetry at 73 cm^{-1}. Decomposition of the normal vibrational mode into the local vibrational mode shows that the CN local stretching mode contributes solely to this normal vibrational mode, confirming the existence of tetrel bonding in **7**. As revealed by the NCI plot analysis, the peripheral H ligands of the T acceptor have an additional weak (attractive) electrostatic interaction with the peripheral F ligands of the T donor, which provides additional stabilization to the complex (see Supporting Information Figure S2).

The TB strength in the series of C donors and neutral T acceptors (**1–7**) shows only a small variation (0.021 for n and 0.87 kcal/mol for ΔE) which is affected by several factors such as the positive charge at the Hs (**5** > **4** > **3** ≈ **6**), the negative electrostatic potential at lp(A) of the T-acceptor (NH$_3$ < OH$_2$ < FH), and the intermonomer distance (e.g., 3.035 Å (**2**) compared to 3.218 (**3**)). It is noteworthy that for X–CH$_3$···NH$_3$ complexes, the TB strength weakens in the order of **3** (X = F) > **4** (X = Cl) > **5** (X = Br) > **6** (X = OH) as the magnitude of the σ-hole decreases (**36** (F–CH$_3$)) > **37** (Cl–CH$_3$) > **38** (Br–CH$_3$) > **39** (OH–CH$_3$) (see Figure 3).

TB involving Si donors: The strength of complexes **8–14** can be understood mostly on the basis of the extreme values of the electrostatic potential of the monomers. First, by varying the T-acceptor in FSiH$_3$···A, where A = FH (**8**), OH$_2$ (**9**), NH$_3$ (**10**), the TB strength trends follows the increase in magnitude of the negative potential at the lp(A) of the Lewis base. It is noticeable that $V_{s,min}$(A) and $V_{s,max}$(T) in **9** are not aligned as in **2**, indicating that even in this case, the stabilization brought by electron delocalization involving the highest occupied orbital HOMO of H$_2$O (the lp(O) orbital of B$_1$ symmetry, see Supporting Information Figure S3) and the lowest unoccupied orbital of FSiH$_3$ (σ^*(SiF) orbital) can influence the geometry of the complex (see Supporting Information Figure S4). Second, by varying the T donors in the XSiH$_3$···NH$_3$ complexes, the TB strength decreases in the order **10** (X = F) > **11** (X = Cl) > **12** (X = Br) ≈ **13** (X = OH) > **14** (X = H) as the Vs at the σ-hole decreases (**41** (F–SiH$_3$) > **42** (Cl–SiH$_3$) > **43** (Br–SiH$_3$) > **44** (HO–SiH$_3$) > **45** (SiH$_4$)). The only exception is **44**, which has a more positive NBO charge at Si (1277 me compared to 852 me in **43**) but a less positive $V_{s,max}$ (1.14 (**44**) compared to 1.46 (**43**), see Table 2 and Supporting Information Figure S1). This could be caused by the stronger σ(O-Si) orbital contraction, whereby the Si atom in HOSiH$_3$ is more electron-deficient than the Si atom in BrSiH$_3$, but due to the higher electronegativity of O, the σ(O-Si) is more compact than σ(Br-Si), resulting in a better shielded Si nucleus, reflected in the less positive potential at the σ-hole region (given by $V_{s,max}$). Substituting F in FSiH$_3$ would lead to an even stronger σ(F-Si) orbital contraction, but this effect would be overcome by the higher electron deficiency of Si.

Stepwise fluorination of SiH$_4$: The successive fluorination of SiH$_4$ (complexes **10** and **15–17**) impacts both the strength and the nature of the TB. Substituting the H collinear to the TB in complex **14** results in complex **10**, which has a stronger TB interaction due to the higher V_s at the σ-hole region and is due to the partial covalent character of the interaction ($H_b = -0.033$ Hartree/Å for complex **10**), which can be understood on the basis of molecular orbital interactions, as the electron delocalization from the highest occupied molecular orbital (HOMO) of the NH$_3$ (lone pair orbital of N, lp(N)) into the lowest unoccupied molecular orbital (LUMO) of FSiH$_3$ (a σ^*(FSi) orbital see Supporting Information Figure S5).

A second fluorine substituent (complex **15**) shows three different electronic effects: (i) the second fluorine withdraws the electron density from Si, decreasing its covalent radius and thus increasing the V_s at the σ-hole region, resulting in a stronger σ-hole-lp(N) electrostatic attraction; (ii) the lp(N)→ σ^*(SiF) electron delocalization is not restricted to the σ^* orbital of the Si–F bond that is collinear to the lp(A) but can also take place to the σ^* orbital of the second Si–F bond (see Table 3); (iii) the orbital effect of the exchange-repulsion between the lone pair orbitals of the peripheral fluorine lp(F) and lp(N) orbital. Effects (i) and (ii) are responsible for the 0.123 Å shorter TB in **15** compared to **10**. However, due to effect (iii), complexes **10** and **15** have similar TB strengths (n = 0.116 for **10**, 0.103 for **15**; ΔE = 6.8 kcal/mol for **10**, and 7.0 kcal/mol for **15**).

The addition of a third fluorine substituent (complexes **16a** and **16b**) leads to shorter and stronger TBs (BSO n = 0.191 (**16a**) and 0.280 (**16b**) compared to 0.103 (**15**)). However, this great increase in the TB strength (especially for **16b**) is not reflected by the binding energies of **16a** and **16b** (ΔE = −7.7 kcal/mol for **16a** and −6.3 kcal/mol for **16b**). The reason for the unexpectedly low ΔE values of **16a** and **16b** is due to the energetic cost associated with the geometric deformation of the monomers upon complexation (E_{def} = 11.8 kcal/mol for **16a**, 21.2 kcal/mol for **16b**). Monomer deformation is mostly caused by the lp(F)–lp(N) exchange-repulsion (effect (iii)), which pushes the peripheral ligands towards the bond collinear to the TB. For example, there is a decrease of 12.3° in the H–Si–F bond angle of HSiF$_3$ upon the formation of **16b**. Monomer deformation and the steric effect on TB complexes were also topics of a recent study carried by Scheiner [148,149]. It is noteworthy that the strongest TB between SiF$_3$H and NH$_3$ is collinear to the V_s at the σ-hole of the Si–H bond (complex **16b**), instead of the most positive potential at the σ-hole of the Si–F bond as one would expect from the σ-hole model or from steric considerations. The stronger and more covalent bond in **16b** is due to the higher stabilization energies brought by electron delocalization from lp(N) into σ^*(FSi) and into the σ^*(HSi) unoccupied orbital, as shown in Table 3. Even if the TB of **16b** is elongated to match the TB distance of **16a**, the NBO second-order delocalization energies of **16b** are still higher than those of **16a** (see Table 3).

Table 3. Natural Bond Orbital (NBO) electron delocalization energies involving the lone pair of NH_3 *.

#	Complex	σ^*(X-Si)	σ^*(Si-R)	σ^*(Si-R')	σ^*(Si-R'')
10	$FSiH_3 \cdots NH_3$	15.7	2.4	2.4	2.4
15	$SiF_2H_2 \cdots NH_3$	12.7	6.2	3.1	3.1
16a	$SiF_3H \cdots NH_3$	16.3	11.9	11.9	7.5
16b	$SiF_3H \cdots NH_3$	11.4	23.5	23.5	23.5
16b[a]	$SiHF_3 \cdots NH_3$	7.9	16.8	16.8	16.8
17	$SiF_4 \cdots NH_3$	20.7	19.5	19.5	19.5

* NBO electron delocalization energies from the second-order perturbation analysis referent to the interaction involving the lp(N) orbital of NH_3 and the σ^*(X-Si) (collinear to the TB), the σ^*(Si-R), σ^*(Si-R') and the σ^*(Si-R'') (peripheral to the TB, see Figure 1) of selected tetrel donors (see Supporting Information Figure S5). Values are in kcal/mol. Calculated with ωB97XD/aug-cc-pVTZ. [a] Complex **16b** with an elongated tetrel bond (TB) to match the TB distance of **16a**.

The addition of a fourth fluorine (complex **17**) makes the TB even stronger (n = 0.335 for **17**), compared to 0.280 for **16**), which is a consequence of the more positive V_s and of the higher electron delocalization that occurs from lp(N) to σ^*(FSi) compared to σ^*(H-Si) (see Table 3 and Supporting Information Figure S5). The substitution of the H collinear to the TB in **16b** by a fluorine does not increase the steric repulsion between the monomer (E_{def} of **17** is almost the same of **16b**), and as a result, ΔE is 5.1 kcal/mol more stable.

TB involving Ge donors: The germanium electron density is more easily polarized by an electronegative substituent than silicon. As a result, V_s at the σ-hole of Ge-donors (**49–52**) are higher than Si donors (see Table 2), the only exception being GeH_4 (**53**), which is a consequence of the higher electronegativity of Ge (χ(Si): 1.74 compared to χ(Ge): 2.02). Due to the stronger lp(N)-σ-hole electrostatic attraction, mono-substituted Ge-donors (**18, 19, 20, 21**) form stronger TBs than mono-substituted Si donors (**10, 11, 12, 13**) when paired with the NH_3 T acceptor. Conversely, the covalent component of this interaction is slightly reduced because of the more diffuse nature of Ge orbitals (see CT and H_b values on Table 1). Similar to C donors and Si donors, for X–$GeH_3 \cdots NH_3$ complexes, the TB strength decreases in the order of **18** (X = F) > **19** (X = Cl) > **20** (X = Br) > **21** (X = OH), as the magnitude of the σ-hole decreases (**49** (F–GeH_3)) > **50** (Cl–GeH_3) > **51** (Br–GeH_3) > **52** (OH–GeH_3) (see Table 2 and Figure 3).

TB in double bonded C and Si donors: When a carbon atom forms a π-bond, density is moved from a p-orbital of the carbon into the π-bond, resulting in a depletion of electron density at the C and the formation of a region of positive potential called a π-hole [150,151]. As noted previously by various authors [152–154], the lp(A)-π-hole electrostatic attraction is an important component of the TB involving double bonded C, Si donors. Another important characteristic of these T-donors is the existence of a low lying empty π^*(XT) orbital which is capable of accepting electron density from the lp(A) orbital of the T-acceptor. In order to evaluate strategies to strengthen TB involving double bonded C and Si donors, complexes **23–27** were investigated. Due to the $D_{\infty h}$ and $C_{\infty v}$ symmetry of CO_2 (**54**) and SCO (**55**), respectively, the π-bond density of these monomers has a central constriction with a negative V_s, leaving a belt-shaped π-hole around the C atom (see Figure 3). In SCO (**55**), a chalcogen bond is also possible due to the formation of a positive V_s at the σ-hole of sulfur. The TBs between CO_2 (**23**) and SCO (**24**) T donors and the prototypical T acceptor NH_3 are weak (n = 0.100 (**23**), 0.074 (**24**)) and electrostatic in nature (H_b > 0.002 Hartree/Å).

Substituting a CO double bond by two CF single bonds in **23** and **24** results in complexes $CF_2O \cdots NH_3$ (**25**) and $CF_2S \cdots NH_3$ (**26**), respectively. The T-donors of these complexes are characterized by having a higher V_s at the π-hole (V_s = 1.85 eV (**56**), 1.29 eV (**57**) compared to 1.18 eV (**54**) and 0.64 eV (**55**)), resulting in stronger TBs (n = 0.127 (**25**), 0.105 (**26a**) compared to 0.100 (**23**) and 0.074 (**24**)). The atypically strong (n = 0.508), highly covalent (H_b = −1.339 Hartree/Å^3) and short interactions (r(TA) = 1.7 Å) found in complex **26b** are formed at the expense of breaking the CS π-bond (n(C=S) decreases from 1.214 in **57** to 0.891 in **26b**). The energetic cost involved in the deformation of the monomers of **26b** is 1.45 kcal/mol higher than the stabilization brought by complexation (E_{int}); hence, the **26b** is less stable than the separated monomers. A small energetic barrier in the dissociative direction separates **26b** from the electrostatic TB complex **26a** (see Figure 6). An even stronger (n = 0.589) but less covalent interaction (H_b = −0.224 Hartree/Å^3) is formed between SiF_2O and NH_3 (complex **27**). In this complex, the SiO double bond is kept almost unaltered (n = 1.465 (**58**); 1.425 (**27**)); consequently, the deformation energy is relatively low compared to the stabilization energy brought by the complexation, resulting in a binding energy of −44.14 kcal/mol.

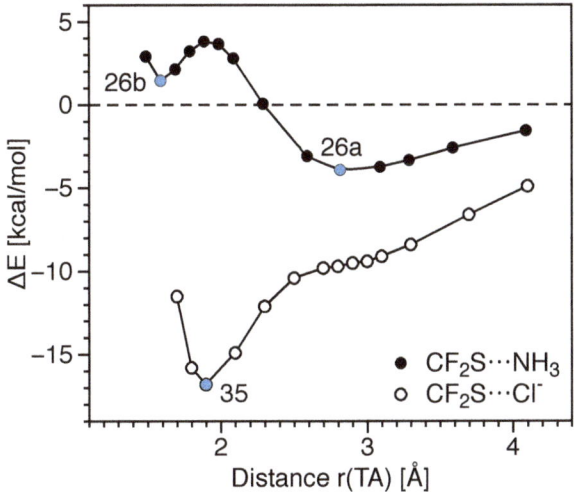

Figure 6. Relationship between the binding energy and interactomic distance computed at the CCSD(T)/aug-cc-pVTZ level of theory. All geometric parameters were optimized at each point of the curves for fixed r(TA) values. The blue dots represent the binding energy at the minima of complex **26** and the minimum of **35**; the black lines connecting points were used to improve interpretation.

3.2. Charge-Assisted Tetrel Bonds

Charge-assisted interactions: Similar to other NCIs [30,34,116,118,155], TBs can be strengthened by having a positively-charged T donor or a negatively-charged T acceptor **30–35**. The simplest positively charged C donor is CH_3^+ (isoelectronic to BH_3). However, due to the availability of an empty p-orbital to coordinate with the lone pairs of the NH_3, the C–N bond in **28** clearly differs from the tetrel bonds. This covalent bond in $[CH_3–NH_3]^+$ (**28**) is much stronger (n = 0.882; ΔE = −110.26 kcal/mol) and covalent (H_b = −1.952 Hartree/Å^3) compared to a TB. The existence of a covalent bond in complex **28** is also confirmed by the NCI plot, showing that there is no noncovalent interaction (see Supporting Information Figure S2). On the other hand, the cationic pnicogen donor FNH_3^+, isoelectronic to FCH_3, forms a noncovalent interaction with the NH_3 (complex **29**) which closely resembles the ones formed by neutral C donors (**1–7**) characterized by an electrostatic nature (H_b) and an interaction collinear to the X–T bond. The only difference is the higher V_s at the σ-hole (V_s = 8.58 eV), which results in a stronger electrostatic interaction (n = 0.236; ΔE = −23.14 kcal/mol).

An increase in the TB strength of as much as 105% (**31**) occurs for complexes involving a neutral T donor and a chloride anion as a T acceptor. This increase does not affect TB strength trends, such as $FGeH_3 > FSiH_3 > FCH_3$, and $CO_2 > SCO$, nor the trends in the covalent character of the TB in these series. Conversely, complex CF_2S forms a weaker bond with Cl^- compared to NH_3 (n = 0.441 (**35**) 0.508 (**26b**)), but has a highly negative binding energy (ΔE = −16.81 kcal/mol of **35** compared to the

1.45 kcal/mol of **26b**). The inverse relationships between the bond strength and interaction energies (E_{int}) or binding energies (ΔE) between these complexes indicate that the local C–N or C–Cl stretching is not the determining factor for the complex stabilization. Other components of the interaction which do not contribute to the strength of T-A local stretching also stabilize the complex. Figure 6 shows the dissociation curves for the CF$_2$S···NH$_3$ (**26**) and CF$_2$S···Cl$^-$ (**35**). Only **26** has a minimum geometry with an electrostatic TB (**26a**), whereas the dissociation curve for **35** has a flat region around r(TA) = 2.9Å separating the electrostatic interaction found for long TBs from the strong covalent interactions in **35**. The barrier energy from **26a** to **26b** is about 3 kcal/mol.

3.3. Tetrel Bonds vs. Other Noncovalent Interactions

We also compared TB with other noncovalent interactions, such as halogen (XB), chalcogen (ChB), and pnicogen (PnB), in mono-fluorinated systems involving the third period series FCl, FSH, FPH$_2$ with a medium (OH$_2$) and a strong (NH$_3$) Lewis base (see Table 4). It was clearly shown that the TB formed by FSiH$_3$ tend to be weaker than the other noncovalent interactions, the only exception being FSiH$_3$···Cl$^-$ (**31**) (n = 0.238) which is slightly stronger than FPH$_2$···Cl$^-$ (n = 0.214) but weaker than FSH···Cl$^-$ (n = 0.264) and FCl···Cl$^-$ (n = 0.382). Increasing the polarizability of the T donor moving FSiH$_3$ to FGeH$_3$ does increase the strength of the tetrel bond enough to compete with the halogen bonds formed by FCl. A better strategy for obtaining TBs that are strong enough to compete with halogen bonds and other noncovalent interactions is the substitution of peripheral Hs in FSiH$_3$ by fluorine atoms. SiF$_4$···NH$_3$ (**17**), for example, has a BSO of n = 0.335 compared to n = 0.216 for the FCl···NH$_3$. A clear advantage of tetrel bonding is that even in the absence of a strong polarizing group collinear to the TB, such as in HSiF$_3$···NH$_3$, the TB is still stronger than other noncovalent interactions (n = 0.280 for **16b**). This TB feature should be extensively explored to tune the strength of TB involving not only Si donors but also the heavier and more polarizable Ge and Sn donors.

Table 4. Summary of energetics, geometric and vibrational spectroscopy data for other types of interactions *.

Complex	ΔE	ΔE_{cp}	r TA	CT	ρ TA	H_b TA	k^a TA	n TA	k^a XT	n XT
F$_2$···OH$_2$ (C$_s$)	−1.42	−1.15	2.662	0.005	0.066	0.022	0.057	0.083	4.488	0.974
Cl$_2$···OH$_2$ (C$_s$)	−2.98	−2.62	2.808	0.015	0.098	0.018	0.097	0.112	2.896	0.761
FCl···OH$_2$ (C$_s$)	−5.22	−4.75	2.566	0.032	0.163	0.016	0.170	0.154	3.967	0.909
FSH···OH$_2$ (C$_s$)	−5.69	−5.15	2.659	0.028	0.138	0.010	0.152	0.144	4.011	0.914
FPH$_2$···OH$_2$ (C$_s$)	−4.63	−4.02	2.780	0.021	0.107	0.006	0.118	0.125	4.198	0.938
F$_2$···NH$_3$ (C$_{3v}$)	−2.00	−1.69	2.615	0.017	0.097	0.027	0.062	0.087	3.821	0.890
Cl$_2$···NH$_3$ (C$_{3v}$)	−4.92	−4.43	2.664	0.055	0.172	0.006	0.132	0.133	2.370	0.680
FCl···NH$_3$ (C$_{3v}$)	−10.13	−9.39	2.320	0.145	0.358	−0.058	0.311	0.216	2.687	0.729
FSH···NH$_3$ (C$_s$)	−8.23	−7.58	2.512	0.081	0.235	−0.020	0.194	0.166	3.309	0.820
FPH$_2$···NH$_3$ (C$_s$)	−6.81	−6.10	2.663	0.057	0.171	−0.012	0.144	0.140	3.794	0.886
FCl···Cl$^-$ (C$_{\infty v}$)	−30.07	−28.98	2.316	0.496	0.547	−0.161	0.855	0.382	1.212	0.465
FSH···Cl$^-$ (C$_s$)	−23.46	−22.48	2.493	0.305	0.377	−0.092	0.443	0.264	1.466	0.518
FPH$_2$···Cl$^-$ (C$_s$)	−19.62	−18.62	2.649	0.208	0.266	−0.058	0.307	0.214	2.136	0.641

* Binding energies (ΔE) and conterpoise corrected binding energies ΔE_{cp} in kcal/mol. Intermonomer bond distance r(TA) in Å. Density at the TA critical point ρ_b in e/Å^3, energy density at the TA critical point H_b in Hartree/Å^3. NPA charge transfer (CT) in e. TA and XT local stretching force constant (k^a) in mdyn/Å and BSO n values. Computed at the CCSD(T)/aug-cc-pVTZ level of theory.

4. Conclusions

In the present work, we investigated a set of 35 representative tetrel complexes (ΔE = −1.4 to −26 kcal/mol) with the objective of finding the factors that influence the strengths of neutral and charged tetrel bonds involving C donors, Si donors, Ge donors, and double bonded C or Si donors. The strength of a tetrel bond is affected by the complex interplay of several factors, such as the magnitude of the σ-hole in the tetrel atom, the negative electrostatic potential at the lp(A) of the T acceptors, the positive charge at the peripheral hydrogen (Hs) of the T donors, exchange-repulsion between the lone pair orbitals of the peripheral atoms of the T donor, and the covalent character which can be rationalized on the basis of electron delocalization from the highest occupied molecular orbital (HOMO) of the T acceptor into the lowest unoccupied orbitals (LUMOs) of the T donor, which is not limited to $\sigma^*(X-T)$ orbital but can also involve the peripheral substituents (orbital of $\sigma^*(R-Si)$ character), allowing the formation of strong tetrel bonds, even in the absence of an electronegative X substituent collinear to the TB. This clearly shows that focusing on just one or two of these factors, in particular, the σ-hole description, can only lead to an incomplete picture [93,94,156,157]. In this work, we derived, for the first time, the intrinsic bond strength of tetrel bonds from calculated vibrational spectroscopy data, which, combined with NBO charges, charge transfer, dipole moments, electrostatic potentials, electron and energy density distributions, difference density distributions, and noncovalent interaction plots calculated at the CCSD(T)/aug-cc-pVTZ level of theory, led to a complete insight into how different electronic effects influence the intrinsic strength of the tetrel bonding.

- Tetrel bonding becomes stronger as the atomic mass of the tetrel center increases as a consequence of increasing the polarizability.
- For $X-TH_3 \cdots NH_3$ complexes, the tetrel bond strength weakens in the order (X = F) > (X = Cl) > (X = Br) ≥ (X = OH) as the magnitude of the σ-hole decreases in the order of $F-TH_3 > Cl-TH_3 > Br-TH_3 \geq OH-TH_3$.
- Successive fluorination of SiH_4 impacts both the strength and the nature of the tetrel bond. The successive fluorinations result in stronger tetrel bonding as a consequence of (i) higher V_s at the σ-hole region; (ii) the partial covalent character of the interaction; (iii) higher electron delocalization that occurs from the highest occupied molecular orbital (HOMO) of the T acceptor to the lowest unoccupied molecular orbital (LUMO) of the T donor. In this series, the binding energy trend deviates from BSO n values due to the high energetic cost associated with the geometric deformation of the monomers upon complexation (E_{def}) which is a consequence of the exchange-repulsion between the lone pair orbitals of the peripheral atoms of the T donor.
- Tetrel bonds in double bonded C donors, e.g., CO_2 with NH_3, are weak and electrostatic in nature. Substituting a C=O double bond with an electron withdrawing group (F atoms) strengthens the tetrel bond.
- A positively-charged Tdonor or negatively-charged T-acceptor strengthens the tetrel bond. It creates higher V_s at the σ-hole, resulting in a stronger electrostatic interaction.

We suggest that future materials based on strong tetrel bonding should be based on Si or heavier tetrel atoms, such as Ge and Sn, combined with peripheral fluorine ligands. Due to the larger size of Ge and Sn, the deformation energy in $XGeF_3 \cdots NH_3$ or $XSnF_3 \cdots NH_3$ should be smaller than $XSiF_3 \cdots NH_3$, making these complexes substantially more stable than $XSiF_3 \cdots NH_3$.

Although all complexes discussed in this paper represent the most stable tetrel-bonded complexes, not all of them represent the most stable structure possible (global minimum). For example, the hydrogen bonded complexes $FH \cdots CFH_3$ (**1**), $OH_2 \cdots CF_3H$ (**2**), $NH_3 \cdots CF_3H$ (**3**) are more stable than the tetrel-bonded complexes. However, a detailed analysis of the competition between tetrel bonds and other noncovalent interactions will be studied in the near future.

Supplementary Materials: The following are available online, Figure S1: Schematic representation of monomers (**36–63**) with atomic charges from the natural population analysis, Figure S2: Noncovalent interactions (NCIs) plot of complexes **1–35**, Figure S3: Selected molecular orbitals of the T-acceptors, Figure S4: Electron difference density distributions $\Delta\rho(r)$ for complexes **1–35**, Figure S5: Combination of donor and acceptor NBO orbitals involved in the electron delocalization of selected complexes; Table S1: Comparison between DLPNO-CCSD(T)/aug-cc-pV5Z and CCSD(T)/aug-cc-pVTZ energies, Table S2: Deviation from DLPNO-CCSD(T)/aug-cc-pV5Z interaction energies, Table S3: Atomic Cartesian coordinates in Å of complexes **1–35**.

Author Contributions: Conceptualization, D.S., V.O., and E.K.; Methodology, E.K.; Formal Analysis, D.S. and V.O.; Investigation, D.S. and V.O.; Data Curation, D.S. and V.O.; Writing—Original Draft Preparation, D.S. and V.O.; Writing—Review & Editing, E.K.; Visualization, D.S. and V.O.; Supervision, E.K.

Funding: This research was funded by National Science Foundation grant number CHE 1464906.

Acknowledgments: We thank Alan Humason for proof-reading and also SMU's HPC for providing excellent computational resources.

Conflicts of Interest: The authors declare no conflict of interest.

Abbreviations

The following abbreviations are used in this manuscript:

ACS	Adibatic connection scheme
BSO n	Bond strength order
CCSD(T)	Coupled cluster theory with singles, doubles, and perturbative triples
CT	Intermonomer charge transfer
EDG	Electron donating group
EWG	Electron withdrawing group
HOMO	Highest occupied molecular orbital
LUMO	Lowest unoccupied molecular orbital
NBO	Natural bond orbital
NCI	Noncovalent interaction
NPA	Natural population analysis
TB	Tetrel bond

References

1. Schneider, H.J. Binding Mechanisms in Supramolecular Complexes. *Angew. Chem. Int. Ed.* **2009**, *48*, 3924–3977. [CrossRef] [PubMed]
2. Bene, J.E.D.; Alkorta, I.; Elguero, J. Exploring the $(H_2C=PH_2)^+$:N-Base Potential Surfaces: Complexes Stabilized by Pnicogen, Hydrogen, and Tetrel Bonds. *J. Phys. Chem. A* **2015**, *119*, 11701–11710. [CrossRef] [PubMed]
3. Politzer, P.; Murray, J.S. Analysis of Halogen and Other σ-Hole Bonds in Crystals. *Crystals* **2018**, *8*, 42. [CrossRef]
4. Dubecký, M.; Mitas, L.; Jurečka, P. Noncovalent Interactions by Quantum Monte Carlo. *Chem. Rev.* **2016**, *116*, 5188–5215. [CrossRef] [PubMed]
5. Scheiner, S. *Hydrogen Bonding: A Theoretical Perspective*; Oxford University Press: New York, NY, USA, 1997.
6. Johnson, E.R.; Keinan, S.; Mori-Sánchez, P.; Contreras-García, J.; Cohen, A.J.; Yang, W. Revealing Noncovalent Interactions. *J. Am. Chem. Soc.* **2010**, *132*, 6498–6506. [CrossRef] [PubMed]
7. Riley, K.E.; Hobza, P. Noncovalent interactions in biochemistry. *WIREs: Comput. Mol. Sci.* **2011**, *1*, 3–17. [CrossRef]
8. Alkorta, I.; Legon, A. Nucleophilicities of Lewis Bases B and Electrophilicities of Lewis Acids A Determined from the Dissociation Energies of Complexes B···A Involving Hydrogen Bonds, Tetrel Bonds, Pnictogen Bonds, Chalcogen Bonds and Halogen Bonds. *Molecules* **2017**, *22*, 1786. [CrossRef] [PubMed]
9. Gholipour, A. Mutual interplay between pnicogen-π and tetrel bond in $PF_3 \perp X$-Pyr···SiH_3CN complexes: NMR, SAPT, AIM, NBO, and MEP analysis. *Struct. Chem.* **2018**. [CrossRef]
10. Christensen, A.S.; Kubař, T.; Cui, Q.; Elstner, M. Semiempirical Quantum Mechanical Methods for Noncovalent Interactions for Chemical and Biochemical Applications. *Chem. Rev.* **2016**, *116*, 5301–5337. [CrossRef] [PubMed]
11. Sessions, R.B.; Gibbs, N.; Dempsey, C.E. Hydrogen Bonding in Helical Polypeptides from Molecular Dynamics Simulations and Amide Hydrogen Exchange Analysis: Alamethicin and Melittin in Methanol. *Biophys. J.* **1998**, *74*, 138–152. [CrossRef]
12. Bene, J.E.D.; Alkorta, I.; Elguero, J. Anionic complexes of F^- and Cl^- with substituted methanes: Hydrogen, halogen, and tetrel bonds. *Chem. Phys. Lett.* **2016**, *655–656*, 115–119. [CrossRef]
13. Priimagi, A.; Cavallo, G.; Metrangolo, P.; Resnati, G. The Halogen Bond in the Design of Functional Supramolecular Materials: Recent Advances. *Acc. Chem. Res.* **2013**, *46*, 2686–2695. [CrossRef] [PubMed]

14. Shi, F.Q.; Li, X.; Xia, Y.; Zhang, L.; Yu, Z.X. DFT Study of the Mechanisms of In Water Au(I)-Catalyzed Tandem [3,3]-Rearrangement/Nazarov Reaction/[1,2]-Hydrogen Shift of Enynyl Acetates: A Proton-Transport Catalysis Strategy in the Water-Catalyzed [1,2]-Hydrogen Shift. *J. Am. Chem. Soc.* **2007**, *129*, 15503–15512. [CrossRef] [PubMed]
15. Shen, Q.J.; Jin, W.J. Strong halogen bonding of 1,2-diiodoperfluoroethane and 1,6-diiodoperfluorohexane with halide anions revealed by UV-Vis, FT-IR, NMR spectroscopes and crystallography. *Phys. Chem. Chem. Phys.* **2011**, *13*, 13721–13729. [CrossRef] [PubMed]
16. Arunan, E.; Desiraju, G.R.; Klein, R.A.; Sadlej, J.; Scheiner, S.; Alkorta, I.; Clary, D.C.; Crabtree, R.H.; Dannenberg, J.J.; Hobza, P.; et al. Definition of the hydrogen bond (IUPAC Recommendations 2011). *Pure Appl. Chem.* **2011**, *83*, 1637–1641. [CrossRef]
17. Arunan, E.; Desiraju, G.R.; Klein, R.A.; Sadlej, J.; Scheiner, S.; Alkorta, I.; Clary, D.C.; Crabtree, R.H.; Dannenberg, J.J.; Hobza, P.; et al. Defining the hydrogen bond: An account (IUPAC Technical Report). *Pure Appl. Chem.* **2011**, *83*, 1619–1636. [CrossRef]
18. Freindorf, M.; Kraka, E.; Cremer, D. A comprehensive analysis of hydrogen bond interactions based on local vibrational modes. *Int. J. Quant. Chem.* **2012**, *112*, 3174–3187. [CrossRef]
19. Kalescky, R.; Zou, W.; Kraka, E.; Cremer, D. Local vibrational modes of the water dimer—Comparison of theory and experiment. *Chem. Phys. Lett.* **2012**, *554*, 243–247. [CrossRef]
20. Kalescky, R.; Kraka, E.; Cremer, D. Local vibrational modes of the formic acid dimer—The strength of the double hydrogen bond. *Mol. Phys.* **2013**, *111*, 1497–1510. [CrossRef]
21. Kraka, E.; Freindorf, M.; Cremer, D. Chiral Discrimination by Vibrational Spectroscopy Utilizing Local Modes. *Chirality* **2013**, *25*, 185–196. [CrossRef] [PubMed]
22. Tao, Y.; Zou, W.; Jia, J.; Li, W.; Cremer, D. Different Ways of Hydrogen Bonding in Water—Why Does Warm Water Freeze Faster than Cold Water? *J. Theory Comp. Chem.* **2016**, *13*, 55–76. [CrossRef] [PubMed]
23. Bauzá, A.; Frontera, A. Aerogen Bonding Interaction: A New Supramolecular Force? *Angew. Chem. Int. Ed.* **2015**, *54*, 7340–7343. [CrossRef] [PubMed]
24. Bauzá, A.; Frontera, A. Theoretical Study on the Dual Behavior of XeO_3 and XeF_4 toward Aromatic Rings: Lone Pair-π versus Aerogen-π Interactions. *ChemPhysChem* **2015**, *16*, 3625–3630. [CrossRef] [PubMed]
25. Bauzá, A.; Frontera, A. π-Hole aerogen bonding interactions. *Phys. Chem. Chem. Phys.* **2015**, *17*, 24748–24753. [CrossRef] [PubMed]
26. Frontera, A.; Bauzá, A. Concurrent aerogen bonding and lone pair/anion-π interactions in the stability of organoxenon derivatives: A combined CSD and ab initio study. *Phys. Chem. Chem. Phys.* **2017**, *19*, 30063–30068. [CrossRef] [PubMed]
27. Desiraju, G.R.; Ho, P.S.; Kloo, L.; Legon, A.C.; Marquardt, R.; Metrangolo, P.; Politzer, P.; Resnati, G.; Rissanen, K. Definition of the halogen bond (IUPAC Recommendations 2013). *Pure Appl. Chem.* **2013**, *85*, 1711–1713. [CrossRef]
28. Politzer, P.; Murray, J.S. Halogen Bonding: An Interim Discussion. *ChemPhysChem* **2013**, *14*, 278–294. [CrossRef] [PubMed]
29. Cavallo, G.; Metrangolo, P.; Milani, R.; Pilati, T.; Priimagi, A.; Resnati, G.; Terraneo, G. The Halogen Bond. *Chem. Rev.* **2016**, *116*, 2478–2601. [CrossRef] [PubMed]
30. Oliveira, V.; Kraka, E.; Cremer, D. The intrinsic strength of the halogen bond: Electrostatic and covalent contributions described by coupled cluster theory. *Phys. Chem. Chem. Phys.* **2016**, *18*, 33031–33046. [CrossRef] [PubMed]
31. Gilday, L.C.; Robinson, S.W.; Barendt, T.A.; Langton, M.J.; Mullaney, B.R.; Beer, P.D. Halogen Bonding in Supramolecular Chemistry. *Chem. Rev.* **2015**, *115*, 7118–7195. [CrossRef] [PubMed]
32. Wolters, L.P.; Schyman, P.; Pavan, M.J.; Jorgensen, W.L.; Bickelhaupt, F.M.; Kozuch, S. The many faces of halogen bonding: A review of theoretical models and methods. *WIREs Comput. Mol. Sci.* **2014**, *4*, 523–540. [CrossRef]
33. Alikhani, E.; Fuster, F.; Madebene, B.; Grabowski, S.J. Topological reaction sites—Very strong chalcogen bonds. *Phys. Chem. Chem. Phys.* **2014**, *16*, 2430–2442. [CrossRef] [PubMed]
34. Oliveira, V.; Cremer, D.; Kraka, E. The Many Facets of Chalcogen Bonding: Described by Vibrational Spectroscopy. *J. Phys. Chem. A* **2017**, *121*, 6845–6862. [CrossRef] [PubMed]

35. Gleiter, R.; Haberhauer, G.; Werz, D.B.; Rominger, F.; Bleiholder, C. From Noncovalent Chalcogen-Chalcogen Interactions to Supramolecular Aggregates: Experiments and Calculations. *Chem. Rev.* **2018**, *118*, 2010–2041. [CrossRef] [PubMed]
36. Mahmudov, K.T.; Kopylovich, M.N.; da Silva, M.F.C.G.; Pombeiro, A.J.L. Chalcogen bonding in synthesis, catalysis and design of materials. *Dalton Trans.* **2017**, *46*, 10121–10138. [CrossRef] [PubMed]
37. Alkorta, I.; Elguero, J.; Bene, J.E.D. Complexes of O=C=S with Nitrogen Bases: Chalcogen Bonds, Tetrel Bonds, and Other Secondary Interactions. *ChemPhysChem* **2018**, *19*, 1886–1894. [CrossRef] [PubMed]
38. Scheiner, S. The Pnicogen Bond: Its Relation to Hydrogen, Halogen, and Other Noncovalent Bonds. *Acc. Chem. Res.* **2012**, *46*, 280–288. [CrossRef] [PubMed]
39. Sarkar, S.; Pavan, M.S.; Row, T.N.G. Experimental validation of 'pnicogen bonding' in nitrogen by charge density analysis. *Phys. Chem. Chem. Phys.* **2015**, *17*, 2330–2334. [CrossRef] [PubMed]
40. Setiawan, D.; Kraka, E.; Cremer, D. Strength of the Pnicogen Bond in Complexes Involving Group 5A Elements N, P, and As. *J. Phys. Chem. A* **2014**, *119*, 1642–1656. [CrossRef] [PubMed]
41. Setiawan, D.; Kraka, E.; Cremer, D. Description of pnicogen bonding with the help of vibrational spectroscopy—The missing link between theory and experiment. *Chem. Phys. Lett.* **2014**, *614*, 136–142. [CrossRef]
42. Setiawan, D.; Cremer, D. Super-pnicogen bonding in the radical anion of the fluorophosphine dimer. *Chem. Phys. Lett.* **2016**, *662*, 182–187. [CrossRef]
43. Thomas, S.P.; Pavan, M.S.; Row, T.N.G. Experimental evidence for 'carbon bonding' in the solid state from charge density analysis. *Chem. Commun.* **2014**, *50*, 49–51. [CrossRef] [PubMed]
44. Bauzá, A.; Mooibroek, T.J.; Frontera, A. Tetrel-Bonding Interaction: Rediscovered Supramolecular Force? *Angew. Chem. Int. Ed.* **2013**, *52*, 12317–12321. [CrossRef] [PubMed]
45. Bauzá, A.; Mooibroek, T.J.; Frontera, A. Tetrel Bonding Interactions. *Chem. Rec.* **2016**, *16*, 473–487. [CrossRef] [PubMed]
46. Bene, J.E.D.; Alkorta, I.; Elguero, J. Carbenes as Electron-Pair Donors To CO_2 for C···C Tetrel Bonds and C–C Covalent Bonds. *J. Phys. Chem. A* **2017**, *121*, 4039–4047. [CrossRef] [PubMed]
47. Alkorta, I.; Elguero, J.; Bene, J.E.D. Azines as Electron-Pair Donors to CO_2 for N···C Tetrel Bonds. *J. Phys. Chem. A* **2017**, *121*, 8017–8025. [CrossRef] [PubMed]
48. Bene, J.E.D.; Alkorta, I.; Elguero, J. Carbon-Carbon Bonding between Nitrogen Heterocyclic Carbenes and CO_2. *J. Phys. Chem. A* **2017**, *121*, 8136–8146. [CrossRef] [PubMed]
49. Bene, J.D.; Elguero, J.; Alkorta, I. Complexes of CO_2 with the Azoles: Tetrel Bonds, Hydrogen Bonds and Other Secondary Interactions. *Molecules* **2018**, *23*, 906. [CrossRef] [PubMed]
50. Grabowski, S.J. Triel Bonds, π-Hole-π-Electrons Interactions in Complexes of Boron and Aluminium Trihalides and Trihydrides with Acetylene and Ethylene. *Molecules* **2015**, *20*, 11297–11316. [CrossRef] [PubMed]
51. Bauzá, A.; Frontera, A. On the Versatility of BH_2X (X=F, Cl, Br, and I) Compounds as Halogen-, Hydrogen-, and Triel-Bond Donors: An Ab Initio Study. *ChemPhysChem* **2016**, *17*, 3181–3186. [CrossRef] [PubMed]
52. Esrafili, M.D.; Asadollahi, S.; Mousavian, P. Anionic tetrel bonds: An ab initio study. *Chem. Phys. Lett.* **2018**, *691*, 394–400. [CrossRef]
53. Li, Q.Z.; Zhuo, H.Y.; Li, H.B.; Liu, Z.B.; Li, W.Z.; Cheng, J.B. Tetrel-Hydride Interaction between XH_3F (X = C, Si, Ge, Sn) and HM (M = Li, Na, BeH, MgH). *J. Phys. Chem. A* **2014**, *119*, 2217–2224. [CrossRef] [PubMed]
54. Lu, Y.; Wang, Y.; Zhu, W. Nonbonding interactions of organic halogens in biological systems: Implications for drug discovery and biomolecular design. *Phys. Chem. Chem. Phys.* **2010**, *12*, 4543–4551. [CrossRef] [PubMed]
55. Mani, D.; Arunan, E. The X–C···π (X = F, Cl, Br, CN) Carbon Bond. *J. Phys. Chem. A* **2014**, *118*, 10081–10089. [CrossRef] [PubMed]
56. Xu, H.; Cheng, J.; Yang, X.; Liu, Z.; Li, W.; Li, Q. Comparison of σ-Hole and π-Hole Tetrel Bonds Formed by Pyrazine and 1,4-Dicyanobenzene: The Interplay between Anion-π and Tetrel Bonds. *ChemPhysChem* **2017**, *18*, 2442–2450. [CrossRef] [PubMed]
57. Grabowski, S.J. Tetrel bond-σ-hole bond as a preliminary stage of the SN_2 reaction. *Phys. Chem. Chem. Phys.* **2014**, *16*, 1824–1834. [CrossRef] [PubMed]
58. Murray, J.S.; Lane, P.; Politzer, P. Expansion of the σ-hole concept. *J. Mol. Model.* **2009**, *15*, 723–729. [CrossRef] [PubMed]

59. Scilabra, P.; Kumar, V.; Ursini, M.; Resnati, G. Close contacts involving germanium and tin in crystal structures: Experimental evidence of tetrel bonds. *J. Mol. Model.* **2018**, *24*, 37. [CrossRef] [PubMed]
60. Donald, K.J.; Tawfik, M. The Weak Helps the Strong: Sigma-Holes and the Stability of MF_4 · Base Complexes. *J. Phys. Chem. A* **2013**, *117*, 14176–14183. [CrossRef] [PubMed]
61. Bundhun, A.; Ramasami, P.; Murray, J.S.; Politzer, P. Trends in σ-hole strengths and interactions of F_3MX molecules (M = C, Si, Ge and X = F, Cl, Br, I). *J. Mol. Model.* **2013**, *19*, 2739–2746. [CrossRef] [PubMed]
62. Scheiner, S. Comparison of CH···O, SH···O, Chalcogen, and Tetrel Bonds Formed by Neutral and Cationic Sulfur-Containing Compounds. *J. Phys. Chem. A* **2015**, *119*, 9189–9199. [CrossRef] [PubMed]
63. Azofra, L.M.; Scheiner, S. Tetrel, chalcogen, and CH···O hydrogen bonds in complexes pairing carbonyl-containing molecules with 1, 2, and 3 molecules of CO_2. *J. Chem. Phys.* **2015**, *142*, 034307. [CrossRef] [PubMed]
64. Wei, Y.; Li, Q.; Scheiner, S. The π-Tetrel Bond and its Influence on Hydrogen Bonding and Proton Transfer. *ChemPhysChem* **2018**, *19*, 736–743. [CrossRef] [PubMed]
65. McDowell, S.A.C.; Joseph, J.A. The effect of atomic ions on model σ-hole bonded complexes of AH_3Y (A = C, Si, Ge; Y = F, Cl, Br). *Phys. Chem. Chem. Phys.* **2014**, *16*, 10854. [CrossRef] [PubMed]
66. Politzer, P.; Murray, J.S.; Clark, T. Halogen bonding and other σ-hole interactions: A perspective. *Phys. Chem. Chem. Phys.* **2013**, *15*, 11178–11189. [CrossRef] [PubMed]
67. Politzer, P.; Murray, J.S. σ-Hole Interactions: Perspectives and Misconceptions. *Crystals* **2017**, *7*, 212. [CrossRef]
68. Politzer, P.; Murray, J.S.; Clark, T.; Resnati, G. The σ-hole revisited. *Phys. Chem. Chem. Phys.* **2017**, *19*, 32166–32178. [CrossRef] [PubMed]
69. Clark, T.; Hennemann, M.; Murray, J.S.; Politzer, P. Halogen bonding: The σ-hole. *J. Mol. Model.* **2007**, *13*, 291–296. [CrossRef] [PubMed]
70. Jönsson, B.; Karlström, G.; Wennerström, H. Ab initio molecular orbital calculations on the water-carbon dioxide system: Molecular complexes. *Chem. Phys. Lett.* **1975**, *30*, 58–59. [CrossRef]
71. Peterson, K.I.; Klemperer, W. Structure and internal rotation of H_2O-CO_2, $HDO-CO_2$, and D_2O-CO_2 van der Waals complexes. *J. Chem. Phys.* **1984**, *80*, 2439–2445. [CrossRef]
72. Mitzel, N.W.; Blake, A.J.; Rankin, D.W.H. β-Donor Bonds in SiON Units: An Inherent Structure- Determining Property Leading to (4+4)-Coordination in Tetrakis-(N,N-dimethylhydroxylamido)silane. *J. Am. Chem. Soc.* **1997**, *119*, 4143–4148. [CrossRef]
73. Mitzel, N.W.; Losehand, U. β-Donorbindungen in Molekülen mit SiON-Einheiten. *Angew. Chem.* **1997**, *109*, 2897–2899. [CrossRef]
74. Southern, S.A.; Bryce, D.L. NMR Investigations of Noncovalent Carbon Tetrel Bonds. Computational Assessment and Initial Experimental Observation. *J. Phys. Chem. A* **2015**, *119*, 11891–11899. [CrossRef] [PubMed]
75. Brammer, L. Halogen bonding, chalcogen bonding, pnictogen bonding, tetrel bonding: Origins, current status and discussion. *Faraday Discuss.* **2017**, *203*, 485–507. [CrossRef] [PubMed]
76. Li, R.; Li, Q.; Cheng, J.; Liu, Z.; Li, W. The Prominent Enhancing Effect of the Cation-π Interaction on the Halogen-Hydride Halogen Bond in $M^1···C_6H_5X···HM^2$. *ChemPhysChem* **2011**, *12*, 2289–2295. [CrossRef] [PubMed]
77. Li, Q.Z.; Sun, L.; Liu, X.F.; Li, W.Z.; Cheng, J.B.; Zeng, Y.L. Enhancement of Iodine-Hydride Interaction by Substitution and Cooperative Effects in NCX-NCI-HMY Complexes. *ChemPhysChem* **2012**, *13*, 3997–4002. [CrossRef] [PubMed]
78. George, J.; Dronskowski, R. Tetrel Bonds in Infinite Molecular Chains by Electronic Structure Theory and Their Role for Crystal Stabilization. *J. Phys. Chem. A* **2017**, *121*, 1381–1387. [CrossRef] [PubMed]
79. Tang, Q.; Li, Q. Interplay between tetrel bonding and hydrogen bonding interactions in complexes involving F_2XO (X=C and Si) and HCN. *Comput. Theor. Chem.* **2014**, *1050*, 51–57. [CrossRef]
80. Liu, M.; Li, Q.; Li, W.; Cheng, J.; McDowell, S.A.C. Comparison of hydrogen, halogen, and tetrel bonds in the complexes of HArF with YH_3X (X = halogen, Y = C and Si). *RSC Adv.* **2016**, *6*, 19136–19143. [CrossRef]
81. Guo, X.; Liu, Y.W.; Li, Q.Z.; Li, W.Z.; Cheng, J.B. Competition and cooperativity between tetrel bond and chalcogen bond in complexes involving F_2CX (X = Se and Te). *Chem. Phys. Lett.* **2015**, *620*, 7–12. [CrossRef]
82. Marín-Luna, M.; Alkorta, I.; Elguero, J. Cooperativity in Tetrel Bonds. *J. Phys. Chem. A* **2016**, *120*, 648–656. [CrossRef] [PubMed]

83. Solimannejad, M.; Orojloo, M.; Amani, S. Effect of cooperativity in lithium bonding on the strength of halogen bonding and tetrel bonding: (LiCN)$_n$···ClYF$_3$ and (LiCN)$_n$···YF$_3$Cl (Y= C, Si and n=1-5) complexes as a working model. *J. Mol. Model.* **2015**, *21*. [CrossRef] [PubMed]
84. Esrafili, M.D.; Mohammadirad, N.; Solimannejad, M. Tetrel bond cooperativity in open-chain (CH$_3$CN)$_n$ and (CH$_3$NC)$_n$ clusters (n = 2–7): An ab initio study. *Chem. Phys. Lett.* **2015**, *628*, 16–20. [CrossRef]
85. Liu, M.; Li, Q.; Li, W.; Cheng, J. Tetrel bonds between PySiX$_3$ and some nitrogenated bases: Hybridization, substitution, and cooperativity. *J. Mol. Graphics Modell.* **2016**, *65*, 35–42. [CrossRef] [PubMed]
86. Wei, Y.; Li, Q.; Yang, X.; McDowell, S.A.C. Intramolecular Si···O Tetrel Bonding: Tuning of Substituents and Cooperativity. *ChemistrySelect* **2017**, *2*, 11104–11112. [CrossRef]
87. Wei, Y.; Cheng, J.; Li, W.; Li, Q. Regulation of coin metal substituents and cooperativity on the strength and nature of tetrel bonds. *RSC Adv.* **2017**, *7*, 46321–46328. [CrossRef]
88. Mani, D.; Arunan, E. The X–C···Y Carbon Bond. In *Noncovalent Forces*; Springer International Publishing: Cham, Switzerland, 2015; pp. 323–356.
89. Mahadevi, A.S.; Sastry, G.N. Cooperativity in Noncovalent Interactions. *Chem. Rev.* **2016**, *116*, 2775–2825. [CrossRef] [PubMed]
90. Politzer, P.; Murray, J.S. The fundamental nature and role of the electrostatic potential in atoms and molecules. *Theor. Chem. Acc.* **2002**, *108*, 134–142. [CrossRef]
91. Murray, J.S.; Politzer, P. The electrostatic potential: An overview. *WIREs Comput. Mol. Sci.* **2011**, *1*, 153–163. [CrossRef]
92. Mani, D.; Arunan, E. The X–C···Y (X = O/F, Y = O/S/F/Cl/Br/N/P) 'carbon bond' and hydrophobic interactions. *Phys. Chem. Chem. Phys.* **2013**, *15*, 14377–14383. [CrossRef] [PubMed]
93. Zierkiewicz, W.; Michalczyk, M.; Scheiner, S. Comparison between Tetrel Bonded Complexes Stabilized by σ and π Hole Interactions. *Molecules* **2018**, *23*, 1416. [CrossRef] [PubMed]
94. Zierkiewicz, W.; Michalczyk, M.; Scheiner, S. Implications of monomer deformation for tetrel and pnicogen bonds. *Phys. Chem. Chem. Phys.* **2018**, *20*, 8832–8841. [CrossRef] [PubMed]
95. Cremer, D.; Kraka, E. From Molecular Vibrations to Bonding, Chemical Reactions, and Reaction Mechanism. *Curr. Org. Chem.* **2010**, *14*, 1524–1560. [CrossRef]
96. Kraka, E.; Setiawan, D.; Cremer, D. Re-evaluation of the Bond Length-Bond Strength Rule: The Stronger Bond Is not Always the Shorter Bond. *J. Comp. Chem.* **2016**, *37*, 130–142. [CrossRef] [PubMed]
97. Setiawan, D.; Kraka, E.; Cremer, D. Hidden Bond Anomalies: The Peculiar Case of the Fluorinated Amine Chalcogenides. *J. Phys. Chem. A* **2015**, *119*, 9541–9556. [CrossRef] [PubMed]
98. Setiawan, D.; Sethio, D.; Cremer, D.; Kraka, E. From strong to weak NF bonds: On the design of a new class of fluorinating agents. *Phys. Chem. Chem. Phys.* **2018**, *20*, 23913–23927. [CrossRef] [PubMed]
99. Lin, C.Y.; Gilbert, A.T.B.; Gill, P.M.W. Calculating molecular vibrational spectra beyond the harmonic approximation. *Theor. Chem. Acc.* **2007**, *120*, 23–35. [CrossRef]
100. Roy, T.K.; Gerber, R.B. Vibrational self-consistent field calculations for spectroscopy of biological molecules: New algorithmic developments and applications. *Phys. Chem. Chem. Phys.* **2013**, *15*, 9468–9492. [CrossRef] [PubMed]
101. Panek, P.T.; Jacob, C.R. Anharmonic Theoretical Vibrational Spectroscopy of Polypeptides. *J. Phys. Chem. Lett.* **2016**, *7*, 3084–3090. [CrossRef] [PubMed]
102. Konkoli, Z.; Cremer, D. A New Way of Analyzing Vibrational Spectra I. Derivation of Adiabatic Internal Modes. *Int. J. Quant. Chem.* **1998**, *67*, 1–9. [CrossRef]
103. Konkoli, Z.; Larsson, J.A.; Cremer, D. A new way of analyzing vibrational spectra. IV. Application and testing of adiabatic modes within the concept of the characterization of normal modes. *Int. J. Quant. Chem.* **1998**, *67*, 41–55. [CrossRef]
104. Zou, W.; Kalescky, R.; Kraka, E.; Cremer, D. Relating Normal Vibrational Modes to Local Vibrational Modes with the Help of an Adiabatic Connection Scheme. *J. Chem. Phys.* **2012**, *137*, 084114. [CrossRef] [PubMed]
105. Kalescky, R.; Kraka, E.; Cremer, D. Description of Aromaticity with the Help of Vibrational Spectroscopy: Anthracene and Phenanthrene. *J. Phys. Chem. A* **2014**, *118*, 223–237. [CrossRef] [PubMed]
106. Freindorf, M.; Tao, Y.; Sethio, D.; Cremer, D.; Kraka, E. New Mechanistic Insights into the Claisen Rearrangement of Chorismate—A Unified Reaction Valley Approach Study. *Mol. Phys.* **2018**, in press. [CrossRef]

107. Cremer, D.; Larsson, J.A.; Kraka, E. New developments in the analysis of vibrational spectra On the use of adiabatic internal vibrational modes. In *Theoretical and Computational Chemistry*; Parkanyi, C., Ed.; Elsevier: Amsterdam, The Netherlands, 1998; pp. 259–327.
108. Zou, W.; Kalescky, R.; Kraka, E.; Cremer, D. Relating normal vibrational modes to local vibrational modes: Benzene and naphthalene. *J. Mol. Model.* **2012**, *19*, 2865–2877. [CrossRef] [PubMed]
109. Humason, A.; Zou, W.; Cremer, D. 11,11-Dimethyl-1,6-methano[10]annulene—An Annulene with an Ultralong CC Bond or a Fluxional Molecule? *J. Phys. Chem. A* **2015**, *119*, 1666–1682. [CrossRef] [PubMed]
110. Kalescky, R.; Kraka, E.; Cremer, D. Identification of the Strongest Bonds in Chemistry. *J. Phys. Chem. A* **2013**, *117*, 8981–8995. [CrossRef] [PubMed]
111. Kalescky, R.; Kraka, E.; Cremer, D. New Approach to Tolman's Electronic Parameter Based on Local Vibrational Modes. *Inorg. Chem.* **2014**, *53*, 478–495. [CrossRef] [PubMed]
112. Kraka, E.; Cremer, D. Characterization of CF Bonds with Multiple-Bond Character: Bond Lengths, Stretching Force Constants, and Bond Dissociation Energies. *ChemPhysChem* **2009**, *10*, 686–698. [CrossRef] [PubMed]
113. Kalescky, R.; Kraka, E.; Cremer, D. Are carbon-halogen double and triple bonds possible? *Int. J. Quantum Chem.* **2014**, *114*, 1060–1072. [CrossRef]
114. Kalescky, R.; Zou, W.; Kraka, E.; Cremer, D. Quantitative Assessment of the Multiplicity of Carbon-Halogen Bonds: Carbenium and Halonium Ions with F, Cl, Br, and I. *J. Phys. Chem. A* **2014**, *118*, 1948–1963. [CrossRef] [PubMed]
115. Oomens, J.; Kraka, E.; Nguyen, M.K.; Morton, T.H. Structure, Vibrational Spectra, and Unimolecular Dissociation of Gaseous 1-Fluoro-1-phenethyl Cations. *J. Phys. Chem. A* **2008**, *112*, 10774–10783. [CrossRef] [PubMed]
116. Oliveira, V.; Kraka, E.; Cremer, D. Quantitative Assessment of Halogen Bonding Utilizing Vibrational Spectroscopy. *Inorg. Chem.* **2016**, *56*, 488–502. [CrossRef] [PubMed]
117. Oliveira, V.; Kraka, E. Systematic Coupled Cluster Study of Noncovalent Interactions Involving Halogens, Chalcogens, and Pnicogens. *J. Phys. Chem. A* **2017**, *121*, 9544–9556. [CrossRef] [PubMed]
118. Oliveira, V.; Cremer, D. Transition from metal-ligand bonding to halogen bonding involving a metal as halogen acceptor a study of Cu, Ag, Au, Pt, and Hg complexes. *Chem. Phys. Lett.* **2017**, *681*, 56–63. [CrossRef]
119. Zhang, X.; Dai, H.; Yan, H.; Zou, W.; Cremer, D. B−H···π Interaction: A New Type of Nonclassical Hydrogen Bonding. *J. Am. Chem. Soc.* **2016**, *138*, 4334–4337. [CrossRef] [PubMed]
120. Zou, W.; Zhang, X.; Dai, H.; Yan, H.; Cremer, D.; Kraka, E. Description of an unusual hydrogen bond between carborane and a phenyl group. *J. Organ. Chem.* **2018**, *865*, 114–127. [CrossRef]
121. Purvis, G.D.; Bartlett, R.J. A full coupled-cluster singles and doubles model: The inclusion of disconnected triples. *J. Chem. Phys.* **1982**, *76*, 1910–1918. [CrossRef]
122. Pople, J.A.; Head-Gordon, M.; Raghavachari, K. Quadratic configuration interaction. A general technique for determining electron correlation energies. *J. Chem. Phys.* **1987**, *87*, 5968–5975. [CrossRef]
123. Dunning, T.H. Gaussian basis sets for use in correlated molecular calculations. I. The atoms boron through neon and hydrogen. *J. Chem. Phys.* **1989**, *90*, 1007–1023. [CrossRef]
124. Woon, D.E.; Dunning, T.H. Gaussian basis sets for use in correlated molecular calculations. III. The atoms aluminum through argon. *J. Chem. Phys.* **1993**, *98*, 1358–1371. [CrossRef]
125. Woon, D.E.; Dunning, T.H. Gaussian basis sets for use in correlated molecular calculations. IV. Calculation of static electrical response properties. *J. Chem. Phys.* **1994**, *100*, 2975–2988. [CrossRef]
126. Wilson, E.B.; Decius, J.C.; Cross, P.C. *Molecular Vibrations. The Theory of Infrared and Raman Vibrational Spectra*; McGraw-Hill: New York, NY, USA, 1955.
127. Kraka, E.; Larsson, J.A.; Cremer, D. Generalization of the Badger Rule Based on the Use of Adiabatic Vibrational Modes. In *Computational Spectroscopy*, Grunenberg, J., Ed.; Wiley: New York, NY, USA, 2010; pp. 105–149.
128. Boys, S.F.; Bernardi, F. The calculation of small molecular interactions by the differences of separate total energies. Some procedures with reduced errors. *Mol. Phys.* **1970**, *19*, 553–566. [CrossRef]
129. Mentel, Ł.M.; Baerends, E.J. Can the Counterpoise Correction for Basis Set Superposition Effect Be Justified? *J. Chem. Theory Comput.* **2013**, *10*, 252–267. [CrossRef] [PubMed]
130. Riplinger, C.; Neese, F. An efficient and near linear scaling pair natural orbital based local coupled cluster method. *J. Chem. Phys.* **2013**, *138*, 034106. [CrossRef] [PubMed]

131. Riplinger, C.; Sandhoefer, B.; Hansen, A.; Neese, F. Natural triple excitations in local coupled cluster calculations with pair natural orbitals. *J. Chem. Phys.* **2013**, *139*, 134101. [CrossRef] [PubMed]
132. Stanton, J.F.; Gauss, J.; Cheng, L.; Harding, M.E.; Matthews, D.A.; Szalay, P.G. CFOUR, Coupled-Cluster techniques for Computational Chemistry, a Quantum-Chemical Program Package. Available online: http://www.cfour.de (accessed on 1 October 2018).
133. Harding, M.; Mezroth, T.; Gauss, J.; Auer, A. Parallel Calculation of CCSD and CCSD(T) Analytic First and Second Derivatives. *J. Chem. Theory Comput.* **2008**, *4*, 64–74. [CrossRef] [PubMed]
134. Neese, F. The ORCA program system. *WIREs Comput. Mol. Sci.* **2011**, *2*, 73–78. [CrossRef]
135. Weinhold, F.; Landis, C.R. *Valency and Bonding: A Natural Bond Orbital Donor-Acceptor Perspective*; Cambridge University Press: Cambridge, UK, 2003.
136. Reed, A.; Curtiss, L.; Weinhold, F. Intermolecular Interactions from A Natural Bond Orbital, Donor-Acceptor Viewpoint. *Chem. Rev.* **1988**, *88*, 899–926. [CrossRef]
137. Glendening, E.D.; Badenhoop, J.K.; Reed, A.E.; Carpenter, J.E.; Bohmann, J.A.; Morales, C.M.; Landis, C.R.; Weinhold, F. NBO6. In *Theoretical Chemistry Institute*; University of Wisconsin: Madison, WI, USA, 2013.
138. Glendening, E.D.; Landis, C.R.; Weinhold, F. NBO 6.0: Natural bond orbital analysis program. *J. Comput. Chem.* **2013**, *34*, 1429–1437. [CrossRef] [PubMed]
139. Keith, T. TK Gristmill Software. Overland Park, KS, USA. Available online: http//aim.tkgristmill.com (accessed on 1 October 2018).
140. Lu, T.; Chen, F. Multiwfn: A multifunctional wavefunction analyzer. *J. Comput. Chem.* **2011**, *33*, 580–592. [CrossRef] [PubMed]
141. Contreras-García, J.; Johnson, E.R.; Keinan, S.; Chaudret, R.; Piquemal, J.P.; Beratan, D.N.; Yang, W. NCIPLOT: A Program for Plotting Noncovalent Interaction Regions. *J. Chem. Theory Comput.* **2011**, *7*, 625–632. [CrossRef] [PubMed]
142. Zou, W.; Nori-Shargh, D.; Boggs, J.E. On the Covalent Character of Rare Gas Bonding Interactions: A New Kind of Weak Interaction. *J. Phys. Chem. A* **2012**, *117*, 207–212. [CrossRef] [PubMed]
143. Kraka, E.; Zou, W.; Filatov, M.; Gräfenstein, J.; Izotov, D.; Gauss, J.; He, Y.; Wu, A.; Konkoli, Z.; Polo, V.; et al. COLOGNE, 2018. Available online: http://www.smu.edu/catco (accessed on 1 October 2018).
144. Li, Y.; Oliveira, V.; Tang, C.; Cremer, D.; Liu, C.; Ma, J. The Peculiar Role of the Au_3 Unit in Au_m Clusters: σ-Aromaticity of the Au_5Zn^+ Ion. *Inorg. Chem.* **2017**, *56*, 5793–5803. [CrossRef] [PubMed]
145. Cremer, D.; Kraka, E. Chemical Bonds without Bonding Electron Density? Does the Difference Electron-Density Analysis Suffice for a Description of the Chemical Bond? *Angew. Chem. Int. Ed.* **1984**, *23*, 627–628. [CrossRef]
146. Cremer, D.; Kraka, E. A Description of the Chemical Bond in Terms of Local Properties of Electron Density and Energy. *Croatica Chem. Acta* **1984**, *57*, 1259–1281.
147. Grabowski, S. Lewis Acid Properties of Tetrel Tetrafluorides—The Coincidence of the σ-Hole Concept with the QTAIM Approach. *Crystals* **2017**, *7*, 43. [CrossRef]
148. Scheiner, S. Systematic Elucidation of Factors That Influence the Strength of Tetrel Bonds. *J. Phys. Chem. A* **2017**, *121*, 5561–5568. [CrossRef] [PubMed]
149. Scheiner, S. Steric Crowding in Tetrel Bonds. *J. Phys. Chem. A* **2018**, *122*, 2550–2562. [CrossRef] [PubMed]
150. Angarov, V.; Kozuch, S. On the σ, π and δ hole interactions: A molecular orbital overview. *New J. Chem.* **2018**, *42*, 1413–1422. [CrossRef]
151. Wang, H.; Wang, W.; Jin, W.J. σ-Hole Bond vs π-Hole Bond: A Comparison Based on Halogen Bond. *Chem. Rev.* **2016**, *116*, 5072–5104. [CrossRef] [PubMed]
152. Dong, W.; Wang, Y.; Cheng, J.; Yang, X.; Li, Q. Competition between σ-hole pnicogen bond and π-hole tetrel bond in complexes of $CF_2=CFZH_2$ (Z= P, As, and Sb). *Mol. Phys.* **2018**, 1–9. [CrossRef]
153. Grabowski, S.J. Hydrogen bonds, and σ-hole and π-hole bonds—Mechanisms protecting doublet and octet electron structures. *Phys. Chem. Chem. Phys.* **2017**, *19*, 29742–29759. [CrossRef] [PubMed]
154. Shen, S.; Zeng, Y.; Li, X.; Meng, L.; Zhang, X. Insight into the π-holebond...π-electrons tetrel bonds between F_2ZO (Z = C, Si, Ge) and unsaturated hydrocarbons. *Int. J. Quantum Chem.* **2017**, *118*, e25521. [CrossRef]
155. Liu, M.; Li, Q.; Li, W.; Cheng, J. Carbene tetrel-bonded complexes. *Struct. Chem.* **2017**, *28*, 823–831. [CrossRef]

156. Xu, H.; Cheng, J.; Yu, X.; Li, Q. Abnormal Tetrel Bonds between Formamidine and TH$_3$F: Substituent Effects. *Chem. Sel.* **2018**, *3*, 2842–2849.
157. Zierkiewicz, W.; Michalczyk, M. On the opposite trends of correlations between interaction energies and electrostatic potentials of chlorinated and methylated amine complexes stabilized by halogen bond. *Theor. Chem. Acc.* **2017**, *136*. [CrossRef]

© 2018 by the authors. Licensee MDPI, Basel, Switzerland. This article is an open access article distributed under the terms and conditions of the Creative Commons Attribution (CC BY) license (http://creativecommons.org/licenses/by/4.0/).

Communication

On the Power of Geometry over Tetrel Bonds

Ephrath Solel and Sebastian Kozuch *

Department of Chemistry, Ben-Gurion University of the Negev, Beer-Sheva 841051, Israel; ephrath@post.bgu.ac.il
* Correspondence: kozuch@bgu.ac.il; Tel.: +972-8-64-61192

Received: 29 September 2018; Accepted: 23 October 2018; Published: 24 October 2018

Abstract: Tetrel bonds are noncovalent interactions formed by tetrel atoms (as σ-hole carriers) with a Lewis base. Here, we present a computational and molecular orbital study on the effect of the geometry of the substituents around the tetrel atom on the σ-hole and on the binding strengths. We show that changing the angles between substituents can dramatically increase bond strength. In addition, our findings suggest that the established Sn > Ge > Si order of binding strength can be changed in sufficiently distorted molecules due to the enhancement of the charge transfer component, making silicon the strongest tetrel donor.

Keywords: tetrel bond; σ-hole; DFT

1. Introduction

Hole interactions [1] are a relatively newly coined term that unites all noncovalent interactions in which a region of positive electrostatic potential on one atom, the hole, interacts with an electron donor. These can be based on σ, π, or δ holes depending on their type of covalent orbital origin [2]. σ-holes are formed at approximately 180° to a σ covalent bond, with the magnitude of the positive electrostatic potential depending on the electronegativity of the neighboring atoms. These interactions are further classified according to the σ-hole-bearing atom: the most studied interaction is hydrogen bonding, but there is also the widely researched halogen bonding [3–5], chalcogen bonding [6–8] (for Group VI atoms), pnictogen bonding [9–11] (Group V), tetrel bonding [12,13] (Group IV), and even aerogen bonding [14] (Group VIII). Better understanding of such noncovalent interactions can help in the study and future design of novel supramolecular complexes, catalysts, and crystal engineering.

Herein, we focus on the effect that the angles around the atom have on the binding strengths of tetrel bonds. We analyze this by examining the effect on the electrostatic hole and on the frontier orbitals in order to explain the dramatic changes in complexation energies. These effects were previously observed in a survey of the Cambridge Structural Database [12]: carbon demonstrates almost nonexistent σ-holes relative to Si, Ge, and Sn [13], with the only found crystal structures exhibiting σ-hole interactions with C based on three-membered rings [15] or cubanes [16], in which the angles around the carbon are far from the optimal tetrahedral angle. The effect of the angles between covalent bonds on interaction strength was also computationally explained by showing that smaller rings cause the σ-hole to be more exposed, increasing its electrostatic potential [12].

It should be noted that upon binding with a Lewis base, there is geometrical deformation around the tetrel atom as substituents move to make more room for the electron donor [17], a distortion that was computed to be more energetically costly for smaller atoms. Freezing the monomer in the complex's distorted geometry eliminates the deformation energy and results in an increase in the tetrel interaction energy. Our aim in this study is to understand the effect of the molecular geometry on bond energy beyond such binding-caused distortions by applying molecular orbital theory on the bonding patterns.

2. Results and Discussion

To estimate the effect the substituent angles have on the σ-hole and on the shape of the frontier orbitals, we examined the TH$_3$F systems (1_T, with T = C, Si, Ge, and Sn, Scheme 1). A weak σ-hole may be formed at the extension of the T-H bonds, but evidently the σ-hole corresponding to the T-F bond is the dominant one. We considered the optimized structures (C_{3V}) with no constrains or at three different fixed F-T-H angles (α = 109°, 100°, and 90°). All computations were performed at the MN15/Def2-TZVPD [18,19] level of theory with Gaussian16 [20] (see the Methods section).

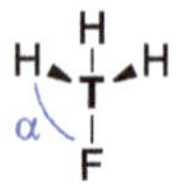

Scheme 1. 1_T model systems.

As can be seen in Figure 1, the LUMO for 1_C is an antibonding F-C σ*. This orbital mostly resides at the extension of the F-T bond (as in all 1_T molecules, see Figures S1–S3) and forms interactions with Lewis bases by charge transfer. For all 1_T molecules, the LUMO shows a larger lobe at the extension of the T-F bond as the angle decreases, which, in principle, aids the orbital interaction with the nucleophile.

The σ$_{F-T}$ orbitals (typically, the HOMO-2) are expected to match the areas with higher and lower electron density [2,21]. In 1_C, the electrostatic hole did not match this criterion. At smaller α, the outer lobe on C was slightly larger, although the electrostatic potential was more positive (Figure 1 and Table 1). For 1_{Si}, 1_{Ge}, and 1_{Sn}, due to the larger and more electropositive tetrel atom, the σ$_{F-T}$ was more localized on the F and more affected by the hydrogens. Thus, there was somewhat less electron density at the outer lobe of σ$_{F-T}$ at smaller angles, which, in principle, matches the trend in the $V_{s,max}$ (the maximum positive potential on the electrostatic potential (ESP) isosurface; see Figure 1 and Table 1). However, it seems that the HOMOs (σ$_{H-T}$ orbitals) are the ones mostly responsible of taking out the electron density, enhancing the σ-hole as the angle decreases by moving the T-H away from the F-T axis (matching the $V_{s,max}$ trend—see Table 1—and the ESP maps—Figure 1 and Figures S1–S3).

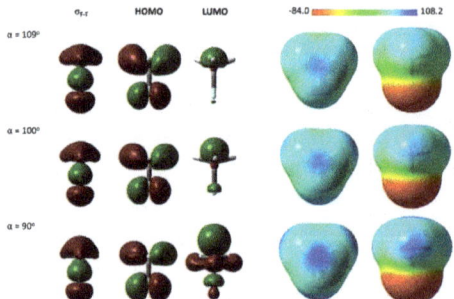

Figure 1. Chosen MOs and electrostatic potential (ESP) maps for 1_C with three different F-C-H angles (α). ESP maps are on the 0.001 density isosurface. The color scale is in kJ mol^{-1}. σ$_{F-T}$ corresponds to the bonding F-T σ orbital, irrespective of its position compared to other orbitals. The HOMO is doubly degenerate.

As can be expected, the F-T-H angle modifies the degree of *sp* hybridization of different orbitals. In NH$_3$ [22], the frontier MOs of the planar geometry exhibit pure *s* or *p* orbitals on the nitrogen, which then mix as the angle of pyramidalization increases. Similarly, according to the NBO analysis in 1_T, the tetrel component of the σ$_{F-T}$ MO has a higher *p* character when the angle is 90°, which decreases when the α angle increases (opposite to the *s* character; see Table 1). This causes stronger σ$_{F-T}$ (shorter F-T bond length) with larger α by focalizing the lobe into the fluorine's direction (Table 1), while also marginally reducing the outer lobe in **1_C**, as explained above (Figure 1).

Table 1. Properties of **1_T** with different σ angles: σ_{F-T} and LUMO energies (kJ mol^{-1}), F-T bond length (Å), %s and %p on T in the NBO F-T σ bond, and the maximal positive electrostatic potential at the σ hole (kJ mol^{-1}).

T	α	σ_{F-T}	LUMO	d_{F-T}	%s [b]	%p [b]	$V_{s,max}$
C	Opt. (109.1°) [a]	−1413.9	154.7	1.376	21.52	78.28	81.2
	109°	−1413.2	154.7	1.376	21.48	78.32	81.2
	100°	−1328.6	147.6	1.426	16.09	83.71	84.1
	90°	−1232.6	84.6	1.511	8.53	91.25	95.2
Si	Opt. (108.3°) [a]	−1328.8	27.2	1.598	21.40	76.03	142.5
	109°	−1332.2	28.7	1.596	21.62	75.81	137.3
	100°	−1282.5	−22.8	1.617	18.39	78.95	200.6
	90°	−1212.8	−123.7	1.655	13.54	83.79	256.5
Ge	Opt. (106.2°) [a]	−1250.8	19.4	1.737	19.96	79.06	164.5
	109°	−1267.7	21.4	1.731	21.20	77.83	148.8
	100°	−1212.4	−18.6	1.751	17.05	81.98	197.0
	90°	−1145.6	−120.0	1.788	11.39	87.66	238.0
Sn	Opt. (104.4°) [a]	−1158.6	−32.6	1.927	18.66	80.49	196.6
	109°	−1182.5	−46.3	1.920	20.75	78.41	170.9
	100°	−1134.9	−58.2	1.935	16.56	82.57	218.6
	90°	−1079.2	−158.9	1.962	11.20	87.94	255.7

[a] Fully optimized molecule, with no angle restrictions. [b] %s and %p are the same for both the bonding and antibonding orbitals.

We computed the complexes of **1_T** with HCN, a prototypical Lewis base for hole interactions that minimizes the influences coming from atoms and bonds other than the tetrel bond (ammonia, for example, exhibits attraction between its partially positive hydrogens and the partially negative hydrogens on the tetrels). The geometry parameters, dissociation energies ($D_e = E_{1_T} + E_{HCN} - E_{1_T\cdots NCH}$), and NBO charge transfer energies (i.e., the $n \to \sigma^*$ perturbational stabilization energy, E^2) are presented in Table 2.

Table 2. Properties of the complexes of **1_T** with HCN at different α angles: (distances in Å, energies in kJ mol^{-1}).

T	α	d_{T-F}	$d_{T\cdots N}$	% Cov. Rad. [b]	D_e [c]	$E^2_{n\to\sigma^*}$ [d]
C	Opt. (109.3°) [a]	1.380	3.154	208	9.3	2.5
	109°	1.381	3.155	208	9.2	2.5
	100°	1.432	3.116	205	8.7	3.0
	90°	1.521	3.013	198	9.5	5.3
Si	Opt. (106.3°) [a]	1.608	2.847	153	18.7	18.0
	109°	1.602	2.944	158	17.0	13.6
	100°	1.625	2.576	138	29.8	37.7
	90°	1.670	2.162	116	56.7	84.1 [e]
Ge	Opt. (104.6°) [a]	1.749	2.931	149	20.4	24.6
	109°	1.738	3.043	154	17.8	18.4
	100°	1.763	2.804	142	25.4	33.9
	90°	1.808	2.532	128	37.8	68.1
Sn	Opt. (102.2°) [a]	1.945	2.934	136	25.7	27.0
	109°	1.930	3.086	143	20.8	17.4
	100°	1.950	2.887	134	28.9	31.0
	90°	1.982	2.703	125	39.4	54.3

[a] Fully optimized molecule with no angle restrictions. [b] Ratio between tetrel bond and the sum of covalent radii of T and N. [c] Tetrel bond dissociation energies. [d] Perturbational stabilization energy according to NBO analysis corresponding to charge transfer. [e] $E^2_{n\to\sigma^*}$ for this complex was calculation by extrapolation, see Supplementary Information.

Table 2 shows that the T-F bonds are all longer compared to the free molecules, as expected upon interaction of a Lewis base with the σ* orbital. Bond strength, due to the higher polarizabilities of the heavier tetrel atoms, is $1_C \ll 1_{Si} < 1_{Ge} < 1_{Sn}$ for the fully optimized molecules or for α = 109°. For the smaller α angles, the T···N distance is shorter and the binding energies larger compared to the unconstrained systems (except for 1_C, which at any rate exhibits very weak binding), with the largest changes with respect to the angle observed with 1_{Si} (see Figure 2A). If we check the effect of changing the tetrel atom at each fixed α angle (Figure 2B), we can see that Ge and Sn show stronger binding than Si only for the fully optimized geometry and for α = 109°. However, upon reduction of the angle to 100° and 90°, the Si shows higher binding than the other tetrel atoms, with significantly shorter T···N distances. As can be seen in Table 2, for α = 90° the T···N distances for 1_{Si}, 1_{Ge}, and 1_{Sn} come close to the sum of the covalent radii of T and nitrogen, pointing to a more covalent character (in the extreme case of 1_{Si}, bond length is only 116% compared to the sum of the covalent radii). In addition, the NBO $n \rightarrow \sigma^*$ component grows as Si > Ge > Sn for the smaller α angles (Figure 2C). This suggests that there are two competing factors affecting binding strength: polarizability, which increases upon descending the column, increasing electrostatic interactions; and orbital interactions, which are stronger for smaller atoms, except for C, and become more dominant at shorter distances and smaller α angles.

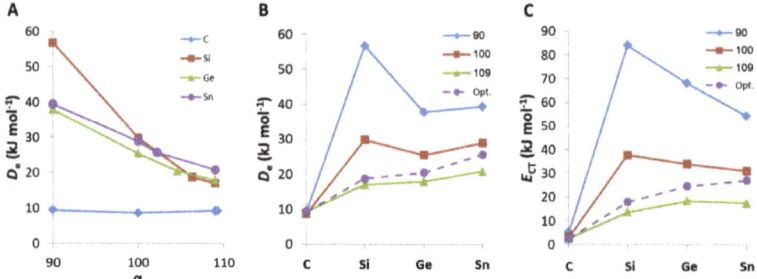

Figure 2. Complexation of 1_T with HCN: (**A**) dissociation energy as a function of the α angle; (**B**) dissociation energy as a function of the tetrel atom; (**C**) NBO $n \rightarrow \sigma^*$ charge transfer energy as a function of the tetrel atom.

We plotted the dissociation energy of the $1_T \cdots$ NCH complexes as a function of both the NBO $n \rightarrow \sigma^*$ charge transfer energy and the $V_{s,max}$ of the uncomplexed tetrel molecules (see Figure S4), which correlate, respectively, with the orbital and electrostatic interactions. The graphs show a linear relationship, indicating that the charge transfer component and the electrostatic interaction (connected with the virtual and occupied MOs of the hole bearer, respectively) go hand in hand in hole interactions [2,21]. However, there is one clear outlier in the $V_{s,max}$ graph (Figure S4B) corresponding to 1_{Si} at 90°, for which the electrostatic potential is an insufficient descriptor. This would suggest that, for complexes with stronger binding energies and smaller intermolecular distances, orbital interaction is more significant—a sign of an incipient covalent bond.

Our results clearly show that at small angles there is a departure from the expected binding order of Sn > Ge > Si > C, as the 1_{Si} shows strongest binding for angles of 100° and 90°. This comes as a result of the better interaction between silicon and nitrogen orbitals compared to the larger Ge and Sn. However, an alternative way to look at this is to check the energy needed to distort the molecules before and after complexation [11]. For α = 100° and 90°, the distortion energy of the monomer is almost always larger than for the complex (Table S1). The difference in distortion energies (ΔE_{dist}) is particularly large for Si (38 kJ mol^{-1} at α = 90°). As the difference in distortion energies equals the difference in dissociation energies ($\Delta E_{dist} = D_e$ constrained $- D_e$ optimized), this can also explain the dramatic increase of complexation energy for 1_{Si} at these angles.

So far, previous observations suggest that the geometry around the tetrel atom can have significant influence on the strength of the tetrel bonds, leading to unusually strong bonds at small angles, especially for T = Si. In order to check this trend in more realistic molecular models, we studied the 2_{Tn} molecules (Scheme 2, with a C$_3$ symmetry). Here, each α angle depends on the varying size of the rings determined by the number of carbon links (n). In 2_{Tn}, the σ-hole is at the extension of the T-C bond and not of the T-O bonds, but the magnitude of the central hole is enlarged by the oxygens (with methylenes instead of oxygens, the σ-hole was smaller and similar in magnitude to the holes on the methylene hydrogens). Many alkoxysilanes and alkoxygermanes are known, and there is also an experimental example similar to 2_{Si2} in which the Si forms strong interactions with electron donors [23].

As can be seen from Table 3, 2_{Cn} does not show binding when n = 2 or 3, and only a decrease in the C-C-O angle to 97° (n = 1) produces some very weak binding. For heavier tetrels, as the number of links (and, correspondingly, α) decreases, the dissociation energy shows significant increase. For n = 3 (close to the unconstrained angles), the dissociation energy has the expected Sn > Ge > Si order. Si and Ge show very similar binding energies for the different n's, with an unfavorable binding for n = 3, but strong binding for n = 1 (D_e > 100 kJ mol^{-1}). The difference in dissociation energies between 2_{Si1} and 2_{Ge1} is negligible, but α is smaller for Ge. This points at the same trend we saw for 1_T—the most dramatic increase in binding with smaller angles occurs with silicon (Figure S5). However, this does not actually make 2_{Si1} the stronger binder due to the smaller α in 2_{Ge1} (if both species had the same α angles, then 2_{Si1} would probably have the higher D_e).

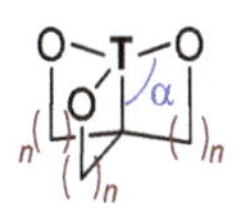

2_{Cn} T = C
2_{Sin} T = Si
2_{Gen} T = Ge
2_{Snn} T = Sn

Scheme 2. 2_{Tn} systems.

Table 3. Bond distances, angles, dissociation energies, and $V_{s,max}$ of 2_{Tn}, and their binding complexes with HCN (distances in Å, energies in kJ mol^{-1}).

T	n	d_{T-C} Monomer	d_{T-C} Complex	α Monomer	α Complex	$d_{T \cdots N}$	% Cov. Rad. [c]	D_e	$V_{s,max}$ [d]
C	1	1.481	1.486	97.6	97.1	2.953	194.3	0.9	42.1
	2	1.561	-	107.8	-	-	-	NB	−107.2
	3	1.542	-	111.4	-	-	-	NB	−153.5
Si	1	1.884	1.954	88.7	83.8	1.931	103.8	111.3	260.3
	2	1.830	1.879	101.0	95.6	2.084	112.0	42.3	147.6
	3 [a]	1.854	1.860	108.5	107.7	3.549	190.8	−3.2	−28.7
Ge	1	1.974	2.043	83.2	79.8	2.047	103.9	112.6	339.2
	2	1.899	1.935	98.9	95.4	2.220	112.7	47.9	186.6
	3	1.934	1.944	109.4	107.8	3.198	162.3	−1.2	9.8
Sn [b]	2	2.077	2.104	93.0	91.0	2.318	107.3	73.9	264.5
	3	2.118	2.147	107.9	104.1	2.470	114.4	30.3	140.2

[a] The complex with HCN interacting with the σ-hole is actually a maximum in energy as attraction between the oxygens and the positive charge of HCN are significant. [b] 2_{Sn1} is unstable. [c] Ratio of the tetrel bond and the sum of covalent radii of T and N. [d] Measured at the extension of the C-T bond.

3. Conclusions

Besides the classical enhancement of the tetrel bond brought by having heavier tetrels, geometry can be an important factor in bond strength. In addition to the release of strain energy [17], a smaller angle between the substituents not only favors the bond by geometrically exposing the tetrel atom, but there is also an electronic effect that boosts the σ-hole and aids the charge transfer. These effects cannot appear in regular halogen bonding due to a lack of side substituents, but can be a feature in pnictogen and chalcogen bonds, or in hypercoordinated halogens [21]. Our findings suggest that, in designing new tetrel bonded complexes, focusing on the geometry around the tetrel atom could allow the use of the more abundant silicone compared to the heavier elements without significantly sacrificing binding strength.

4. Methods

All density functional theory (DFT) computations were done at the MN15/Def2-TZVPD [18,24] level with Gaussian16 [20]. All energies reported do not include ZPE correction. All minima were confirmed with frequency computations. In order to check the validity of the used DFT method, the dissociation energies for 1_{Si} and 1_{Ge} were computed with CCSD(T)/CBS (complete basis set extrapolation from aug-cc-pvtz/aug-cc-pvqz, carried out in ORCA [25]) at the geometries found by MN15/Def2-TZVPD (see Table S2). The CCSD(T)/CBS results showed values very close to those of DFT, with a maximum difference of 8.2 kJ mol^{-1}, and displayed the same trends (stronger binding for 1_{Ge} at 109° and for 1_{Si} at 90°). NBO analyses were done with NBO3.1 [26] as appears in Gaussian16.

Supplementary Materials: The following are available online: Figures S1–S3: Chosen MOs and ESP maps for 1_C, 1_{Ge}, 1_{Sn} with three different F-T-H angles; Figure S4: graphs for the complexation of 1_T with HCN; Figure S5: graphs for the complexation of 2_{Tn} with HCN.

Author Contributions: E.S. and S.K. both contributed to the conceptualization, computation and writing of this project.

Funding: This research was funded by the Israeli Science Foundation (grant No. 631/15) and the German–Israeli Foundation (grant I-2481-302.5/2017). E.S. acknowledges funding from the Israeli Ministry of Science and Technology through the Shulamit Aloni Fellowship for Promoting Women in Science, and from the Kreitman Foundation Postdoctoral Fellowship.

Conflicts of Interest: The authors declare no conflict of interest. The funders had no role in the design of the study; in the collection, analyses, or interpretation of the data; in the writing of the manuscript; and in the decision to publish the results.

References

1. Clark, T.; Hennemann, M.; Murray, J.S.; Politzer, P. Halogen bonding: The σ-hole. *J. Mol. Model.* **2007**, *13*, 291–296. [CrossRef] [PubMed]
2. Angarov, V.; Kozuch, S. On the σ, π and δ hole interactions: A molecular orbital overview. *New J. Chem.* **2018**, *42*, 1413–1422. [CrossRef]
3. Politzer, P.; Lane, P.; Concha, M.C.; Ma, Y.; Murray, J.S. An overview of halogen bonding. *J. Mol. Model.* **2007**, *13*, 305–311. [CrossRef] [PubMed]
4. Politzer, P.; Murray, J.S.; Clark, T. Halogen bonding and other σ-hole interactions: A perspective. *Phys. Chem. Chem. Phys.* **2013**, *15*, 11178–11189. [CrossRef] [PubMed]
5. Auffinger, P.; Hays, F.A.; Westhof, E.; Ho, P.S. Halogen bonds in biological molecules. *Proc. Natl. Acad. Sci. USA* **2004**, *101*, 16789–16794. [CrossRef] [PubMed]
6. Wang, W.; Ji, B.; Zhang, Y. Chalcogen bond: A sister noncovalent bond to halogen bond. *J. Phys. Chem. A* **2009**, *113*, 8132–8135. [CrossRef] [PubMed]
7. Mahmudov, K.T.; Kopylovich, M.N.; Guedes da Silva, M.F.C.; Pombeiro, A.J.L. Chalcogen bonding in synthesis, catalysis and design of materials. *Dalton Trans.* **2017**, *46*, 10121–10138. [CrossRef] [PubMed]
8. Pascoe, D.J.; Ling, K.B.; Cockroft, S.L. The origin of chalcogen-bonding interactions. *J. Am. Chem. Soc.* **2017**, *139*, 15160–15167. [CrossRef] [PubMed]
9. Scheiner, S. A new noncovalent force: Comparison of P···N interaction with hydrogen and halogen bonds. *J. Chem. Phys.* **2011**, *134*, 094315. [CrossRef] [PubMed]
10. Del Bene, J.E.; Alkorta, I.; Elguero, J. The pnicogen bond in review: Structures, binding energies, bonding properties, and spin-spin coupling constants of complexes stabilized by pnicogen bonds. In *Noncovalent Forces*; Scheiner, S., Ed.; Springer International Publishing: New York, NY, USA, 2015.
11. Adhikari, U.; Scheiner, S. Comparison of P···D (D = P,N) with other noncovalent bonds in molecular aggregates. *J. Chem. Phys.* **2011**, *135*, 184306. [CrossRef] [PubMed]
12. Bauzá, A.; Mooibroek, T.J.; Frontera, A. Tetrel bonding interactions. *Chem. Rec.* **2016**, *16*, 473–487. [CrossRef] [PubMed]

13. Murray, J.S.; Lane, P.; Politzer, P. Expansion of the σ-hole concept. *J. Mol. Model.* **2009**, *15*, 723–729. [CrossRef] [PubMed]
14. Bauzá, A.; Frontera, A. Aerogen bonding interaction: A new supramolecular force? *Angew. Chem. Int. Ed.* **2015**, *54*, 7340–7343. [CrossRef]
15. Anisimov, V.M.; Zolotoi, A.B.; Antipin, M.Y.; Lukin, P.M.; Nasakin, O.E.; Struchkov, Y.T. Hexacyanocyclopropane. Synthesis and structure. *Mendeleev Commun.* **1992**, *2*, 24–25. [CrossRef]
16. Bauzá, A.; Mooibroek, T.J.; Frontera, A. Influence of ring size on the strength of carbon bonding complexes between anions and perfluorocycloalkanes. *Phys. Chem. Chem. Phys.* **2014**, *16*, 19192–19197. [CrossRef] [PubMed]
17. Zierkiewicz, W.; Michalczyk, M.; Scheiner, S. Implications of monomer deformation for tetrel and pnicogen bonds. *Phys. Chem. Chem. Phys.* **2018**, *20*, 8832–8841. [CrossRef] [PubMed]
18. Yu, H.S.; He, X.; Li, S.L.; Truhlar, D.G. MN15: A Kohn–Sham global-hybrid exchange–correlation density functional with broad accuracy for multi-reference and single-reference systems and noncovalent interactions. *Chem. Sci.* **2016**, *7*, 5032–5051. [CrossRef] [PubMed]
19. Kozuch, S.; Martin, J.M.L. Halogen bonds: Benchmarks and theoretical analysis. *J. Chem. Theory Comput.* **2013**, *9*, 1918–1931. [CrossRef] [PubMed]
20. Frisch, M.J.; Trucks, G.W.; Schlegel, H.B.; Scuseria, G.E.; Robb, M.A.; Cheeseman, J.R.; Scalmani, G.; Barone, V.; Petersson, G.A.; Nakatsuji, H.; et al. *Gaussian 16 (Revision A.03)*; Gaussian, Inc.: Wallingford, UK, 2016.
21. Kirshenboim, O.; Kozuch, S. How to twist, split and warp a σ-hole with hypervalent halogens. *J. Phys. Chem. A* **2016**, *120*, 9431–9445. [CrossRef] [PubMed]
22. Walsh, A.D. 469. The electronic orbitals, shapes, and spectra of polyatomic molecules. Part IV. Tetratomic hydride molecules, AH3. *J. Chem. Soc. (Resumed)* **1953**, 2296–2301. [CrossRef]
23. Kobayashi, J.; Kawaguchi, K.; Kawashima, T. Water-coordinated neutral silane complex: A frozen intermediate of hydrolysis of alkoxysilanes. *J. Am. Chem. Soc.* **2004**, *126*, 16318–16319. [CrossRef] [PubMed]
24. Rappoport, D.; Furche, F. Property-optimized Gaussian basis sets for molecular response calculations. *J. Chem. Phys.* **2010**, *133*, 134105. [CrossRef] [PubMed]
25. Neese, F. The ORCA program system. *Wiley Interdiscip. Rev. Comput. Mol. Sci.* **2012**, *2*, 73–78. [CrossRef]
26. Glendening, E.D.; Reed, A.E.; Carpenter, J.E.; Weinhold, F. *NBO*, Version 3.1; TCI: Madison, WI, USA, 1998.

Sample Availability: Samples of the compounds are not available from the authors.

© 2018 by the authors. Licensee MDPI, Basel, Switzerland. This article is an open access article distributed under the terms and conditions of the Creative Commons Attribution (CC BY) license (http://creativecommons.org/licenses/by/4.0/).

Article

Tetrel Bonding Interactions in Perchlorinated Cyclopenta- and Cyclohexatetrelanes: A Combined DFT and CSD Study

Antonio Bauzá * and Antonio Frontera *

Department of Chemistry, Universitat de les Illes Balears, Crta de Valldemossa km 7.5, 07122 Palma de Mallorca (Baleares), Spain
* Correspondence: antonio.bauza@uib.es (A.B.); toni.frontera@uib.es (A.F.); Fax.: +34-971-173426 (A.F.)

Academic Editor: Steve Scheiner
Received: 3 July 2018; Accepted: 15 July 2018; Published: 19 July 2018

Abstract: In this manuscript, we combined DFT calculations (PBE0-D3/def2-TZVP level of theory) and a Cambridge Structural Database (CSD) survey to evaluate the ability of perchlorinated cyclopenta- and cyclohexatetrelanes in establishing tetrel bonding interactions. For this purpose, we used Tr_5Cl_{10} and Tr_6Cl_{12} (Tr = Si and Ge) and HCN, HF, OH^- and Cl^- as electron donor entities. Furthermore, we performed an Atoms in Molecules (AIM) analysis to further describe and characterize the interactions studied herein. A survey of crystal structures in the CSD reveals that close contacts between Si and lone-pair-possessing atoms are quite common and oriented along the extension of the covalent bond formed by the silicon with the halogen atom.

Keywords: tetrel bonding interactions; CSD search; DFT calculations; AIM analysis

1. Introduction

The fascinating progress achieved in modern chemistry during the last decade has been supported by an in-depth understanding of noncovalent interactions, which are the pillars of supramolecular chemistry [1,2]. Therefore, their proper comprehension is key for chemists working in this area of research, since many chemical and biological processes are regulated by a precise combination of noncovalent forces, which often dictate the pathway of highly specific recognition mechanisms. For instance, the formation process of novel supramolecular assemblies is usually governed by an intricate combination of interactions between hosts and guests, presenting high affinities, even in highly competitive media [3–6]. For this reason, it is necessary to adequately describe and understand noncovalent interactions between molecules to achieve progress in this field of research. In this context, hydrogen bonding interactions are known as a classical supramolecular force present in many chemical and biological environments [7]. Similarly, halogen bonding interactions [8] have been found to share both strength and directionality features with hydrogen bonds. Consequently, the Cambridge Structural Database (CSD) was inspected in a series of studies in order to gain some insights into the impact of this interaction in solid state chemistry [9,10]. The scientific interest regarding this interaction has expanded exponentially due to its recognition as a prominent player in biological media and the design of new materials; leading to a wide amount of theoretical and experimental studies [11–14]. In addition, it has been widely recognized that σ-holes (and more recently π-holes [15]) can also appear in positive electrostatic potential regions involving covalently bond atoms of groups III to VIII [16–22]. Besides, several theoretical studies have focused on the study of their physical nature [23–27], concluding that it is basically sustained by the interaction of an electron-rich entity (electron donor) with a σ-hole (electron acceptor), in a close way to hydrogen and halogen-bonding interactions [7,12].

In this regard, the recognition of tetrel-bonding interactions [28] (i.e., an attractive noncovalent force between a σ-/π-hole present in a group IV atom and a Lewis base) has increased among the scientific community over the past years. In particular, both experimental [24,29] and theoretical [30,31] chemists have contributed to expanding current knowledge by evaluating their impact on solid state, ref. [32] biological systems [33] and chemical reactivity [34]. Of particular interest among the scientific community is perhalogenated cyclohexasilanes, due to its ability to act as a multiple tetrel bond donor using the twelve available σ-holes. In fact, crystallographic studies [35–37] have shown that perhalogenated cyclohexasilanes can strongly bind electron-rich moieties, such as halide anions or organocyanides (such as acetonitrile). Several theoretical studies [38–41] have explored this possibility by theoretically analyzing a series of anion/lone pair-Si inverted sandwiched complexes and confirming their ability to behave as efficient ditopic anion receptors.

In this context, we wondered about the possibility of (i) expanding current knowledge to cyclopentatetrelanes (Si and Ge) and (ii) exploring the effect of Ge in cyclohexa-derivatives. In order to achieve this goal, we used Tr_5Cl_{10} and Tr_6Cl_{12} molecules, where Tr = Si and Ge, and HCN, HF, OH$^-$ and Cl$^-$ moieties, as neutral and anionic electron donors, respectively (see Figure 1). In addition, we performed an Atoms in Molecules (AIM) analysis to further characterize the interactions described herein. Finally, we carried out a CSD survey in order to find experimental evidence of the importance of tetrel bonding interactions in the solid state involving perhalogenated cyclopenta- and cyclohexatetrelanes.

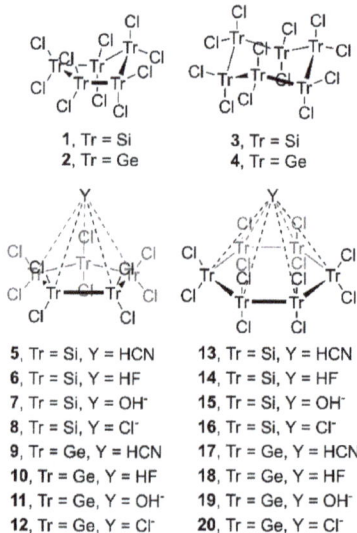

Figure 1. Compounds and complexes 1–20 studied in this work.

2. Results and Discussion

2.1. Preliminary MEP Analysis

We firstly computed the molecular electrostatic potential (MEP) mapped onto the van der Waals surface for compounds **1** and **3** in their respective envelope and chair conformations (Figure 2A). As noted, both molecules show areas of positive electrostatic potential on extension of the Si-Cl and Si-Si bonds, named σ-holes. Particularly, in case of **1**, the most positive MEP region is located at one face of the molecule (the face opposite to the axial Cl atom bonded to the *endo* carbon atom). This region of positive MEP is formed by the superposition of four Si-Cl σ-holes (see Figure 2A, left).

On the other hand, in case of compound **3**, six small σ-holes at the extension of the six Si–Cl axial bonds can be observed. The MEP value at these symmetrically distributed σ-holes is significantly smaller (12.6 kcal/mol) that that at the s-hole of the five membered ring, because only one Si–Cl bond is involved. In addition, both molecules present a low σ-hole accessibility, since they are closely surrounded by four (in **1**) and three (in **3**) negative belts belonging to the chlorine substituents, which disfavor the interaction with electron rich species, owing to both electrostatic and steric repulsive effects. However, when a planar disposition is imposed (see Figure 2B), the σ-holes gain in both magnitude size as well as become more accessible, thus enhancing the interaction with electron rich guests from both electrostatic and steric perspectives.

As noted in Figure 2B, in all cases, a positive electrostatic potential region can be located on the center of the ring, as a consequence of the combination of five (in **1** and **2**) and six (in **3** and **4**) Cl-Tr σ-holes (Tr = Si, Ge). The presence of this region ensures an attractive interaction with an electron-rich entity. In addition, the MEP values at the center of the ring are more positive for Ge derivatives (compounds **2** and **4**) than for their Si analogous (compounds **1** and **3**), thus expecting more favorable interaction energy values for complexes involving the former, as it is known for other σ-hole interactions [16]. It is also worthy to note than the MEP values are more positive for six membered rings (compounds **3** and **4**), due to the participation of an additional Cl-Si σ-hole, thus anticipating larger interaction energy values from an electrostatic point of view.

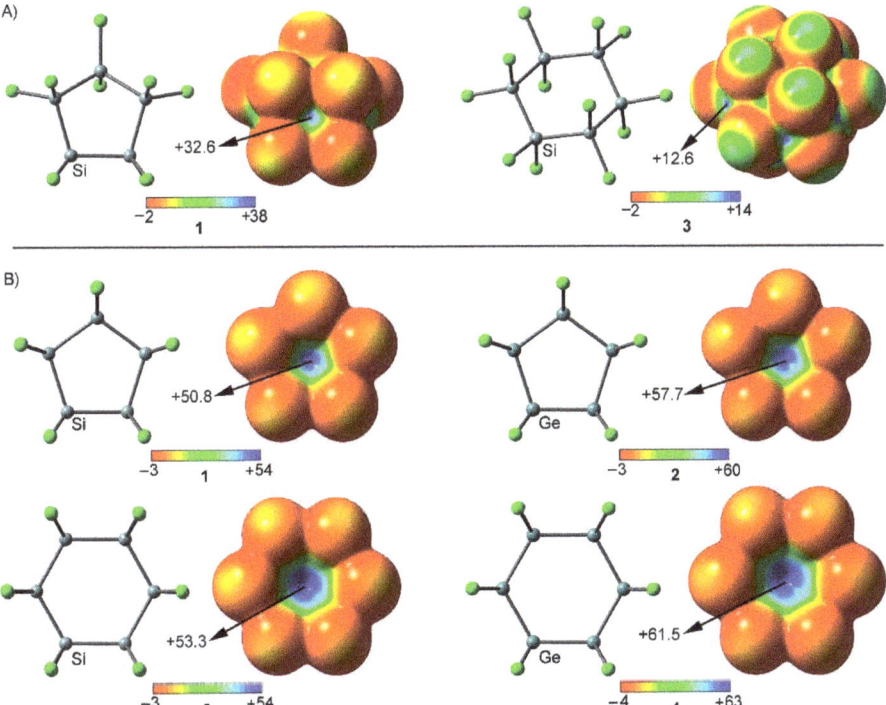

Figure 2. (**A**) MEP surfaces of compounds **1** and **3** in envelope and chair conformations, respectively. (**B**) MEP surfaces of compounds **1** to **4** in a planar disposition. Energies at selected points of the surface (0.001 a.u.) are given in kcal/mol.

2.2. Energetic and Geometric Results

Table 1 gathers the interaction energies and equilibrium distances of optimized complexes **5** to **20** (see Figure 3), computed at the PBE0-D3/def2-TZVP level of theory. From analysis of the results, several points arise. First, in all cases with the exception of complexes **13** and **14**, the interaction energy values are favorable and vary from moderately strong (in case of neutral donors) to strong (in case of charged donors), ranging between −107.5 and −3.7 kcal/mol. Second, complex **11** involving OH$^-$ obtained the most favorable interaction energy value of the study, while complex **14** involving HF obtained the poorest binding energy value of the study. Finally, complexes involving Ge (**9** to **12** and **17** to **20**) achieved larger interaction energy values than those involving Si (**5** to **8** and **13** to **16**), in agreement with the MEP analysis discussed above.

For complexes involving perchlorinated cyclopentatetrelanes (**5** to **12**), complexes **9** and **11** involving HCN and OH$^-$ obtained the largest interaction energy values of their respective series (−11.4 and −107.5 kcal/mol). On the other hand, complexes **6** and **10** achieved the poorest binding energy values of the series, owing to the low basicity of the HF molecule (−3.7 and −7.7 kcal/mol, respectively). Finally, complexes **8** and **12** involving Cl$^-$ obtained a lower interaction energy value than their OH$^-$ analogous (−61.4 and −68 kcal/mol, respectively), due to the higher basicity of the latter.

Among complexes **13** to **20** involving perchlorinated cyclohexatetrelanes, a similar behavior is observed in case of charged complexes **15**, **16**, **19** and **20**, where those involving OH$^-$ (**15** and **19**) obtained a larger interaction energy value (−95.3 and −106.3 kcal/mol) than those involving Cl$^-$ (complex **16**, −59.3 kcal/mol and complex **20**, −70 kcal/mol). On the other hand, in case of neutral complexes (**13**, **14**, **17** and **18**), those involving HCN as electron donor (**13** and **17**) obtained a more favorable binding energy value (−0.4 and −8.3 kcal/mol, respectively) than those involving HF (**14** and **18**), in agreement to that observed for complexes involving cyclopentatetrelanes. It is also worth noting that the magnitude of the interaction energy is almost negligible in case of complex **13** and repulsive in case of complex **14** (+3.2 kcal/mol). For these complexes, we computed the 1:2 assemblies (one cyclohexasilane and two lone pair donor molecules, denoted as **13A** for HCN and **14A** for HF), obtaining favorable interaction energy values of −12.6 kcal/mol for complex **13A** and −5.7 kcal/mol in case of complex **14A** (see Table 1). To further clarify the large difference between 1:1 and 1:2 complexes, we computed the energetic difference between the planar and the chair conformation in compound **3**, which is 12.3 kcal/mol. Therefore, the interaction of compound **3** with HCN (six concurrent tetrel bonds) is just able to compensate the difference between the chair and planar conformation, thus resulting in a negligible binding energy. The binding energy of the 1:2 complex (**13A**) is −12.2 kcal/mol, because six additional tetrel bonds are established (twelve in total, six in each side of the ring). In case of complex **14**, due to the lower basicity of the HF, the formation of the six Si⋯F tetrel bonds (1:1 complex) is not able to compensate the 12.3 kcal/mol required for changing the chair conformation into a planar one. Consequently, the 1:1 complex results to be 3.2 kcal/mol higher in energy than the separated monomers (only compensates around 9.1 kcal/mol). In good agreement, when the 1:2 complex (**14A**) is formed, the interaction energy becomes favorable (−5.7 kcal/mol) thus the additional six tetrel bonding interactions account for −8.9 kcal/mol. For the complexes of compound **4** (Ge instead of Si) all computed interaction energies are favorable because the difference in energy between the chair and planar conformation is only 6.8 kcal/mol.

Finally, it is also somewhat unexpected that complexes involving cyclopentatetrelanes (**5** to **12**) obtained more favorable binding energy values than their corresponding cyclohexatetrelane analogous (**13** to **20**), contrary to that obtained in the MEP analysis shown above for the planar molecules. Among other factors like proximity of the σ-holes and/or the negative belts of the chlorine atoms, the most likely explanation is that the difference in energy between the envelope and planar conformation is of 2.7 kcal/mol in **1** and 1.2 kcal/mol in **2**.

Although the interaction described above resembles lone pair–π (or anion–π) interactions [42], where a positive electrostatic potential region located at the center of the aromatic moiety interacts with an electron rich moiety, we (and other research groups [41]) consider this particular interaction as a σ-hole bonding. That is, the positive electrostatic potential area emerges over the center of the ring as the superposition of six/five σ-holes at the extension of the Si/Ge–Cl covalent bonds.

Table 1. Interaction energies without and with BSSE correction (ΔE and ΔE$_{BSSE}$, respectively, kcal/mol), equilibrium distances (R, Å) and value of the density at the bond CP ($10^2 \times \rho$, a.u.) for complexes **5–20** at the PBE0-D3/def2-TZVP level of theory.

Complex	ΔE [a]	ΔE$_{BSSE}$	R [b]	$10^2 \times \rho$
5	−7.8	−7.1	2.350	1.17
6	−5.0	−3.7	2.271	0.94
7	−116.3	−102.4	1.401	3.67
8	−67.3	−61.4	2.067	2.72
9	−12.3	−11.4	2.340	0.97
10	−9.1	−7.6	2.237	0.81
11	−121.1	−107.5	1.430	3.36
12	−73.8	−68.0	2.121	2.39
13	−1.2	−0.5	2.145	1.09
13A [a]	−14.1	−12.6	2.157	-
14	+1.9	+3.2	2.051	0.91
14A [a]	−8.3	−5.7	2.062	-
15	−109.1	−95.3	1.223	2.70
16	−65.4	−59.3	1.849	2.23
17	−9.2	−8.3	2.118	0.90
18	−5.6	−4.1	2.021	0.75
19	−120.0	−106.3	1.196	2.46
20	−76.2	−70.0	1.870	1.97

[a] **13A** and **14A** are 1:2 complexes where two HCN and HF molecules are located above and below the Si$_n$ molecular plane. [b] Distances measured from the electron rich atom to the ring centroid.

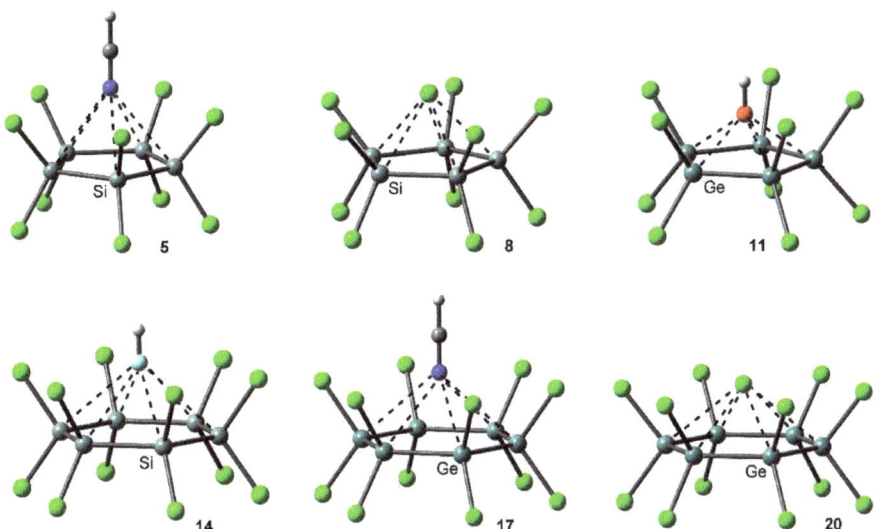

Figure 3. PBE0-D3/def2-TZVP optimized geometries of complexes **5, 8, 11, 14, 17** and **20**.

2.3. AIM and NCI Analyses

We have used the Bader's theory of "atoms in molecules" [43] (AIM) to characterize the noncovalent interactions shown in complexes **5–20**. A bond critical point (CP) and a bond path connecting two atoms is an unambiguous evidence of interaction. The AIM distribution of critical points and bond paths computed for some representative examples are shown in Figure 4. As noted, for complexes involving cyclopentatetrelanes (**7**, **9** and **12**) five symmetrically distributed bond CPs interconnect the electron donor and tetrel atoms, thus characterizing five simultaneous tetrel bonding interactions.

On the other hand, in case of complexes **15**, **17** and **18** involving cyclohexatetrelanes, six symmetrically distributed bond CPs interconnect the electron donor atom and the tetrel atoms, which characterize six simultaneous tetrel bonding interactions. Furthermore, in all cases, several ring CPs emerge (five for complexes **7**, **9** and **12** and six for complexes **15**, **17** and **18**), due to the formation of several supramolecular rings, which further describe the interaction. It is also worthy to mention that in case of complex **9**, a cage CP is observed, which also describes the interaction. Curiously, in case of the Si compounds, the bond path connects the Si–Si bond CP to the electron rich atom. Finally, the value of the laplacian in all cases is positive, as is common in closed shell calculations.

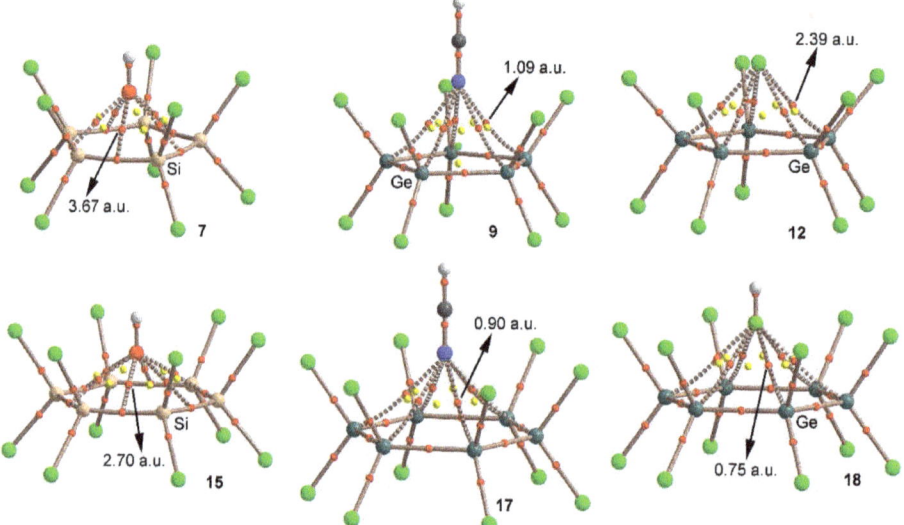

Figure 4. Distribution of critical points (red spheres) and bond paths for complexes **7**, **9**, **12**, **15**, **17** and **18** at the PBE0/def2-TZVP level of theory. Bond, ring and cage CPs are represented by red, yellow and green spheres, respectively. The values of the charge density (ρ) at the bond critical points that emerge upon complexation are indicated in a.u.

We have also carried out an Non Covalent Interactions (NCI) plot [44] of some representative examples to further analyze the tetrel bonding complexes discussed above (see Figure 5). The NCI visualization index enables the identification and characterization of non-covalent interactions in an efficient way. The NCI plot allows an assessment of host–guest assembly complementarity and the extent to which weak interactions stabilize a complex. The information provided is basically qualitative, that is, which molecular regions are involved in the interaction.

As noted, in case of complexes involving neutral donors (**5** and **10**), a green isosurface covers the entire cyclopentatetrelane moiety and characterizes the five simultaneous tetrel bonds. On the other hand, in case of anionic complexes **15** and **20**, the color of the isosurface is blue due to the existence of a strong electrostatic contribution to the interaction. Particularly, in case of complex **15**, the isosurface shows a more-pronounced blue region, in agreement with the strong interaction energy of complex **15** (see Table 1). In both complexes the isosurface is extended among all six σ-holes from the Si and Ge atoms. The nonexistence of surface at the center of the ring is in good agreement with the proposed σ-hole nature of the interaction instead of anion–π.

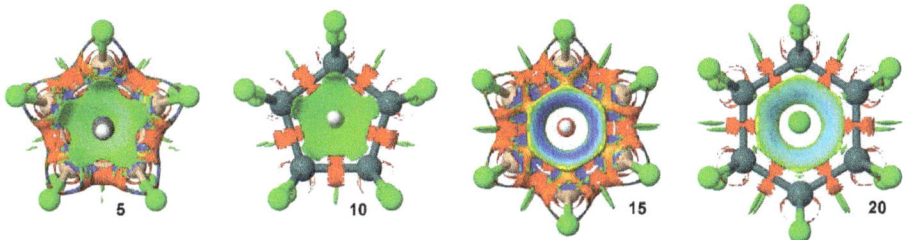

Figure 5. NCI plots of complexes **5**, **10**, **15** and **20**. The gradient cut-off is s = 0.35 au, and the color scale is −0.04 < ρ < 0.04 au.

2.4. CSD Search

We have explored CSD [45] to find evidence of the importance of tetrel bonding interactions involving perhalogenated cyclopenta- and cyclohexatetrelanes. During the search, we considered any sp³ tetrel atom apart from C (from Si to Pb) and any type of substituent in five and six membered rings. We found 11 structures containing cyclopentasilanes and 23 structures containing cyclohexasilanes (see Supplementary Materials—ESI for the complete list of structures). No structures were found involving other tetrel atoms. In addition, among these structures, 4 belonging to cyclopentasilane moieties and 19 structures involving cyclohexasilanes exhibit tetrel bonding interactions. Some examples are shown in Figure 6. In detail, in DUDSUS [46], the crystal packing consists of discrete [$Si_{32}Cl_{45}$]⁻ cluster units formed by a Si_{20} dodecahedral core bearing an endohedral Cl⁻ ion. Moreover, each Si_{20} core carries eight chloro and twelve trichlorosilyl substituents that fulfill all silicon cluster atom valencies. In addition, these electron withdrawing groups ensure the presence of Si σ-holes pointing inside of the cavity, leading to the establishment of multiple tetrel bonding interactions that act as a stabilizing source of the Cl⁻ ions. On the other hand, in ELAFIH [47] and AHASEJ [48] structures, the solid state architecture is governed by the formation of 2:1 dimers involving a perchlorinated cyclopenta- and cyclohexasilane rings and two acetonitrile and chloride molecules, respectively, in a 2:1 inverted sandwich fashion. It is also worthy to remark that experimentally only the 2:1 complexes are observed, in line with the energetic results obtained for complexes **13** and **14**. Finally, the distance values obtained are also within the range of the ones retrieved from the solid state, giving reliability to the theoretical results and highlighting the importance of these interactions in the solid state architecture of cyclopenta- and cyclohexasilanes.

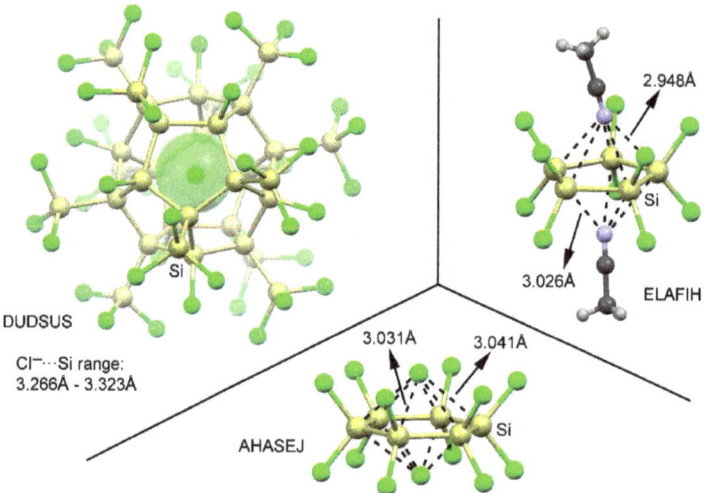

Figure 6. Partial views of the X-ray structure of some perchlorinated cyclopenta- and cyclohexasilanes establishing tetrel bonding interactions. The CSD codes are indicated.

3. Theoretical Methods

The geometries of the complexes studied herein have been fully optimized at the PBE0-D3/def2-TZVP level of theory. The calculations were performed by using the program TURBOMOLE version 7.0 (University of Karlsruhe, Karlsruhe, Germany) [49]. The calculation of the interaction energy values was performed using the formula $E_{int} = E_{AB} - E_A - E_B$, where E_{AB} corresponds to the energy of the optimized complex, while E_A and E_B refer to the energies of the optimized isolated monomers. The C_{5v} or C_{6v} symmetry point groups were used in the optimization of the anionic complexes and no symmetry constrains were imposed in the neutral complexes. It should be mentioned that the geometries of the neutral complexes (stationary points) converge to C_s and C_{2v} point groups for the five-membered and six-membered rings, respectively. For some anionic complexes, we carried out optimization without imposing symmetry constraints and the final optimized geometries (stationary points) adopted either C_{5v} or C_{6v} symmetry. The interaction energies were calculated with correction for the basis set superposition error (BSSE) by using the Boys–Bernardi counterpoise [50]. The Bader's "Atoms in molecules" theory was used to study the interactions discussed herein by means of the AIMAll calculation package (version 17.11.14, TK Gristmill Software, Overland Park, KS, USA) [51]. The calculations for the wavefunction analysis were performed by means of the Gaussian 09 calculation package (version B.01, Gaussian inc., Wallingford, CT, USA) [52]. The NCI plot is a visualization index based on electron density and its derivatives, and enables identification and visualization of non-covalent interactions. The isosurfaces correspond to both favorable and unfavorable interactions, as differentiated by the sign of the second density Hessian eigenvalue and defined by the isosurface color. The color scheme is a red-yellow-green-blue scale with red for ρ^+_{cut} (repulsive) and blue for ρ^-_{cut} (attractive). Yellow and green surfaces correspond to weak repulsive and weak attractive interactions, respectively. The models were designed based on previous theoretical studies that analyzed the ability of cyclohexasilanes to properly accommodate two anionic donor entities (mostly halogen ions, see [41]). We also included an OH$^-$ ion and two neutral electron donors (HCN and HF molecules) to obtain a more representative set of complexes. In addition, we chose cyclohexagermanane and the cyclopenta- derivatives of Si and Ge as tetrel bond donors to gain further insights into the behavior of this family of compounds.

4. Conclusions

In this manuscript, we analyzed the ability of perchlorinated cyclopenta- and cyclohexatetrelanes to establish tetrel bonding interactions with both neutral and charged electron donors. We used Tr_5Cl_{10} and Tr_6Cl_{12} (Tr = Si and Ge) and HCN, HF, OH$^-$ and Cl$^-$ moieties as electron donor molecules. In relation to this, complex **11** involving cyclopentagermanane and OH$^-$ as a Lewis base obtained the largest binding energy value of the study. On the other hand, complex **14** involving HF and cyclohexasililane achieved the poorest interaction energy value of the study. We also demonstrated that the energetic cost of forming the 2:1 complex compensates the energy penalty of passing from chair to planar conformation in cyclohexasilane complexes **13** and **14**. In addition, we have observed a reinforcement of the interaction strength ongoing from Si to Ge in both cyclopenta- and cyclohexatetrelane systems, as it is commonly observed for other σ-hole interactions. Furthermore, we performed Atoms in Molecules (AIM) analysis to further characterize the interactions described above. Finally, several experimental examples retrieved from the Cambridge Structural Database (CSD) were shown in order to provide reliability to the results and to highlight the importance of these interactions in the solid state of cyclopenta- and cyclohexatetrelanes.

Supplementary Materials: Supplementary Materials are available online, cartesian coordinates of the complexes and results from the CSD search.

Author Contributions: A.B. and A.F. conceived and designed the calculations; A.B. and A.F. analyzed the data; A.B. wrote the paper.

Funding: We thank the MINECO/AEI (projects CTQ2014-57393-C2-1-P and CTQ2017-85821-R FEDER funds) for financial support. We thank the CTI for computational facilities.

Conflicts of Interest: The authors declare no conflict of interest.

Abbreviations

The following abbreviations are used in this manuscript:

AIM	Atoms in molecules
MEP	Molecular electrostatic potential
BSSE	Basis Set Superposition Error
CSD	Cambridge Structural Database
CP	Critical point
NCIplot	Non Covalent Interactions plot
MINECO	Ministerio de Economía y Competitividad
AEI	Agencia Española de Investigación

References

1. Schneider, H.J. Binding mechanisms in supramolecular complexes. *Angew. Chem. Int. Ed.* **2009**, *48*, 3924–3977. [CrossRef] [PubMed]
2. Schneider, H.J.; Yatsimirski, A. *Principles and Methods in Supramolecular Chemistry*; John Wiley: Chichester, UK, 2000.
3. Lehn, J.M. *Supramolecular Chemistry Concepts and Perspectives*; Wiley–VCH: Weinheim, Germany, 1995.
4. Vögtle, F. *Supramolecular Chemistry: An Introduction*; Wiley: New York, NY, USA, 1993.
5. Beer, P.D.; Gale, P.A.; Smith, D.K. *Supramolecular Chemistry*; Oxford University Press: Oxford, UK, 1999.
6. Steed, J.W.; Atwood, J.L. *Supramolecular Chemistry*; Wiley: Chichester, UK, 2000.
7. Grabowski, S.J. What is the covalency of hydrogen bonding? *Chem. Rev.* **2011**, *111*, 2597–2625. [CrossRef] [PubMed]
8. Metrangolo, P.; Resnati, G. Halogen bonding: A paradigm in supramolecular chemistry. *Chem. Eur. J.* **2001**, *7*, 2511–2519. [CrossRef]
9. Murrayrust, P.; Motherwell, W.D.S. Computer retrieval and analysis of molecular geometry. 4. Intermolecular interactions. *J. Am. Chem. Soc.* **1979**, *101*, 4374–4376. [CrossRef]

10. Bauzá, A.; Quiñonero, D.; Deyà, P.M.; Frontera, A. Halogen bonding versus chalcogen and pnicogen bonding: A combined Cambridge structural database and theoretical study. *CrystEngComm* **2013**, *15*, 3137–3144. [CrossRef]
11. Brown, A.; Beer, P.D. Halogen bonding anion recognition. *Chem. Commun.* **2016**, *52*, 8645–8658. [CrossRef] [PubMed]
12. Politzer, P.; Murray, J.S. Halogen bonding: An interim discussion. *ChemPhysChem* **2013**, *14*, 278–294. [CrossRef] [PubMed]
13. Politzer, P.; Murray, J.S.; Clark, T. Halogen bonding and other σ-hole interactions: A perspective. *Phys. Chem. Chem. Phys.* **2013**, *15*, 11178–11189. [CrossRef] [PubMed]
14. Bauzá, A.; Frontera, A. Supramolecular nanotubes based on halogen bonding interactions: Cooperativity and interaction with small guests. *Phys. Chem. Chem. Phys.* **2017**, *19*, 12936–12941. [CrossRef] [PubMed]
15. Murray, J.S.; Lane, P.; Clark, T.; Riley, K.E.; Politzer, P. Σ-holes, π-holes and electrostatically-driven interactions. *J. Mol. Model.* **2012**, *18*, 541–548. [CrossRef] [PubMed]
16. Bauzá, A.; Mooibroek, T.J.; Frontera, A. The Bright Future of Unconventional σ/π-Hole Interactions. *ChemPhysChem* **2015**, *16*, 2496–2517. [CrossRef] [PubMed]
17. Bauzá, A.; Frontera, A. Aerogen Bonding Interaction: A New Supramolecular Force? *Angew. Chem. Int. Ed.* **2015**, *54*, 7340–7343. [CrossRef] [PubMed]
18. Politzer, P.; Murray, J.S.; Clark, T. Halogen bonding: An electrostatically-driven highly directional noncovalent interaction. *Phys. Chem. Chem. Phys.* **2010**, *12*, 7748–7757. [CrossRef] [PubMed]
19. Grabowski, S.J. Triel Bonds, π-Hole-π-Electrons Interactions in Complexes of Boron and Aluminium Trihalides and Trihydrides with Acetylene and Ethylene. *Molecules* **2015**, *20*, 11297–11316. [CrossRef] [PubMed]
20. Bauzá, A.; Mooibroek, T.J.; Frontera, A. Directionality of π-holes in nitro compounds. *Chem. Commun.* **2015**, *51*, 1491–1493. [CrossRef] [PubMed]
21. Grabowski, S.J. Triel bonds-complexes of boron and aluminum trihalides and trihydrides with benzene. *Struct. Chem.* **2017**, *28*, 1163–1171. [CrossRef]
22. Adriaenssens, L.; Gil-Ramírez, G.; Frontera, A.; Quiñonero, D.; Escudero-Adán, E.C.; Ballester, P. Thermodynamic characterization of halide–π interactions in solution using "two-wall" aryl extended calix[4]pyrroles as model system. *J. Am. Chem. Soc.* **2014**, *136*, 3208–3218. [CrossRef] [PubMed]
23. Scheiner, S. The Pnicogen Bond: Its Relation to Hydrogen, Halogen, and Other Noncovalent Bonds. *Acc. Chem. Res.* **2013**, *46*, 280–288. [CrossRef] [PubMed]
24. Marín-Luna, M.; Alkorta, I.; Elguero, J. Cooperativity in Tetrel Bonds. *J. Phys. Chem. A* **2016**, *120*, 648–656. [CrossRef] [PubMed]
25. Wang, W.; Ji, B.; Zhang, Y. Chalcogen Bond: A Sister Noncovalent Bond to Halogen Bond. *J. Phys. Chem. A* **2009**, *113*, 8132–8135. [CrossRef] [PubMed]
26. Bauzá, A.; Frontera, A. Theoretical study on the dual behavior of XeO_3 and XeF_4 toward aromatic rings: Lone pair–π versus aerogen–π interactions. *ChemPhysChem* **2015**, *16*, 3625–3630. [CrossRef] [PubMed]
27. Bauzá, A.; Frontera, A. π-Hole aerogen bonding interactions *Phys. Chem. Chem. Phys.* **2015**, *17*, 24748–24753. [CrossRef] [PubMed]
28. Bauzá, A.; Mooibroek, T.; Frontera, A. Tetrel-bonding interaction: Rediscovered supramolecular force? *Angew. Chem. Int. Ed.* **2013**, *52*, 12317–12321. [CrossRef] [PubMed]
29. Southern, S.A.; Bryce, D.L. NMR investigations of noncovalent carbon tetrel bonds. computational assessment and initial experimental observation. *J. Phys. Chem. A* **2015**, *119*, 11891–11899. [CrossRef] [PubMed]
30. Southern, S.A.; Errulat, D.; Frost, J.M.; Gabidullin, B.; Bryce, D.L. Prospects for (207)Pb solid-state NMR studies of lead tetrel bonds. *Faraday Discuss.* **2017**, *203*, 165–186. [CrossRef] [PubMed]
31. Scheiner, S. Systematic elucidation of factors that influence the strength of tetrel bonds. *J. Phys. Chem. A* **2017**, *121*, 5561–5568. [CrossRef] [PubMed]
32. Shukla, R.; Chopra, D. Characterization of the short O=C···O=C π-hole tetrel bond in the solid state. *CrystEngComm* **2018**, *20*, 3308–3312. [CrossRef]
33. Bauzá, A.; Frontera, A. RCH_3···O interactions in biological systems: Are they trifurcated H-bonds or noncovalent carbon bonds? *Crystals* **2016**, *8*, 26. [CrossRef]
34. Grabowski, S.J. Tetrel bond-σ-hole bond as a preliminary stage of the SN_2 reaction. *Phys. Chem. Chem. Phys.* **2014**, *16*, 1824–1834. [CrossRef] [PubMed]

35. Choi, S.B.; Kim, B.K.; Boudjouk, P.; Grier, D.G. Amine-promoted disproportionation and redistribution of trichlorosilane: formation of tetradecachlorocyclohexasilane dianion. *J. Am. Chem. Soc.* **2001**, *123*, 8117–8118. [CrossRef] [PubMed]
36. Dai, X.; Schulz, D.L.; Braun, C.W.; Ugrinov, A.; Boudjouk, P. "Inverse Sandwich" complexes of perhalogenated cyclohexasilane. *Organometallics* **2010**, *29*, 2203–2205. [CrossRef]
37. Teichmann, J.; Köstler, B.; Tillmann, J.; Moxter, M.; Kupec, R.; Bolte, M.; Lerner, H.-W.; Wagner, M. Halide-ion diadducts of perhalogenated cyclopenta- and cyclohexasilanes. *Z. Anorg. Allg. Chem.* **2018**. [CrossRef]
38. Robertazzi, A.; Platts, J.A.; Gamez, P. Anion···Si interactions in an inverse sandwich complex: A computational study. *ChemPhysChem* **2014**, *15*, 912–917. [CrossRef] [PubMed]
39. Pokhodnya, K.; Anderson, K.; Kilina, S.; Boudjouk, P. Toward the mechanism of perchlorinated cyclopentasilane (Si_5Cl_{10}) ring flattening in the $[Si_5Cl_{10}·2Cl]^{2-}$ dianion. *J. Phys. Chem. A* **2017**, *121*, 3494–3500. [CrossRef] [PubMed]
40. Pokhodnya, K.; Anderson, K.; Kilina, S.; Naveen, D.; Boudjouk, P. Mechanism of charged, neutral, mono-, and polyatomic donor ligand coordination to perchlorinated cyclohexasilane (Si_6Cl_{12}). *J. Phys. Chem. A* **2018**, *122*, 4067–4075. [CrossRef] [PubMed]
41. Vedha, S.A.; Solomon, R.V.; Venuvanalingam, P. On the nature of hypercoordination in dihalogenated perhalocyclohexasilanes. *J. Phys. Chem. A* **2013**, *117*, 3529–3538. [CrossRef] [PubMed]
42. Geboes, Y.; de Proft, F.; Herrebout, W.A. Lone pair···π interactions involving an aromatic π-system: Complexes of hexafluorobenzene with dimethyl ether and trimethylamine. *Chem. Phys. Lett.* **2016**, *647*, 26–30. [CrossRef]
43. Bader, R.F.W. A quantum theory of molecular structure and its applications. *Chem. Rev.* **1991**, *91*, 893–928. [CrossRef]
44. Contreras-García, J.; Johnson, E.R.; Keinan, S.; Chaudret, R.; Piquemal, J.-P.; Beratan, D.N.; Yang, W. NCIPLOT: A program for plotting noncovalent interaction regions. *J. Chem. Theory Comput.* **2011**, *7*, 625–632. [CrossRef] [PubMed]
45. Groom, C.R.; Bruno, I.J.; Lightfoot, M.P.; Ward, S.C. The Cambridge Structural Database. *Acta Cryst.* **2016**, *B72*, 171–179. [CrossRef] [PubMed]
46. Tillmann, J.; Wender, J.H.; Bahr, U.; Bolte, M.; Lerner, H.-W.; Holthausen, M.C.; Wagner, M. One-step synthesis of a [20] silafullerane with an endohedral chloride ion. *Angew. Chem. Int. Ed.* **2015**, *54*, 5429–5433. [CrossRef] [PubMed]
47. Dai, X.; Anderson, K.J.; Schulz, D.L.; Boudjouk, P. Coordination chemistry of Si_5Cl_{10} with organocyanides. *Dalton Trans.* **2010**, *39*, 11188–11192. [CrossRef] [PubMed]
48. Tillmann, J.; Lerner, H.-W.; Bats, J.W. CCDC 1414760: Experimental Crystal Structure Determination. *CSD Commun.* **2015**. [CrossRef]
49. Ahlrichs, R.; Bär, M.; Hacer, M.; Horn, H.; Kömel, C. Electronic structure calculations on workstation computers: The program system Turbomole. *Chem. Phys. Lett.* **1989**, *162*, 165–169. [CrossRef]
50. Boys, S.B.; Bernardi, F. The calculation of small molecular interactions by the differences of separate total energies. Some procedures with reduced errors. *Mol. Phys.* **1970**, *19*, 553–566. [CrossRef]
51. Keith, T.A. *AIMAll (Version 13.05.06)*; TK Gristmill Software: Overland Park, KS, USA, 2013.
52. Frisch, M.J.; Trucks, G.W.; Schlegel, H.B.; Scuseria, G.E.; Robb, M.A.; Cheeseman, J.R.; Scalmani, G.; Barone, V.; Petersson, G.A.; Nakatsuji, H.; et al. *Gaussian 09, Revision B.01*; Gaussian, Inc.: Wallingford, CT, USA, 2009.

Sample Availability: Samples of the compounds are not available from the authors.

© 2018 by the authors. Licensee MDPI, Basel, Switzerland. This article is an open access article distributed under the terms and conditions of the Creative Commons Attribution (CC BY) license (http://creativecommons.org/licenses/by/4.0/).

Article

Intermolecular Non-Covalent Carbon-Bonding Interactions with Methyl Groups: A CSD, PDB and DFT Study

Tiddo J. Mooibroek

van 't Hoff Institute for Molecular Sciences, Universiteit van Amsterdam, Science Park 904, 1098 XH, Amsterdam, The Netherlands; t.j.mooibroek@uva.nl; Tel.: +31(0)205-25-72-08

Received: 28 July 2019; Accepted: 12 September 2019; Published: 16 September 2019

Abstract: A systematic evaluation of the CSD and the PDB in conjunction with DFT calculations reveal that non-covalent Carbon-bonding interactions with X–CH$_3$ can be weakly directional in the solid state ($P \leq 1.5$) when X = N or O. This is comparable to very weak CH hydrogen bonding interactions and is in line with the weak interaction energies calculated (≤ -1.5 kcal·mol^{-1}) of typical charge neutral adducts such as [Me$_3$N-CH$_3$···OH$_2$] (**2a**). The interaction energy is enhanced to ≤ -5 kcal·mol^{-1} when X is more electron withdrawing such as in [O$_2$N-CH$_3$··O=C^{dme}] (**20b**) and to ≤ 18 kcal·mol^{-1} in cationic species like [Me$_3$O$^+$-CH$_3$···OH$_2$]$^+$ (**8a**).

Keywords: intermolecular interactions; non-covalent interactions; carbon-bonding interactions; crystal structure database analysis; density functional theory

1. Introduction

The manner in which molecules interact with one another is largely determined by non-covalent interactions.c [1] So-called 'σ-Hole interactions'c [2–5] like hydrogen bonding are prominent identifiable interactions that bear biological significance [6]. Such σ-hole interactions have also been identified with other non-metals [7–11] like halogen atoms to generate halogen bonding interaction [12,13]. The impact of halogen bonding interactions on molecular biology has come into focus since about 2004 [14]. Indeed, evaluations of the protein data bank (PDB) [15] revealed that halogen bonding is structurally very similar to hydrogen bonding [12,14,16–18] and can be functionally relevant [19–22]. Relatively weak π-hole interactions [4,23–31] involving organic carbonyls, [26,32–36] π-acidic aromatics, [37,38] metal carbonyls [33,34,36,39] and nitro-compounds [40–45] are increasingly acknowledged as relevant drivers of molecular aggregation such as in ligand-protein complexes.

The impact of a novel type of weak interaction on molecular recognition phenomena naturally leads one to speculate that other non-canonical interactions may play a similar role. One interesting candidate are σ-hole interactions involving sp^3-hybridized C-atoms. Such interactions have been studied since about 2013 [7,46] and are particularly interesting because sp^3-C is abundant in living systems. More specifically, the methyl group (X–CH$_3$, where X = any atom or group) is frequently encountered in natural and synthetic compounds and 'non-covalent Carbon bonding' involving methyl groups has thus been studied by various researchers [47–59]. Most of these contributions are computational inquiries, while a small amount of these articles also deals with an analysis of non-covalent Carbon bonding interactions in protein structures present in the Protein Data Bank (PDB) [47,50,56]. Interestingly, none of the studies so far have systematically evaluated the crystal structure data present in the Cambridge Structure Database (CSD) [60,61]. What is more, evaluations of the PDB were largely anecdotal or only considered structures that comply to the (rather strict) geometric criteria of a Carbon bonding geometry. Some also included *intra*molecular contact distances (which are notoriously difficult to evaluate).

In this contribution a combined CSD and PDB evaluation is presented aimed at elucidating whether electron rich entities have a preferential orientation around a methyl group within a rather large envelope, i.e., whether intermolecular non-covalent Carbon bonding interactions with methyl groups are directional. For evaluative purposes, several Density Functional Theory (DFT) computations were conducted as well. This combined database/DFT study reaffirms that non-covalent Carbon-bonding interactions with X-CH$_3$ can be significant, although the interaction is hardly directional, in particular when the methyl group is poorly polarized such as most C–CH$_3$ structures.

2. Materials and Methods

2.1. General Information on Database Analyses

The CSD [60,61] version 5.40 including two updates (until May 2019) was inspected using ConQuest [62] version 2.0.2 (build 246353, 2019). X-ray powder structures were omitted from the searches, which were further limited to structures containing 3D coordinates and those with an R-factor ≤ 0.1. The PDB was queried using Relibase [63] 3.2.3 and restricted to protein and DNA crystal structures where the packing environment was also searched. No other restrictions were imposed on the PDB search. Datasets were obtained using the general query shown in Figure 1a. The methyl groups were split in those connected to a C, N, O, P, or S atom (X in the figure, in the PDB search specified as part of a ligand). The interacting 'electron rich' partners (ElR in the figure) considered were a water, amide or carboxy-O atom, a sulphur atom or the centroid of an aryl ring (in the PDB search always specified as part of the protein). The geometric constraints imposed on the searches were that the *inter*molecular distance d between the methyl C-atom and ElR was ≤5 Å and that the X–CH$_3$···ElR angle (α) was 90°–180°. All the data were thus confined within a hemisphere with a basal radius of 5 Å, centered on the methyl C-atoms as is shown in Figure 1b.

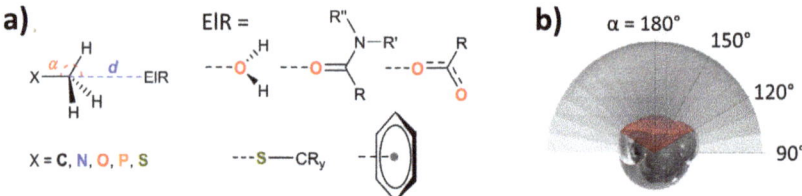

Figure 1. Representation of the method used to retrieve and analyse data from the CSD and the PDB. (**a**) general query to obtain data with $d \leq 5$Å, α = 90°–180°, X = C, N, O, P or S and ElR (electron rich entity) is as indicated. (**b**) Illustration of the method used to assess directionality (see text for details).

2.2. Methodology to Generate P(α) Plots

The datasets obtained as described above (2.1) were analysed to assess whether the distribution of ElR within the methyl-centered hemisphere reflects any directionality. This method has been successfully applied to assess the directional behaviors (in the solid state) of various other weak non-covalent interactions such as anion/lone-pair-π, [29,64] CH-π, [11,65] halogen-π [66,67] and nitro π-hole interactions [42,68]. The method works by first computing the freely accessible volumes at each α-value (α^{free}) by subtracting the volume of a model methyl group from a spherical cone with 5 Å height and a cone angle of 180-α. This can be achieved by using the 3D-drawing program Autodesk® Inventor® Pro [29]. This is illustrated in Figure 1b, where the spherical cones are shown at 10° intervals. The model methyl group was generated by using standard aliphatic C–H bond distances (1.06 Å) [69] and the van der Waals radius of C (1.70 Å) and H (1.09 Å) [70]. The interfering volume between each spherical cone and the model methyl group can be obtained using the 'inspect interference' option in Autodesk® Inventor® Pro; the red part in Figure 1b is the interfering volume involving a spherical cone with a cone angle of 60° (i.e., at α = 120°). The volume differences between such 'free' volumes with increasing values of α thus give the absolute volume distribution of freely accessible volume around a methyl model within the hemisphere, as a function of α: $\Delta\alpha^{free}(\alpha)$. Dividing each volume ($\Delta\alpha^{free}$) in this distribution by the total freely accessible volume (i.e., the volume of a hemisphere minus the interfering volume of the model methyl group in that hemisphere) thus gives the relative volume distribution as a function of α: $\Delta^{rel}\alpha^{free}(\alpha)$. This distribution is the random (or volume) distribution. The data retrieved form the CSD and the PDB can be binned as a function of α. Relating this binned data to all the data in a dataset thus gives the observed relative distribution as a function of α: $\Delta^{rel}\alpha^{data}(\alpha)$. The quotient of this relative data distribution over the random distribution is a measure for the actual probability (P) of finding data at a certain value of α. That is, $P(\alpha)$ is unity for a random distribution of data, while P-values larger than unity reveal a relative concentration of data, which is indicative of attractive interactions.

2.3. Methodology to Generate N(d') Plots

A second analysis involved plotting the hit fraction (N) for a subset with α = 160°–180° as a function of the van der Waals corrected H$_3$C···ElR distance d' (i.e., d − vdW(C) − vdW(ElR)): [70] $N(d')$. Such distributions show how much of the data is involved in van der Waals overlap with the methyl C-atom along the vector of the X–CH$_3$ bond and how such data is distributed. For attractive interacting pairs this distribution is expected to exhibit a peak-like feature, or an S-like curvature when the cumulative hit fraction is used.

2.4. Computational Methods

DFT geometry optimization calculations were performed with Spartan 2016 at the B3LYP [71,72]-D3 [73]/def2-TZVP [74,75] level of theory, which is known to give accurate results at reasonable computational cost and a very low basis set superposition error (BSSE) [74,75]. The typical starting geometry for possible Carbon bonding adducts was set to d' = −0.1 Å and α = 180°, and in the case of dimethylacetamide the C···O=C angle was also set to 180°. The geometry optimizations were performed without any constraints. For other geometries (e.g., a H-bonded geometry), the molecular fragments were manually oriented in a suitable constellation before starting an unconstrained geometry optimization. The Amsterdam Density Functional (ADF) [76] modelling suite at the B3LYP [71,72]-D3 [73]/TZ2P [74,75] level of theory (no frozen cores) was used for energy decomposition and 'atoms in molecules' [77] analyses. Details of the Morokuma-Ziegler inspired energy decomposition scheme used in the ADF-suite have been reported elsewhere [76,78] and the scheme has proven useful to evaluate hydrogen bonding interactions [79].

3. Results and Discussion

3.1. P(α) Plots

A numerical overview of the amount of crystallographic information files (CIFs) and protein data bank files (PDBs) for each search query is given in Table S1, together with the amount of hits found in each dataset (a .cif or .pdb file can contain multiple hits). Shown in Figure 2 are the $P(\alpha)$ plots for the CSD (left) and PDB (right) data plotted at 5° intervals involving X = C, N, O and ElR = water-O (top) or amide-O (bottom). These datasets were chosen because they all contained a large number of hits (>7,500) and thus allow for the most reliable comparison. A complete set of $P(\alpha)$ plots is provided in Figure S1.

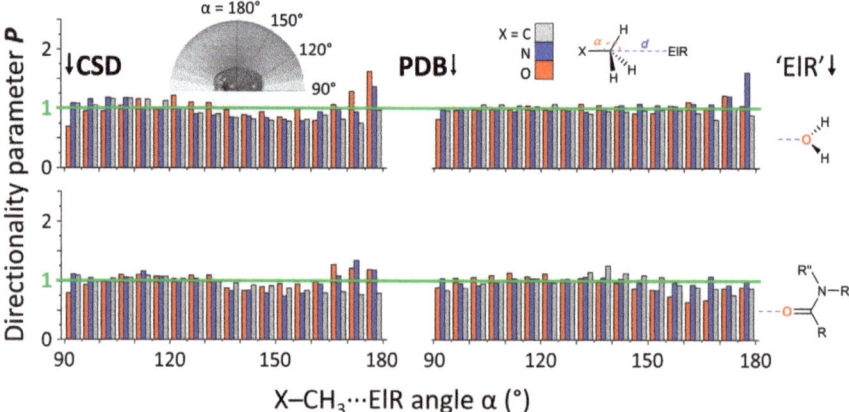

Figure 2. $P(\alpha)$ directionality plots for the data retrieved from the CSD (left) and the PDB (right) using the general query shown in the top-right inset figure for X–CH$_3$···ElR pairs. X can be C, N or O and 'ElR' can be a water or an amide O-atom. The insert figure in the top left is intended as a guide to the eye to interpret the spatial location of data with a certain value of α. Due to the amount of data per dataset (see Table S1 for numerical overview), the plots are given at a 5° resolution for α. A full set of $P(\alpha)$ plots (i.e., for all the X vs ElR pairs in Figure 1) is given in Figure S1. The P value of 1 is highlighted in green and indicates a random distribution of data. N (CSD/PDB) = 46,000/29,508 (C, water); 18,170/17,101 (N, water); 7190/11,392 (O, water); 53,473/22,538 (C, amide); 7663/10,855 (N, amide); 9,158/11,064 (O, amide).

The data plotted in Figure 1 largely trace the line at $P = 1$ (highlighted in green), which is indicative of a random distribution of data. For X = O and N, these values are somewhat above unity around α = 160°–180° for water O-atoms in both databases and for amide O-atoms in the CSD. The maximum P-values are very small at about 1.5, which indicates a very small amount of directionality. Indeed, maximum P-values for several weak inter-molecular interactions are: ~2.5 for CH-π; [11,65] ~3 for π interactions with nitro compounds; [42,68] ~2.5–5 for anion-π and lone-pair-π; [29,64] ~2.5–10 for halogen-π [67] and also about 2.5–10 for halogen bonding with aryl-halogens. [66] Interestingly, the P-values did not peak near α = 90°–120°, an angle congruent with hydrogen bonding. These data thus suggest that the Carbon binding geometry is more directional than a hydrogen bonding geometry, although this directionality is very weak. In all cases for X = C, the P-values around α = 160°–180° are below unity, suggesting that the Carbon bonding geometry is least favored in these instances. The data for ElR = RCO$_2$ and R$_y$CS are very similarly distributed and the data for aryl rings is skewed towards α = 160°–180° only for the CSD data (for X = C, N, O, P and S, see Figure S1). For all other cases where X = P, very few hits were obtained (most numerous was RCO$_2$ in the CSD with N = 1,454). While some datasets with X = S were of a reasonable size, too many were well below N = 7,500 and these data will thus not be discussed in the main text (there is a small discussion in the caption of Figure S1).

3.2. N(d') plots

The data characterized by α = 160°–180° was inspected further by means of $N(d')$ plots as described in the methods and materials section. A numerical overview of the amount of data in each dataset as well as the (relative) amount of van der Waals overlap found in each dataset is given in Table S2. Shown in Figure 3 are plots of the hit fractions (in %) as a function of d' (the van der Waals corrected d) for X = C, N, O and ElR = water or amide. The same data plotted as cumulative hit fractions is shown in Figure S2 and the $N(d')$ plots for all datasets containing >500 hits are shown in Figure S3.

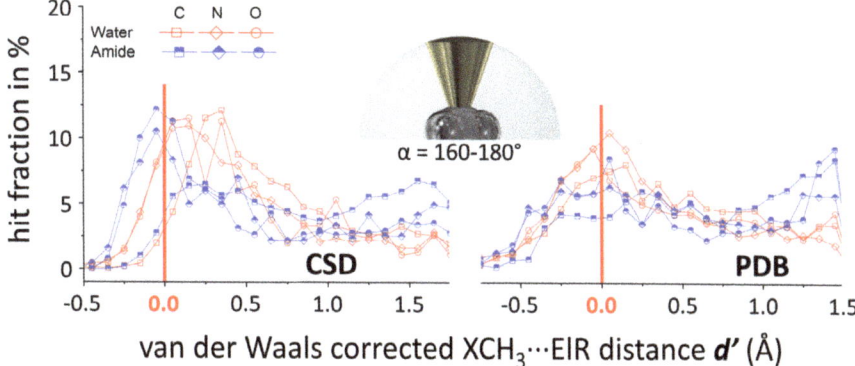

Figure 3. Hit fraction (in %) as a function of the van der Waals corrected $XCH_3\cdots ElR$ distance d' (in Å) for several datasets from Table S2 ($\alpha \geq 160°$, as illustrated by the inset figure). The interacting pairs involve water-O (red, empty) or amide-O (blue, half-filled) with X = C (squares), N (diamonds), or O (circles). See Figure S2 for the same data plotted as a cumulative hit fraction. N (CSD/PDB) = 2,376/1,622 (C, water); 1,089/1,186 (N, water); 452/757 (O, water); 2,599/1,179 (C, amide); 504/650 (N, amide); 640/483 (O, amide).

In the CSD (left), The data involving $N/O-CH_3\cdots O^{amides}$ are very similarly distributed and grouped near $d' \approx 0$ Å with about 30% of all the data involved in van der Waals overlap. Likewise, data involving $N/O-CH_3\cdots O^{waters}$ are also very similarly distributed but group near the larger $d' \approx 0.25$ Å with 15% van der Waals overlap. The $C-CH_3\cdots O^{amides/waters}$ data hardly displays van der Waals overlap (3–4%) and is broadly grouped around $d' = 0.4$ Å for waters and not grouped at all for amides. Similar trends are present in the PDB (right), albeit the features are much less pronounced and the $N/O-CH_3\cdots O$ datasets with amides and waters are very similar.

The data presented in Figure 3 thus imply a somewhat directional nature of $X-CH_3\cdots O^{amides/waters}$ interactions for X = N and O, but not at all for X = C. These findings are in line with the lack of directionality observed in the $P(\alpha)$ plots for X = C and the somewhat directional behavior for X = N or O (see Figure 2). A likely explanation for this is the larger (Pauling) electronegativity of N (3.04) and O (3.44) compared to C (2.55), resulting in a larger degree of polarization of the $X-CH_3$ bond for X = N or O. Another conceivable manner to make a methyl group pore electropositive is to bind it to a cationic fragment such as in protonated or quaternary $R_3N^+-CH_3$ fragments. Thus, an additional dataset was retrieved from the CSD involving $R_3N^+-CH_3\cdots ElR$ pairs that fulfilled the $d \leq 5$ Å and $\alpha = 160°–180°$ criteria (see bottom entries in Tables S1 and S2). Shown in Figure 4 are the $N(d')$ plots of the most numerous datasets involving $R_3N^+-CH_3$ (hexagonals, N^+) together with similar datasets involving all possible $N-CH_3$ fragments (diamonds, N). In all three cases (ElR = water, carboxy or aryl), the distribution is shifted to lower d' distances for cationic $R_3N^+-CH_3$, which means that relatively more van der Waals overlap is present in these datasets. As can be seen especially from the cumulative $N(d')$ plots (left), the grouping is tightest with carboxy O-atoms (green), followed by water O-atoms (red) and aryl rings centroids (grey) are not grouped at all (nearly linear).

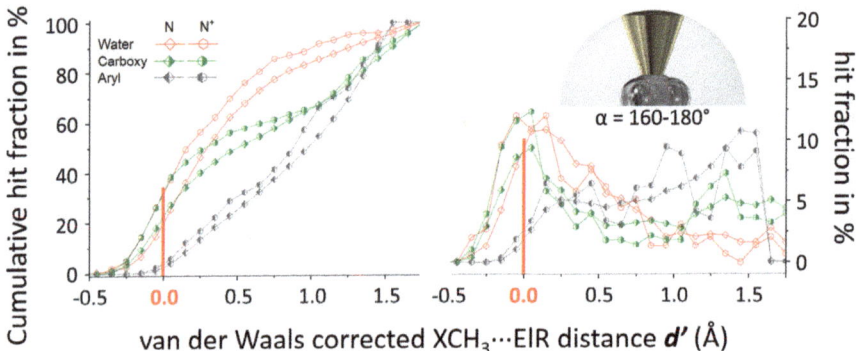

Figure 4. Cumulative (**left**) and regular (**right**) hit fraction (in %) as a function of the van der Waals corrected $XCH_3 \cdots ElR$ distance d' (in Å) for the datasets from Table S2 ($\alpha = 160°–180°$ as illustrated in the inset figure). The interacting pairs involve water-O (red, empty), carboxy-O (green, right-filled) or the centroid of an aryl ring (grey, left-filled). X can be any N (diamond) or a cationic (tetravalent) N^+ (hexagonal). N (N/N^+) = 1,809/290 (water, red); 2,062/274 (carboxy, green); 10,585/527 (aryl, grey).

3.3. Computations

In order to gain insight into the nature and energetics of possible non-covalent Carbon bonding interactions involving various methyl groups and water or amide O-atoms, DFT calculations were performed of $X–CH_3$ adducts with water and with dimethylacetamide (dma, see methods section for details). An overview of these adducts is given in Table 1, together with α of the optimized structures, the total interaction energy of the adducts in kcal·mol^{-1} and the percentages of electrostatic (E), orbital (O), and dispersion (D) interactions that contribute to this total energy [79]. Perspective views and atoms-in-molecules analyses of all converged structures are shown in Figure S4 and several representative examples are shown in Figure 5.

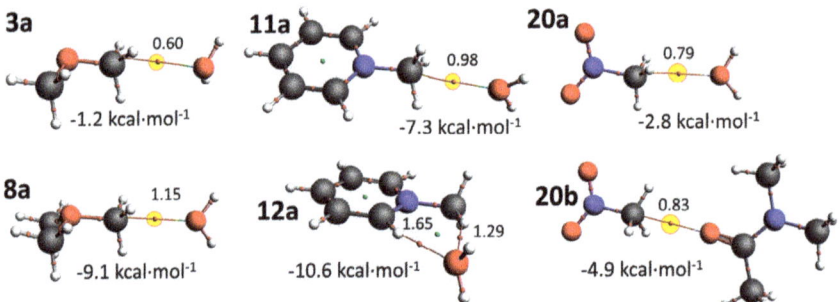

Figure 5. Ball and stick representations of molecular adducts selected from Table 1 that were optimized by DFT (B3LYP-D3/def2-TZVP). The thin lines are bond paths (bp's) and the small red spheres are bond critical points (bcp's) obtained from an 'atoms-in-molecules' analysis. The bond density (ρ) is in arbitrary units $\cdot 10^2$ and bcp's indicative of non-covalent Carbon bonding have been highlighted in yellow.

Table 1. Numerical overview of adducts computed with DFT between an indicated X-CH$_3$ methyls and water (adducts 'a') or dimethylacetamide (dma, adducts 'b'). Using ADF at the B3LYP-D3/def2-TZVP level of theory, interaction energies (in kcal·mol^{-1}) were computed and an energy decomposition analyses is shown as a percentage of the total amount of interaction energies split up as electrostatic (E), orbital (O) and dispersion (D) interactions. Entries in grey are not consistent with a Carbon bonding geometry. See Figure S4 for perspective views and atoms in molecules analyses.

Interacting X-CH$_3$:	adduct	Water–O			adduct	Dimethylacetamide (dma) – C=O		
		α (°)	ΔE	E/O/D in %		α (°)	dE	E/O/D in %
Me–CH$_3$	1a	170[a]	−1.2	32/22/46%	1b	122[f]	−2.8	26/16/58%
Me$_2$N–CH$_3$	2a	174[b]	−0.8	36/18/45%	2b	173[b]	−1.6	33/22/44%
MeO–CH$_3$	3a	175[b]	−1.2	48/17/35%	3b	164[f]	−3.2	27/16/56%
Me$_2$P–CH$_3$	4a	167[a]	−1.2	28/32/41%	4b	91[d]	−4.2	45/21/33%
MeS–CH$_3$	5a	177[b]	−1.0	44/17/38%	5b	100[d]	−3.7	46/21/32%
Me$_2$C$^+$–CH$_3$	6a	75[c]	−12.6	66/22/12%	6b	109[c]	−82.4	43/55/2%
Me$_3$N$^+$–CH$_3$	7a	173[b]	−7.8	73/15/12%	7b	170[b]	−14.7	70/19/10%
Me$_2$O$^+$–CH$_3$	8a	176[b]	−9.1	72/18/10%	8b	175[b]	−17.6	69/22/8%
Me$_3$P$^+$–CH$_3$	9a	55[d]	−8.2	69/14/18%	9b	71[d]	−21.7	64/23/13%
Me$_2$S$^+$–CH$_3$	10a	168[b]	−7.6	73/15/12%	10a	170[b]	−15.0	70/20/10%
^{Py}N$^+$–CH$_3$	11a	170[b]	−7.3	73/15/12%	11b	175[b]	−14.2	71/19/10%
	12a	93[d]	−10.6	70/22/9%	12b	91[d]	−20.3	65/27/8%
	13a	68[e]	−9.0	70/14/16%	13b	74[e]	−18.2	64/21/15%
I–CH$_3$ [g]	14a	175[b]	−1.6	53/18/28%	14b	175[b]	−3.0	52/20/28%
	15a	84[d]	−3.8	55/23/21%				
F$_3$C–CH$_3$ [g]	16a	179[b]	−1.8	59/13/27%	16b	175[b]	−3.4	59/15/26%
	17a	76[d]	−3.8	60/15/24%				
N≡C–CH$_3$ [g]	18a	158[b]	−2.3	65/13/23%	18b	172[b]	−4.3	62/15/23%
	19a	68[d]	−5.1	63/19/18%				
O$_2$N–CH$_3$ [g]	20a	165[b]	−2.8	66/13/21%	20b	179[b]	−4.9	64/15/21%
	21a	86[d]	−6.4	63/25/12%				

[a] One of the water H-atoms and one of the RCH$_3$ C-atoms are closest to each other; [b] Carbon bonding interaction geometry; [c] Interaction with the cationic C; [d] Hydrogen bonding interaction(s); [e] Interaction with cationic N; [f] CH-π interaction; [g] The interaction energy with benzene was also computed, starting from a geometry with X–CH$_3$···benzene centroid = 180°. All adducts converged at a geometry where this angle was about 90° degrees. Interaction energies are about 4–5 kcal·mol^{-1} and dominated by electrostatics (35–40%) and dispersion (40–50%, see Figure S5 for details).

For comparison purposes, adducts with ethane were computed as shown in entries 1 of Table 1. Both structures converged to a hydrogen bonding interaction. ΔE = −1.2 kcal·mol^{-1} in **1a** and a methyl acts as electron donating site; i.e., an O–H···C hydrogen bonding interaction. This can be understood due to the polarization in ethane, where both C's are most electronegative and are polarized by the H-atoms. Adduct **1b** is about twice as stable with ΔE = −2.8 kcal·mol^{-1} and also features a hydrogen bonding interaction, but now between a methyl CH and a π-bond in dma. Both **1a** and **1b** are stabilized mainly by dispersion (46–58%), then electrostatics (32–26%) and least by orbital interactions (22–16%). The neutral water adducts where X = permethylated N, O, P, or S are energetically nearly identical to the ethane adduct (**2a**–**5a**). Like with ethane, **4a** converged in a O–H···C hydrogen bonding interaction, which can be rationalized by the lower (Pauling) electronegativity of P (2.19) compared to C (2.55). **2a**, **3a** and **5a** converged at a geometry consistent with a Carbon bonding interaction. This is illustrated for structure **3a** in Figure 5, where a single bond critical point (bcp) is located between C and O with a bond density of 0.60 · 10^2 a.u.. This can be rationalized by the higher (Pauling) electronegativity of N (3.05), O (3.44) and S (2.58) compared to C (2.55). Electrostatics or dispersion are the main energetic stabilizing factor in adducts **2a**–**5a** which is typical for weak and non-directional interactions like in adduct

1a. A similar series with dma was computed as adducts **2b–5b**. Only **2b** converged in a geometry consistent with a Carbon bonding interaction (with dispersion as main driver) while the others are C–H···O hydrogen bonding interactions. Carbon bonding interactions with regular permethylated main group elements are thus comparable to very weak C–H hydrogen bonding interactions and less than about −1.5 kcal·mol^{-1} in strength. These energies are in line with earlier computations with the adducts [H$_2$N-CH$_3$···OCH$_2$] [47] and [HO-CH$_3$···OH$_2$] [48] of −0.7 and −1.0 kcal·mol^{-1} respectively.

The cationic adducts **6–11a** were computed as well and the most stable of these involved the Me$_3$C$^+$ carbocation in **6** (adducts with pentamethylated Carbon are unstable). The bonding interaction in **6b** is largely covalent, as evidenced by the interaction energy of −82.4 kcal·mol^{-1}, the large orbital contribution (55%), a dense bond critical point (18.4 · 10^2 a.u.) and a clear pyramidalization of the central C-atom (see Figure S4). Of the other adducts, all except **6a** (Me$_3$C$^+$···O interaction) and **9** (with the least electronegative P) converged into an X–CH$_3$···O Carbon bonding geometry. This is illustrated for **8a** and **11a** in Figure 4, where a clear bcp can be seen in between ^{methyl}C and O^{water} with a bond density of 1.15 · 10^2 and 0.98 · 10^2 a.u. for **8a** and **11a** respectively. The bonding energies in **7–11a** are mainly electrostatic in origin (~70%) and about −8 kcal·mol^{-1} for water and −15 kcal·mol^{-1} for dma. The most stable adducts in both series involved the most electronegative O (3.44) in Me$_3$O$^+$ (**8**). Two alternative configurations with N-methylpyridinium were also computed (**12** and **13**). In **12**, the O points in between two CH hydrogens as is illustrated for **12a** in Figure 4. In adducts **13** the O atom is located directly above the cationic N$^+$. Both **12** and **13** are more stable than the Carbon bonding geometry found in **11**, suggesting that hydrogen bonding interactions are most preferred. The interaction energies of Carbon bonding interaction with cationic species is similar to previous data of the adducts: [H$_3$N$^+$-CH$_3$···OCH$_2$] (−9.7 kcal·mol^{-1}); [47] [Me$_3$N$^+$-CH$_3$···OC(H)NH$_2$] (−13 kcal·mol^{-1}); [49] [Me$_2$S$^+$-CH$_3$···OH$_2$/NH$_3$/OCH$_2$] (about −8–9 kcal·mol^{-1}); [47,49] and [R$_2$S$^+$-CH$_3$···various lone-pairs] (about −9.0 kcal·mol^{-1}) [50].

As the calculations with cationic species imply that electron withdrawing substituents amplify the Carbon bonding interaction, it was decided to compute adducts with small molecules that have an electron withdrawing group: Iodomethane (**14**), 1,1,1-trifluoroethane (**16**), acetonitrile (**18**) and nitromethane (**20**). All these adducts converged as a Carbon bonding geometry and are energetically favorable by 1.6–2.8 kcal·mol^{-1} for water and 3.0–4.9 kcal·mol^{-1} for dma. The adducts involving nitromethane (**20**) were most stable and are shown in Figure 5, together with an aim analysis revealing a single C···O bcp (ρ = 0.79 · 10^2 and 0.83 · 10^2 a.u. for **20a** and **20a** respectively). For the water adducts, H-bonding geometries were also optimized: **15a**, **17a**, **19a** and **21a**. All these adducts were about twice as stable and the Carbon bonding geometry. This can be ascribed to the fact that besides a C–H···O hydrogen bonding interaction, another weak hydrogen bonding interaction is present as well (i.e., C–H···I in **15a**, C–H···F in **17a** and C–H···π in **19a**, see also see also Figure S4). For example, a dimer of HCF$_3$ is estimated at −2.6 kcal·mol^{-1} and exhibits two weak C–H···F hydrogen bonding interactions (not shown). The interaction energies with neutral yet polarized methyl groups (**14–20**) liken those reported by others for: [F-CH$_3$···OH$_2$/NH$_3$/PH$_3$] (about −2–3 kcal·mol^{-1}); [48,51–53] [F-CH$_3$···C$_2$H$_2$] (−1.2 kcal·mol^{-1}); [54] [Hlg-CH$_3$···C$_2$H$_4$/NH$_3$/PH$_3$] (−1–4 kcal·mol^{-1}); [53,55] [NC-CH$_3$···C$_2$H$_4$/dma] (−3–5 kcal·mol^{-1}); [55,56] and [O$_2$N-CH$_3$···dma] (−4.9 kcal·mol^{-1}) [56].

3.4. General Discussion

From all the calculations collected in Table 1 it is evident that the interaction energies of Carbon bonding geometries with the sp^2-O in dma (adducts 'b') is consistently about twice as strong as the interaction with sp^3-O in water (adducts 'a'). This is in line with the larger amount of van der Waals overlap observed in the $N(d')$ plots (Figure 2, ~30% for amides vs ~15% for water). The interaction energies of adducts with a Carbon bonding geometry range from very weak (below −1.5 kcal·mol^{-1} in **2**, **3**, **5**), to moderately weak (between −1.5 and −5 kcal·mol^{-1} in **14**, **16**, **18** and **20**) to fairly strong in the cationic adducts (between −7 and −18 kcal·mol^{-1} in **6–11**). ΔE becomes smaller (more stable) in the order **2 < 14 < 16 < 18 < 20 < 7 < 8**. Within this series, the orbital contribution remains constant at about 15–20%, while the electrostatic component increases from 30–35% in **2** to about 65% in **8**. This implies that stronger Carbon bonding interactions are mainly driven by electrostatic interactions and that weaker such adducts are driven by dispersion. These computational results are consistent with recent literature reports [47–59] and the database analyses presented here; neutral adducts are very weak and thus hardly (or non) directional but can be made stronger (and thus presumably more directional) when X in X–CH$_3$ is strongly polarized (see especially Figure 4). The relevance of Carbon bonding interactions with methyl groups is thus likely limited to highly polarized and/or cationic species. While this limits the scope considerably, it is worth pointing out that ligands with methyl groups related to those in adducts **2–10** and **14–21** are abundant within proteins structures and that cationic methyl groups also occur. For example, methylated methionine residues in methyl transferases [80] and nicotinamide derivatives such as nicotinamide adenine dinucleotide [81,82].

4. Summary and Conclusions

The CSD and the PDB were systematically evaluated for potential directional behavior of intermolecular non-covalent Carbon bonding interactions involving X–CH$_3$ and electron rich entities such as O/S atoms or an aryl ring (ElR) within a hemisphere of 5 Å basal radius (centered on C). It was found that X–CH$_3$···ElR interactions can be as directional as very weak hydrogen bonding interaction involving C–H ($P_{max} \leq 1.50$) but not directional at all when X = C. Grouping of data with significant amounts of van der Waals overlap (up to ~30%) was observed in various sub-datasets in the region where the X–CH$_3$···ElR angle α is 160°–180°. These distributions were significantly shifted to shorter distances (i.e., more van der Waals overlap) in the case of cationic R$_3$N$^+$–CH$_3$···O$^{water/amide}$ compared to charge-neutral R$_2$N–CH$_3$···O$^{water/amide}$ interactions.

Model DFT calculations revealed that charge neutral X–CH$_3$···O adducts with water and dimethylacetamide are very weak ($\leq$ −1.5 kcal·mol^{-1} in **2**, **3a**, **5a**) and are often not the energy minima of the adducts (**1**, **3b**, **4**, **5b**). The interaction energies can be increased by deploying a more electron withdrawing X (−1.5 to −5 kcal·mol^{-1} in **14**, **16**, **18** and **20**). Rendering X cationic leads to even more stable adducts (−7.0 to −18 kcal·mol^{-1}) in **7**, **8**, **10** and **11**). Carbon-bonding adducts with dimethylacetamide are consistently twice as stable as those with water. Energy decomposition analyses showed that increased stability is driven by electrostatics and atom-in-molecule analyses regularly gave a clear bond critical point involving the methyl C-atom.

It is thus concluded that this combined database / DFT study reaffirms that intermolecular non-covalent Carbon interactions with X–CH$_3$ is electrostatically driven and can be significant. The interaction can even by mildly directional in the solid state (comparable to weak CH hydrogen bonding interactions), provided X is sufficiently electron withdrawing.

Supplementary Materials: The following are available online, Table S1. Numerical overview of datasets retrieved from the CSD and the PDB. See Figure S1 and Figure 2 for P (α) directionality plots. For data of refined datasets see Table S2 ($\alpha \geq 160°$ and van der Waals overlap). 'n.a.' stands for 'not assessed'. The data for quaternary (cationic) N-atoms is not used in the paper for P(α) plot but they were used in N(d') plots. Table S2. Numerical overview of the amount of data in each dataset from Table S1 characterized by $\alpha \geq 160°$ (N$\alpha \geq 160°$), together with the amount of data within that dataset where van der Waals radii overlap (NΣvdW). 'n.a.' stands for 'not assessed'. Datasets with N$\alpha \geq 160°$ larger than 100 were inspected further by means of the N(d') plots shown in Figures S2–S3. As a guide to the eye, datasets with less than 100 hits are in grey, those between 100 and 500 hits are in blue and those above 500 are in regular black. Figure S1. P (α) directionality plots for the data retrieved from the CSD (left) and the PDB (right) using the general query shown in the top-left inset figure for X–CH3···ElR pairs. X can be C, N, O, P, or S and 'ElR' can be a water, amide or carboxyl O-atom, an RyCS S-atom (y = 2 or 3, R = any non-metal) or the centroid of an aryl ring, as is indicated in the right-hand side of the figure. The insert figure in the top right is intended as a guide to the eye to interpret the spatial location of data with a certain value of α. Due to the amount of data per dataset (see Table S1 for numerical overview), the top four P (α) plots are given at a 5° resolution for α and the bottom six at a 10° resolution. NB: Interestingly, in the P(α) plots for X = S and ElR = water, P is above unity at α = 160°–180° for the CSD data, while P < 1 for the PDB data in this same region. In both databases, the P-values are $\geq$ 1 around α = 105°. This indicates that the H-bonding geometry ($\alpha \approx 105°$) is somewhat directional on both databases but that the carbon bonding geometry is more directional only in the CSD. However, the dataset retrieved from the CSD is much smaller (1,546 hits) than the dataset from the PDB (30,725 hits) and the observed feature α = 160°–180° in the CSD might well be an artefact. Similarly, the P(α) plots with X = S and an amide O-atom reveal that P $\geq$ 1 at α = 90°–105°, again congruent with a hydrogen bonding geometry. P $\geq$ 1 also at α = 170°–180, but only for the PDB data. As this dataset is more voluminous (N = 11,215 vs

2,175 in the CSD), this implies that a carbon bonding geometry is more directional (at least in protein structures). A possible reason for this discrepancy might be that many cysteine and methionine residues are involved in metal coordination or directly methylated thus polarizing the S–C bond. For example, iron-sulphur clusters are held in place by cysteine–Fe coordination bonds [1] and methylated methionine residues are a (crystallographically known) intermediate in methyl transferases. [2]. Figure S2. Cumulative hit fraction (in %) as a function of the van der Waals corrected XCH3···ElR distance d' (in Å) for several datasets from Table S2 ($\alpha \geq 160°$, as illustrated by the inset figure). The interacting pairs involve water-O (red, empty) or amide-O (blue, half-filled) with X = C (squares), N (diamonds), or O (circles). See Figure 2 for the same data plotted as a regular hit fraction. Figure S3. Cumulative (top) and regular (bottom) hit fraction (in %) as a function of the van der Waals corrected XCH3···ElR distance d' (in Å) for all the datasets from Table S2 ($\alpha \geq 160°$, illustrated with the inset figure) that contain ≥ 500 data points. The inset legends show the nature of X (horizontally) and ElR (vertically). NB: In addition to the trends observed and discussed in the main text according to Figure 2, It is interesting to note that in all cases, for X = C the plot is shifted most to longer d' and has the least van der Waals overlap. Moreover, carboxy O-atoms are distributed about the same as amide O-atoms and thio S-atoms about the same as water. Aryls are always randomly distributed with the least amount of van der Waals overlap. Figure S4. Ball and stick representations of perspective views of all molecular adducts listed in Table 1 that were optimized by DFT (B3LYP-D3/def2-TZVP). The thin lines are bond paths (bp's) and the small red spheres are bond critical points (bcp's) obtained from an 'atoms-in-molecules' analysis. The bond density (ρ) is in arbitrary units 102 and bcp's indicative of non-covalent carbon bonding have been highlighted in yellow. Figure S5. Ball and stick representations of molecular adducts selected from Table 1 that were optimized by DFT (B3LYP-D3/def2-TZVP). The thin lines are bond paths (bp's) and the small red spheres are bond critical points (bcp's) obtained from an 'atoms-in-molecules' analysis. The bond density (ρ) is in arbitrary units ·102 and bcp's indicative of non-covalent carbon bonding have been highlighted in yellow.

Funding: This research was funded by the Netherlands Organization for Scientific Research (NWO) grant number 723.015.006.

Conflicts of Interest: The author declares no conflict of interest.

References

1. Lehn, J.M. *Supramolecular Chemistry: Concepts and Perspectives*, 1st ed.; Wiley VCH: Weinheim, Germany, 1995.
2. Clark, T. σ-Holes. *WIREs Comput. Mol. Sci.* **2013**, *3*, 13–20. [CrossRef]
3. Politzer, P.; Murray, J.S.; Clark, T. Halogen bonding and other σ-hole interactions: A perspective. *Phys. Chem. Chem. Phys.* **2013**, *15*, 11178–11189. [CrossRef] [PubMed]
4. Bauza, A.; Mooibroek, T.J.; Frontera, A. The bright future of unconventional σ/π-hole interactions. *Chem. Phys. Chem.* **2015**, *16*, 2496–2517. [CrossRef] [PubMed]
5. Crabtree, R.H. Hypervalency, secondary bonding and hydrogen bonding: Siblings under the skin. *Chem. Soc. Rev.* **2017**, *46*, 1720–1729. [CrossRef] [PubMed]
6. Grabowski, S.J. *Hydrogen Bonding: New Insights*; Springer: Heidelberg, Germany, 2006.
7. Bauza, A.; Mooibroek, T.J.; Frontera, A. Tetrel-bonding interaction: Rediscovered supramolecular force? *Angew. Chem. Int. Ed.* **2013**, *52*, 12317–12321. [CrossRef] [PubMed]
8. Bauza, A.; Mooibroek, T.J.; Frontera, A. Tetrel bonding interactions. *Chem. Rec.* **2016**, *16*, 473–487. [CrossRef] [PubMed]
9. Bauza, A.; Frontera, A.; Mooibroek, T.J. 1,1,2,2-Tetracyanocyclopropane (TCCP) as supramolecular synthon. *Phys. Chem. Chem. Phys.* **2016**, *18*, 1693–1698. [CrossRef]
10. Bauza, A.; Mooibroek, T.J.; Frontera, A. Influence of ring size on the strength of carbon bonding complexes between anions and perfluorocycloalkanes. *Phys. Chem. Chem. Phys.* **2014**, *16*, 19192–19197. [CrossRef]
11. Bauza, A.; Mooibroek, T.J.; Frontera, A. Non-covalent sp^3 carbon bonding with ArCF$_3$ is analogous to CH-π interactions. *Chem. Commun.* **2014**, *50*, 12626–12629. [CrossRef]
12. Scholfield, M.R.; van der Zanden, C.M.; Carter, M.; Ho, P.S. Halogen bonding (X-bonding): A biological perspective. *Protein Sci.* **2013**, *22*, 139–152. [CrossRef]
13. Cavallo, G.; Metrangolo, P.; Milani, R.; Pilati, T.; Priimagi, A.; Resnati, G.; Terraneo, G. The halogen bond. *Chem. Rev.* **2016**, *116*, 2478–2601. [CrossRef] [PubMed]
14. Auffinger, P.; Hays, F.A.; Westhof, E.; Ho, P.S. Halogen bonds in biological molecules. *Proc. Natl. Acad. Sci. USA* **2004**, *101*, 16789–16794. [CrossRef] [PubMed]
15. Berman, H.M.; Westbrook, J.; Feng, Z.; Gilliland, G.; Bhat, T.N.; Weissig, H.; Shindyalov, I.N.; Bourne, P.E. The Protein Data Bank. *Nucleic Acids Res.* **2000**, *28*, 235–242. [CrossRef] [PubMed]

16. Voth, A.R.; Khuu, P.; Oishi, K.; Ho, P.S. Halogen bonds as orthogonal molecular interactions to hydrogen bonds. *Nat. Chem.* **2009**, *1*, 74–79. [CrossRef] [PubMed]
17. Hardegger, L.A.; Kuhn, B.; Spinnler, B.; Anselm, L.; Ecabert, R.; Stihle, M.; Gsell, B.; Thoma, R.; Diez, J.; Benz, J.; et al. Systematic investigation of halogen bonding in protein-ligand interactions. *Angew. Chem. Int. Ed.* **2011**, *50*, 314–318. [CrossRef] [PubMed]
18. Zhang, Q.; Xu, Z.J.; Shi, J.Y.; Zhu, W.L. Underestimated halogen bonds forming with protein backbone in protein data bank. *J. Chem Inf. Model.* **2017**, *57*, 1529–1534. [CrossRef] [PubMed]
19. Ho, P.S. Halogen bonding in medicinal chemistry: From observation to prediction. *Future Med. Chem.* **2017**, *9*, 637–640. [CrossRef] [PubMed]
20. Mendez, L.; Henriquez, G.; Sirimulla, S.; Narayan, M. Looking back, looking forward at halogen bonding in drug discovery. *Molecules* **2017**, *22*, 1397. [CrossRef] [PubMed]
21. Voth, A.R.; Hays, F.A.; Ho, P.S. Directing macromolecular conformation through halogen bonds. *Proc. Natl. Acad. Sci. USA* **2007**, *104*, 6188–6193. [CrossRef] [PubMed]
22. Danelius, E.; Andersson, H.; Jarvoll, P.; Lood, K.; Grafenstein, J.; Erdelyi, M. Halogen bonding: A powerful tool for modulation of peptide conformation. *Biochemistry* **2017**, *56*, 3265–3272. [CrossRef] [PubMed]
23. Murray, J.S.; Lane, P.; Clark, T.; Riley, K.E.; Politzer, P. σ-holes, π-holes and electrostatically-driven interactions. *J. Mol. Model.* **2012**, *18*, 541–548. [CrossRef] [PubMed]
24. Murrayrust, P.; Burgi, H.B.; Dunitz, J.D. Chemical reaction paths. 5. SN1 reaction of tetrahedral molecules. *J. Am. Chem. Soc.* **1975**, *97*, 921–922. [CrossRef]
25. Choudhary, A.; Gandla, D.; Krow, G.R.; Raines, R.T. Nature of Amide Carbonyl-Carbonyl Interactions in Proteins. *J. Am. Chem. Soc.* **2009**, *131*, 7244–7246. [CrossRef] [PubMed]
26. Bartlett, G.J.; Choudhary, A.; Raines, R.T.; Woolfson, D.N. n → π* interactions in proteins. *Nat. Chem. Biol.* **2010**, *6*, 615–620. [CrossRef] [PubMed]
27. Newberry, R.W.; Bartlett, G.J.; VanVeller, B.; Woolfson, D.N.; Raines, R.T. Signatures of n → π* interactions in proteins. *Protein Sci.* **2014**, *23*, 284–288. [CrossRef]
28. Schottel, B.L.; Chifotides, H.T.; Dunbar, K.R. Anion-π interactions. *Chem. Soc. Rev.* **2008**, *37*, 68–83. [CrossRef]
29. Mooibroek, T.J.; Gamez, P. Anion-arene and lone pair-arene interactions are directional. *Cryst. Eng. Comm.* **2012**, *14*, 1027–1030. [CrossRef]
30. Frontera, A.; Gamez, P.; Mascal, M.; Mooibroek, T.J.; Reedijk, J. Putting anion-π interactions into perspective. *Angew. Chem. Int. Ed.* **2011**, *50*, 9564–9583. [CrossRef]
31. Mooibroek, T.J.; Gamez, P.; Reedijk, J. Lone pair-π interactions: A new supramolecular bond? *Cryst. Eng. Comm.* **2008**, *10*, 1501–1515. [CrossRef]
32. Burgi, H.B. Stereochemistry of reaction paths as determined from crystal-structure data - relationship between structure and energy. *Angew. Chem. Int. Ed.* **1975**, *14*, 460–473. [CrossRef]
33. Doppert, M.T.; van Overeem, H.; Mooibroek, T.J. Intermolecular π-Hole / n→π* interactions with carbon monoxide ligands in crystal structures. *Chem. Commun.* **2018**, *54*, 12049–12052. [CrossRef] [PubMed]
34. Ruigrok van der Werve, A.; van Dijk, Y.R.; Mooibroek, T.J. π-Hole / n→π* interactions with acetonitrile in crystal structures. *Chem. Commun.* **2018**, *54*, 10742–10745. [CrossRef] [PubMed]
35. Lewinski, J.; Bury, W.; Justyniak, W. Significance of intermolecular S center dot center dot center dot C(π) interaction involving M-S and -C=O centers in crystal structures of metal thiolate complexes. *Eur. J. Inorg. Chem.* **2005**. [CrossRef]
36. Echeverria, J. The n → π* interaction in metal complexes. *Chem. Commun.* **2018**, *54*, 3061–3064. [CrossRef]
37. Gamez, P.; Mooibroek, T.J.; Teat, S.J.; Reedijk, J. Anion binding involving π-acidic heteroaromatic rings. *Acc. Chem. Res.* **2007**, *40*, 435–444. [CrossRef] [PubMed]
38. Gamez, P. The anion-π interaction: Naissance and establishment of a peculiar supramolecular bond. *Inorg. Chem. Front.* **2014**, *1*, 35–43. [CrossRef]
39. Echeverria, J. Intermolecular Carbonyl ··· Carbonyl Interactions in Transition-Metal Complexes. *Inorg. Chem.* **2018**, *57*, 5429–5437. [CrossRef]
40. Mooibroek, T.J. Coordinated nitrate anions can be directional π-hole donors in the solid state: A CSD study. *Cryst. Eng. Comm.* **2017**, *19*, 4485–4488. [CrossRef]
41. Bauza, A.; Frontera, A.; Mooibroek, T.J. NO_3^- anions can act as Lewis acid in the solid state. *Nat. Commun.* **2017**, *8*, 14522. [CrossRef]

42. Bauza, A.; Frontera, A.; Mooibroek, T.J. π-Hole interactions involving nitro compounds: Directionality of nitrate esters. *Cryst. Growth Des.* **2016**, *16*, 5520–5524. [CrossRef]
43. Franconetti, A.; Frontera, A.; Mooibroek, T.J. Intramolecular π-hole interactions with nitro aromatics. *Cryst. Eng. Comm.* **2019**. [CrossRef]
44. Bauza, A.; Frontera, A.; Mooibroek, T.J. π-Hole Interactions Involving Nitro Aromatic Ligands Within Protein Structures. *Chem. Eur. J.* **2019**. [CrossRef]
45. Li, W.; Spada, L.; Tasinato, N.; Rampino, S.; Evangelisti, L.; Gualandi, A.; Cozzi, P.G.; Melandri, S.; Barone, V.; Puzzarini, C. Theory Meets Experiment for Noncovalent Complexes: The Puzzling Case of Pnicogen Interactions. *Angew. Chem. Int. Ed.* **2018**, *57*, 13853–13857. [CrossRef] [PubMed]
46. Grabowski, S.J. Tetrel bond-sigma-hole bond as a preliminary stage of the S(N)2 reaction. *Phys. Chem. Chem. Phys.* **2014**, *16*, 1824–1834. [CrossRef] [PubMed]
47. Bauza, A.; Frontera, A. RCH_3 center dot center dot center dot O Interactions in Biological Systems: Are They Trifurcated H-Bonds or Noncovalent Carbon Bonds? *Crystals* **2016**, *6*, 26. [CrossRef]
48. Mani, D.; Arunan, E. The X-C ··· Y (X = O/F, Y = O/S/F/Cl/Br/N/P) 'carbon bond' and hydrophobic interactions. *Phys. Chem. Chem. Phys.* **2013**, *15*, 14377–14383. [CrossRef] [PubMed]
49. Scheiner, S. Ability of IR and NMR Spectral Data to Distinguish between a Tetrel Bond and a Hydrogen Bond. *J. Phys. Chem. A* **2018**, *122*, 7852–7862. [CrossRef]
50. Trievel, R.C.; Scheiner, S. Crystallographic and Computational Characterization of Methyl Tetrel Bonding in S-Adenosylmethionine-Dependent Methyltransferases. *Molecules* **2018**, *23*, 2965. [CrossRef]
51. Sethio, D.; Oliveira, V.; Kraka, E. Quantitative Assessment of Tetrel Bonding Utilizing Vibrational Spectroscopy. *Molecules* **2018**, *23*, 2763. [CrossRef]
52. Scheiner, S. Systematic Elucidation of Factors That Influence the Strength of Tetrel Bonds. *J. Phys. Chem. A* **2017**, *121*, 5561–5568. [CrossRef]
53. Laconsay, C.J.; Galbraith, J.M. A valence bond theory treatment of tetrel bonding interactions. *Comput. Theor. Chem.* **2017**, *1116*, 202–206. [CrossRef]
54. Wei, Y.X.; Li, H.B.; Cheng, J.B.; Li, W.Z.; Li, Q.Z. Prominent enhancing effects of substituents on the strength of π···σ-hole tetrel bond. *Int. J. Quantum Chem.* **2017**, *117*, e25448. [CrossRef]
55. Mani, D.; Arunan, E. The X-C ··· π (X = F, Cl, Br, CN) Carbon Bond. *J. Phys. Chem. A* **2014**, *118*, 10081–10089. [CrossRef] [PubMed]
56. Mundlapati, V.R.; Sahoo, D.K.; Bhaumik, S.; Jena, S.; Chandrakar, A.; Biswal, H.S. Noncovalent Carbon-Bonding Interactions in Proteins. *Angew. Chem. Int. Ed.* **2018**, *57*, 16496–16500. [CrossRef] [PubMed]
57. Varadwaj, P.R.; Varadwaj, A.; Jin, B.Y. Significant evidence of C ··· O and C center dot center dot center dot C long-range contacts in several heterodimeric complexes of CO with CH_3-X, should one refer to them as carbon and dicarbon bonds! *Phys. Chem. Chem. Phys.* **2014**, *16*, 17238–17252. [CrossRef] [PubMed]
58. Garcia-Llinas, X.; Bauza, A.; Seth, S.K.; Frontera, A. Importance of R-CF_3 center dot center dot center dot O Tetrel Bonding Interactions in Biological Systems. *J. Phys. Chem. A* **2017**, *121*, 5371–5376. [CrossRef] [PubMed]
59. Thomas, S.P.; Pavan, M.S.; Row, T.N.G. Experimental evidence for 'carbon bonding' in the solid state from charge density analysis. *Chem. Commun.* **2014**, *50*, 49–51. [CrossRef] [PubMed]
60. Allen, F.H. The Cambridge Structural Database: A quarter of a million crystal structures and rising. *Acta Crystallogr. Sect. B-Struct. Sci.* **2002**, *58*, 380–388. [CrossRef] [PubMed]
61. Groom, C.R.; Bruno, I.J.; Lightfoot, M.P.; Ward, S.C. The Cambridge Structural Database. *Acta Crystallogr. Sect. B-Struct. Sci.* **2016**, *72*, 171–179. [CrossRef] [PubMed]
62. Bruno, I.J.; Cole, J.C.; Edgington, P.R.; Kessler, M.; Macrae, C.F.; McCabe, P.; Pearson, J.; Taylor, R. New software for searching the Cambridge Structural Database and visualizing crystal structures. *Acta Crystallogr. Sect. B-Struct. Sci.* **2002**, *58*, 389–397. [CrossRef]
63. Hendlich, M.; Bergner, A.; Gunther, J.; Klebe, G. Relibase: Design and development of a database for comprehensive analysis of protein-ligand interactions. *J. Mol. Biol.* **2003**, *326*, 607–620. [CrossRef]
64. Mooibroek, T.J.; Gamez, P. Directional character of solvent- and anion-pentafluorophenyl supramolecular interactions. *Cryst. Eng. Comm.* **2012**, *14*, 3902–3906. [CrossRef]
65. Mooibroek, T.J.; Gamez, P. How directional are D-H center dot center dot center dot phenyl interactions in the solid state (D = C, N, O)? *CrystEngComm* **2012**, *14*, 8462–8467. [CrossRef]

66. Mooibroek, T.J.; Gamez, P. Halogen ··· phenyl supramolecular interactions in the solid state: Hydrogen versus halogen bonding and directionality. *CrystEngComm* **2013**, *15*, 1802–1805. [CrossRef]
67. Mooibroek, T.J.; Gamez, P. Halogen bonding versus hydrogen bonding: What does the Cambridge Database reveal? *Cryst. Eng. Comm.* **2013**, *15*, 4565–4570. [CrossRef]
68. Bauza, A.; Mooibroek, T.J.; Frontera, A. Directionality of π-holes in nitro compounds. *Chem. Commun.* **2015**, *51*, 1491–1493. [CrossRef] [PubMed]
69. Allen, F.H.; Kennard, O.; Watson, D.G.; Brammer, L.; Orpen, A.G.; Taylor, R. Tables of bond lengths determined by X-ray and neutron-diffraction. 1. Bond lengths in organic compounds. *J. Chem. Soc. Perkin Trans. 2* **1987**, S1–S19. [CrossRef]
70. Bondi, A. van der Waals Volumes and Radii. *J. Phys. Chem.* **1964**, *68*, 441–452. [CrossRef]
71. Lee, C.T.; Yang, W.T.; Parr, R.G. Development of the Colle-Salvetti correlation-energy formula into a functional of the electron density. *Phys. Rev. B* **1988**, *37*, 785–789. [CrossRef]
72. Becke, A.D. Density-functional exchange-energy approximation with correct asymptotic-behaviour. *Phys. Rev. A* **1988**, *38*, 3098–3100. [CrossRef]
73. Grimme, S.; Antony, J.; Ehrlich, S.; Krieg, H. A consistent and accurate ab initio parametrization of density functional dispersion correction (DFT-D) for the 94 elements H-Pu. *J. Chem. Phys.* **2010**, *132*, 154104. [CrossRef]
74. Weigend, F.; Ahlrichs, R. Balanced basis sets of split valence, triple zeta valence and quadruple zeta valence quality for H to Rn: Design and assessment of accuracy. *Phys. Chem. Chem. Phys.* **2005**, *7*, 3297–3305. [CrossRef]
75. Weigend, F. Accurate Coulomb-fitting basis sets for H to Rn. *Phys. Chem. Chem. Phys.* **2006**, *8*, 1057–1065. [CrossRef]
76. te Velde, G.; Bickelhaupt, F.M.; Baerends, E.J.; Guerra, C.F.; van Gisbergen, S.J.A.; Snijders, J.G.; Ziegler, T. Chemistry with ADF. *J. Comput. Chem.* **2001**, *22*, 931–967. [CrossRef]
77. Bader, R.F.W. Atoms in Molecules. *Acc. Chem. Res.* **1985**, *18*, 9–15. [CrossRef]
78. Bickelhaupt, F.M.; Baerends, E.J. Kohn-Sham density functional theory: Predicting and understanding chemistry. In *Reviews in Computational Chemistry, Vol 15*; Lipkowitz, K.B., Boyd, D.B., Eds.; Wiley-Vch, Inc.: New York, NY, USA, 2000; Volume 15, pp. 1–86.
79. van der Lubbe, S.C.C.; Guerra, C.F. The Nature of Hydrogen Bonds: A Delineation of the Role of Different Energy Components on Hydrogen Bond Strengths and Lengths. *Chem. Asian J.* **2019**, *14*, 2760–2769. [CrossRef]
80. Schubert, H.L.; Blumenthal, R.M.; Cheng, X.D. Many paths to methyltransfer: A chronicle of convergence. *Trends Biochem.Sci.* **2003**, *28*, 329–335. [CrossRef]
81. Foster, J.W.; Moat, A.G. Nicotinamide Adenosine-dinucleotide Biosynthesis and Pyridine-nucleotide Cycle Metabilism in Microbial Systems. *Microbiol. Rev.* **1980**, *44*, 83–105.
82. Pollak, N.; Dolle, C.; Ziegler, M. The power to reduce: Pyridine nucleotides - small molecules with a multitude of functions. *Biochem. J.* **2007**, *402*, 205–218. [CrossRef]

© 2019 by the author. Licensee MDPI, Basel, Switzerland. This article is an open access article distributed under the terms and conditions of the Creative Commons Attribution (CC BY) license (http://creativecommons.org/licenses/by/4.0/).

Article

Identification of the Tetrel Bonds between Halide Anions and Carbon Atom of Methyl Groups Using Electronic Criterion

Ekaterina Bartashevich [1,*], Yury Matveychuk [1] and Vladimir Tsirelson [1,2]

1 Research Laboratory of Multiscale Modelling of Multicomponent Functional Materials, REC Nanotechnology, South Ural State University, 454080 Chelyabinsk, Russia; matveichukyv@susu.ru (Y.M.); vtsirelson@yandex.ru (V.T.)
2 Quantum Chemistry Department, D.I. Mendeleev University of Chemical Technology, 125047 Moscow, Russia
* Correspondence: bartashevichev@susu.ru; Tel.: +7-351-267-9564

Received: 26 February 2019; Accepted: 16 March 2019; Published: 19 March 2019

Abstract: The consideration of the disposition of minima of electron density and electrostatic potential along the line between non-covalently bound atoms in systems with $Hal^-\cdots CH_3-Y$ (Hal^- = Cl, Br; Y = N, O) fragments allowed to prove that the carbon atom in methyl group serves as an electrophilic site provider. These interactions between halide anion and carbon in methyl group can be categorized as the typical tetrel bonds. Statistics of geometrical parameters for such tetrel bonds in CSD is analyzed. It is established that the binding energy in molecular complexes with tetrel bonds correlate with the potential acting on an electron in molecule (PAEM). The PAEM barriers for tetrel bonds show a similar behavior for both sets of complexes with Br^- and Cl^- electron donors.

Keywords: tetrel bond; electron density; electrostatic potential; potential acting on an electron in molecule

1. Introduction

The problem of categorizing non-covalent interactions in molecular crystals and complexes is now a focus of attention [1,2]. Nowadays, the systematization of the halogen, chalcogen, pnictogen, and tetrel bonds already exists [3]; however, in most cases, only the simplistic geometrical approach underlies the analysis of such types of interactions. In this context, the types of non-covalent interactions are traditionally discussed in terms of interatomic distances and angles, which specify the mutual orientation of pivotal chemical bonds [4,5]. However, due to the pronounced and specific electrostatic component of such non-covalent bonds [6,7], more careful analysis of the electronic features of the halogen, chalcogen, pnictogen, and tetrel bonds is required. Such analysis needs to focus on the features of valence electron shells and related anisotropy of the electrostatic potential of interacting atoms.

The estimation of the binding energy for molecules in the $Y_4T\cdots Hal^-$ complexes, where a tetrel atom T = C, Si, Ge, Sn, as well as the description of electron density characteristics for tetrel bonds, were presented in References [8–13]. The carbon atom in the CH_3-group is fairly often noted as the owner of σ-hole, and the fact that the oxygen atom can act as an electron-rich center in the $CH_3\cdots O$ tetrel bonding has been confirmed in studies [14,15]. Note that in these early works such non-covalent interaction has been referred as a "carbon bond". Pal et al. [16] described the $CH_3\cdots N$ tetrel bonding in a Co(II) coordination polymeric system using the analysis of calculated electron density. The typical tetrel bonds formed by CH_3-group in crystals have been also been observed by high-precision X-ray diffraction method using the analysis of the experimental electron density [17,18].

The rows of binding energy were calculated for the series of complexes in which the compounds NH$_3$ [19–22], PH$_3$ and AsH$_3$ [23], benzene and unsaturated hydrocarbons [24], HCN and pyrazine [25] acted as the tetrel bond acceptors. According to these results, the tetrel bond strength depends on the tetrel atom, the Lewis base that acts as a tetrel bond acceptor, and the fragments covalently bound with this atom. Therefore, in the complexes with ammonia at transition from CF$_4$ to SnF$_4$, the binding energy increases from −0.82 to −25.53 kcal/mol (MP2/aug-cc-pVDZ) [21], that is, the tetrels of higher periods form stronger interactions in complexes. Decrease of the Lewis base strength in series from X = NH$_3$ to AsH$_3$ reduces the binding energy from −3.23 to −1.28 kcal/mol (MP2/aug-cc-pVDZ) in complexes X⋯CH$_3$Cl [23]. Therefore, the values of binding energy estimated in neutral complexes with tetrel bonds formed by the carbon in a methyl group are commonly very small. The binding energy grows more sharply if the hydrogen atoms are replaced by halogens, which leads to enhancement of electron acceptor properties of the tetrel atom providing the σ-hole for bonding. This fact is illustrated by complexes H$_4$Sn⋯NH$_3$ (−2.44 kcal/mol) and F$_4$Sn⋯NH$_3$ (−25.53 kcal/mol) with the tetrel bond lengths of 3.17 and 2.28 Å (MP2/aug-cc-pVDZ), respectively [21]. If the anion acts as the tetrel bond acceptor, the estimated binding energy can be greater by an order of magnitude as compared to the neutral complexes [9–11]. The highest value of binding energy, −93.58 kcal/mol, has been recorded for the SnF$_4$⋯F$^-$ complex (MP2/aug-ccpVTZ) [11]; replacing the F$^-$ ion with Cl$^-$ and Br$^-$ reduces this energy by 2–3 times.

Therefore, a variety of functional groups can act as the tetrel bond donors in molecular complexes and a methyl group as an electrophilic site deliverer is quite often in the focus of attention for the study of tetrel bond properties. The tetrel bonds formed by the methyl carbon atom occur in crystalline systems as well. There are at least two studies [17,26] based on high-precision X-ray diffraction data, which established the participation of the carbon atom of a methyl group in non-covalent interactions. These facts motivate us to focus on the search for evidential electronic criterion for recognizing the type of non-covalent bonding and for systematization of the electrophilic site features for the carbon atom of a methyl group in molecular complexes and crystals.

Quantum Theory of Atoms in Molecules (QTAIM) [27] suggests zero-flux conditions for electron-density gradient [28] and electrostatic-potential gradient [29] to determine the boundaries of chemically bonded atoms and electrically neutral atomic fragments, respectively. For non-covalent interactions with a significant electrostatic component, the boundaries of the electron density basins (ϱ-basins) do not coincide with the boundaries of the electrostatic potential (ESP) basins (φ-basins). It means that electrons formally belonging to the electron donor atom can be electrostatically attracted to the nucleus of an electron acceptor atom along a specific direction. The superposition of the ϱ- and φ-basins has been discussed in the literature [30–32], and the features of zero-flux surface in electrostatic potential in solids have been studied using the experimental charge density [29,33].

Previously we have proposed the following electronic criterion for recognizing the atom that prescribes the name of non-covalent bonding [34,35]. The minimum of electrostatic potential along the interaction line is located at the side of the atom that donates electrons; the minimum of electron density is closer to the atom that delivers its electrophilic site for bonding. More explicitly, the latter atom prescribes the name of bonding. This observation opens up the broad possibilities for identifying the role of atoms involved in non-covalent interactions. For example, atoms of the Group 14 of the Periodic Table are able to deliver their electrophilic sites for non-covalent interactions with electron donors for the tetrel bond formation, and the suggested electronic criterion can specify the electron acceptor role of these atoms. Along the Hal$^-$⋯C interatomic line, the 1D minimum of ESP should be located closer to the electron donor atom, while the minimum of the electron density will be found closer to the atom that is an electron acceptor. Such disposition of minima indicates that the fraction of electron density from the atomic ϱ-basin of an electron donor is electrostatically attracted to the nucleus of an electron acceptor atom. In this case, the disposition of the 1D minima of both functions can unambiguously indicate that only one of the pair of atoms provides its electrophilic site. Namely, the minimum of electron density on the interatomic line is always closer to the atom that has delivered its own electrophilic site for bonding.

We have recently proposed the potential acting on an electron in molecule (PAEM [36]) [37] as a function that not only characterizes the properties of non-covalent bonding with a significant electrostatic component, but also allows us to observe the quantitative relationship with the interaction energy in complexes. Unlike ESP, the PAEM contains both Coulomb and exchange components. The first of them has a classic nature and the second one is the two-electron contribution of the quantum exchange-correlation potential. PAEM was examined [38] for the halogen and chalcogen bonding characterization and its usefulness was confirmed.

The aim of the present study is to demonstrate the efficiency and productivity of the above-mentioned electronic criterion and the PAEM for analysis of tetrel bonds between the carbon atom of methyl groups and halide anions, which sometimes occur in molecular crystals. We also try to understand to what extent the characteristics of the gas-phase complexes are suitable for describing $Hal^-\cdots CH_3Y$ (Hal^- = Cl, Br; Y = N, O) tetrel bonds in crystals.

2. Results and Discussion

2.1. Population of $Hal^-\cdots CH_3Y$ Tetrel Bonds in Crystals

The search of short contacts between a halide anion and the methyl carbon, $Hal^-\cdots CH_3-Y$, in Cambridge Structural Database (CSD) v.5.39 (The Cambridge Crystallographic Data Centre, Cambridge, UK) [39] was performed with the following restrictions: organic derivatives without disordered, polymeric, powder, organometallic and repeating structures have been considered. The main condition was to choose interactions in which the halide anion, Hal^- = Cl, Br, I, was placed on the extension of the C–Y covalent bond, where Y = C, N, O. We set that condition using the angle θ (Hal^-–C–Y), the value of which was in the range from 160° to 180°. At the same time, we selected the structures with interatomic distances, d(Hal, C), falling into the range ($r_{vdw}(C) + r_{vdw}(Hal) \pm 0.2$ Å), where r_{vdw} is the Bondi atomic radius [40]. The total number of selected structures that satisfied those conditions was 164. The analysis of the obtained sample has shown that the Y atom covalently bound with the CH_3 group is nitrogen in most cases. We have found 43 cases of $Cl^-\cdots CH_3-N$ interactions, 36 cases of $Br^-\cdots CH_3-N$ interactions and 53 cases of $I^-\cdots CH_3-N$ interactions. Oxygen and carbon are involved in such covalent bonds much less frequently and approximately equally. In these cases, the methyl group forms more interactions with the Cl^- anion than with the Br^- and I^- anions taken together. It can be concluded that the polarity of the covalent bond CH_3-Y affects the probability and strength of the tetrel bond formation. It should be noted that the distances d(Hal, C) for the cases of $Hal^-\cdots CH_3-C$ interactions everywhere exceed the sum of van der Waals radii (Figure 1). If the CH_3-group is bound with oxygen ($Hal^-\cdots CH_3-O$), the distances d(Hal, C) can be less than the sum of van der Waals radii, though all of these distances are more than this sum for I^- cases.

Therefore, we conclude that the studied type of interactions, $Hal^-\cdots CH_3-Y$, are not widely spread within crystals listed in CSD, but they are not exceptional.

2.2. Evidence of Electrophilic Sites for the CH_3-Groups Bound in Tetrel Bonds

Let us now look at the same examples of halide crystal structures containing tetrel bonds (Figure 2). All the results considered in this chapter were obtained for crystal structures by the calculations with the periodic boundary conditions. In the crystalline N,N,N',N'-tetramethylchloroformamide chloride, LONGEB [41], the Cl^- anion forms non-covalent interactions with the Cl, H, C atoms, which are characterized by interatomic distances smaller than the sums of van der Waals radii. The $Cl^-\cdots Cl-C$ non-covalent interaction with a distance of 3.122 Å refers to a typical charge-assisted halogen bond; the next five hydrogen bonds, $Cl^-\cdots H-C$, are characterized by interatomic distances ranging from 2.924 to 2.664 Å. Finally, the $Cl^-\cdots CH_3$ interaction of 3.425 Å can be called a tetrel bond. In the crystalline dimethylmethyleneammonium chloride VAPREJ [42], the chloride anion forms eight

Cl⁻···H–C interactions, which are shorter than the sum of van der Waals radii and two Cl⁻···CH$_3$ tetrel bonds. In the dimethylmethylenimine bromide crystal, LILLOH [43], in addition to multiple Br⁻···H–C interactions, there are two tetrel bonds: Br⁻···CH$_3$ (3.533 Å) and Br⁻···CH$_2$ (3.503 Å). Quantum-topological analysis of the electron density in all considered crystals have confirmed the presence of the Hal⁻···C bond path and bcp of electron density (Table 1). Our series of tetrel bonds in the considered crystals does not vary much, and we have observed the small changes in electron density at the bond critical points, $\varrho(r_{bcp})$, which are in the range 0.0042–0.0070 a.u. for Cl⁻···CH$_3$–Y and 0.0060–0.0068 a.u. for Br⁻···CH$_3$–Y.

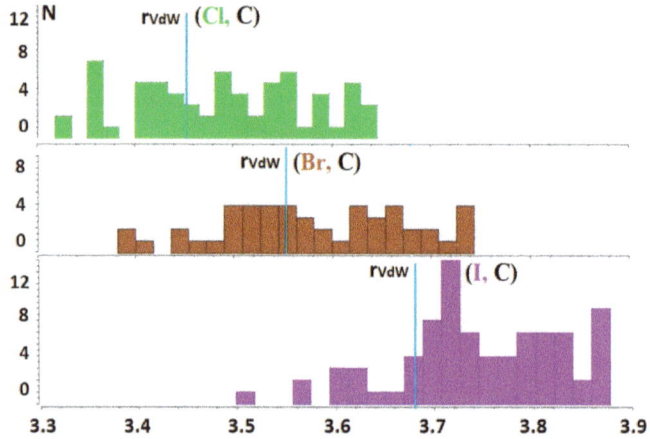

Figure 1. The distribution of the Hal⁻···CH$_3$–Y interactions in crystals for Cl⁻ (green), Br⁻ (brown), I⁻ (violet); the blue lines mark the sum of van der Waals radii.

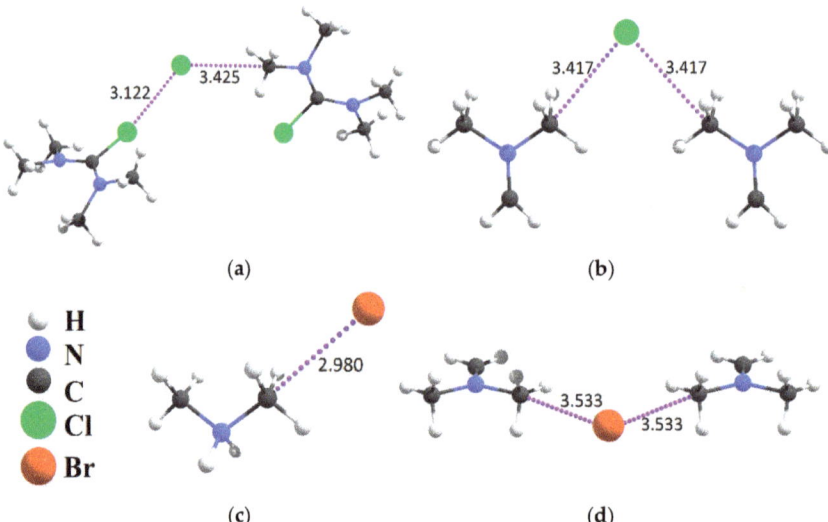

Figure 2. Fragments of structures with tetrel bonds and other non-covalent interactions in crystals LONGEB (a), VAPREJ (b), POSTUM02 (c), LILLOH (d). The interatomic distances are given in Angstroms.

Table 1. Experimental and calculated tetrel and C–Y bond lengths, D (Å), angles Hal⁻···C–Y, θ and calculated electron density, $\varrho(r_{bcp})$ (a.u.) at bond critical points for considered crystals.

Crystal	Bond	Crystal D_{exp}, θ (Hal⁻···C–N)$_{exp}$	Crystal D_{calc}, θ (Hal⁻···C–N)$_{calc}$	$\varrho(r_{bcp})$, Crystal
GETQIF	Cl(3)⁻···C(2)	3.4584 169.08	3.4260 163.91	0.0056
	C(2)–N(1)	1.4815	1.4958	0.2441
LONGEB	Cl(1)⁻···C(4)	3.4251 175.28	3.4087 166.64	0.0068
	C(4)–N(2)	1.4722	1.4740	0.2458
VAPREJ	Cl(1)⁻···C(1)	3.417 164.88	3.4385 164.49	0.0064
	C(1)–N(1)	1.466	1.4747	0.2480
TMHYZC	Cl(1)⁻···C(2)	3.4374 174.96	3.4280 176.34	0.0062
	C(2)···N(1)	1.4976	1.5080	0.2406
ZENJAD	Cl(1)⁻···C(7)	3.5111 170.58	3.4644 171.72	0.0056
	C(7)–O(2)	1.4471	1.4468	0.2306
LILLOH	Br(1)⁻···C(2)	3.533 167.25	3.5664 166.86	0.0061
	C(2)–N(1)	1.474	1.4735	0.2490
FADXIR	Br(1)⁻···C(6)	3.6014 170.87	3.5722 173.15	0.0058
	C(6)–N(1)	1.5025	1.5082	0.2371
POSTUM02	Br(1)⁻···C(1)	3.7012 175.21	3.6667 174.15	0.0048
	C(1)–N(1)	1.4852	1.4928	0.2432
ZZZGVM01	Br(1)⁻···C(2)	3.742 168.65	3.7283 169.04	0.0042
	C(2)–N(1)	1.474	1.4968	0.2437
ZZZUQO03	Br(1)⁻···C(1)	3.685 171.12	3.6819 171.11	0.0049
	C(1)–N	1.487	1.5039	0.2411

It is possible to demonstrate the electrophilic site on a carbon atom using the electrostatic potential (ESP) mapped on the isosurface of electron density or the distribution of Electron Localization Function (ELF) [44] for CH$_3$-group, which participate in a tetrel bond. For example, relatively higher positive values of ESP on the isosurface of electron density (0.003 a.u.) we can see in the region of the σ-hole, which belongs to the C atomic basin in trimethylammonium cation (Figure 3a,b).

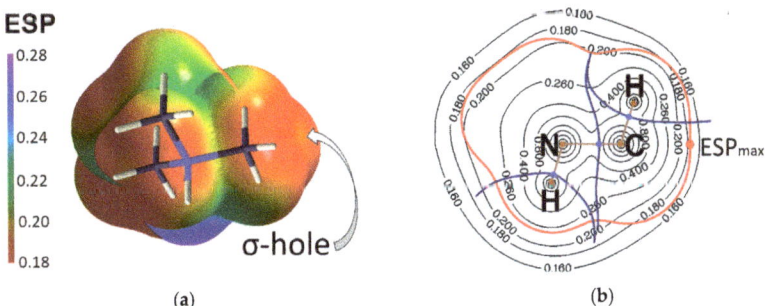

Figure 3. (a) ESP in the trimethylammonium cation on the isosurface of electron density of 0.003 a.u.; (b) contour map of ESP in the plane N-C-H, red point indicates the maximum of ESP on the van der Waals surface (red line) and belongs to C atomic basin.

Moving strictly along the line linking the Cl(1)⁻ and C(4) atoms of CH₃-group (Figure 4a), we reach the extension of the covalent bond formed by CH₃-group. Along this line the ELF is less than 0.5 near the carbon atom, showing the region of the reduced probability of electron pairing. It can be considered as the manifestation of the carbon atom σ-hole. The values of ELF (r_{bcp}) at the bond critical points do not exceed 0.05. This excludes the hypothesis about significant covalent character of the Hal⁻···CH₃Y tetrel bonds. At about 0.8 Å from the chloride anion nucleus the maximum values of ELF are distributed around the circumference. Figure 4b, depicting the tetrel bonds formed by the bromide anion in LILLOH crystal, shows a similar ELF distribution. Near the carbon atom and along the Br(1)⁻···C(2) line the ELF does not attain high values, but it increases sharply, affecting the basins of hydrogen atoms, if we slightly deviate from this line.

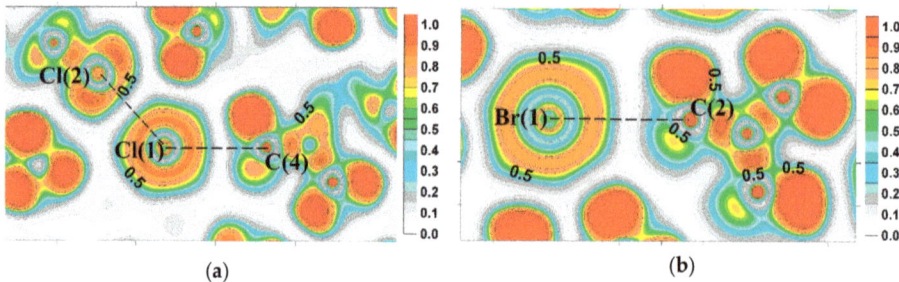

Figure 4. The ELF for (a) the halogen and tetrel bonds Cl(2)···Cl(1)⁻···C(4) in LONGEB crystal; (b) the tetrel bond in LILLOH crystal.

Now let us consider and evaluate how the electronic criterion works for the cases of non-covalent interactions formed by the carbon atoms of methyl groups in halide crystals. In Figure 5a it can be seen that in the LONGEB crystal, the Cl⁻ anion forms two non-covalent interactions at least, as follows from the presence of corresponding bcp. In both cases, Cl(2)···Cl(1)⁻···C(4), the one-dimensional ESP minimum is closer to electron donating anion, Cl(1)⁻ in a crystal. The electron density minima along the Cl(2)···Cl(1)⁻ and Cl(1)⁻···C(4) lines are located on the side of the Cl(2) and C(4) atoms. They indicate the electrophilic site providers and dictate the name of the non-covalent bonding.

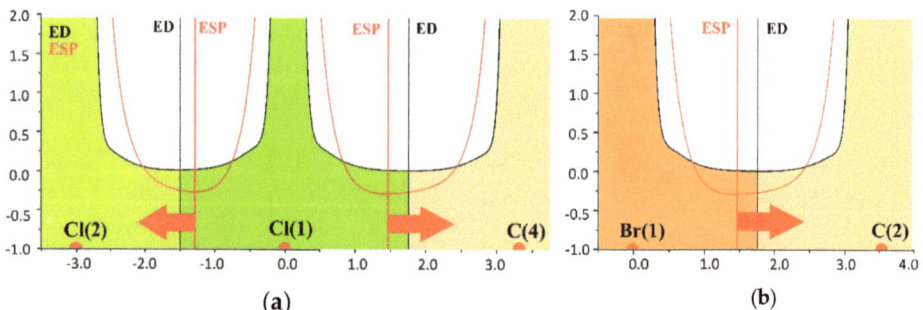

Figure 5. The disposition of electron density and electrostatic potential minima (a.u.) (a) along two interatomic lines Cl(2)···Cl(1)⁻···C(4), (Å), in LONGEB; (b) along the tetrel bond in LILLOH. The arrows point to the electrophilic site provider.

According to the proposed electronic criterion, the first interaction, Cl(2)···Cl(1)⁻, can be categorized as a charge-assisted halogen bond, and the second one, Cl(1)⁻···C(4), is a tetrel bond enhanced by charges. In Figure 5b, the minimum of electron density along the Br(1)⁻···C(2) line is located on the side of the C(4) atom, while the minimum of ESP is closer to Br(1)⁻. Such disposition of minima shows that the carbon atom accepts electrons along the Br(1)⁻···C(2) line and that interaction can be called a tetrel bond.

2.3. Binding Energy in Molecular Complexes with the Hal⁻···CH3 Tetrel Bonds

The determination of the equilibrium geometry for ion pairs "halide anion–cation" extracted from the crystal environment is not a straightforward procedure. In general, the retention of the halide anion position strictly on the extension of a covalent bond of CH_3-group is rather difficult in the gas phase state. This task requires us to dwell on the level of gas phase calculations, different from those used for crystal structures. Nevertheless, this step allows us to obtain the stationary state for the maximal number of complexes, for which the tetrel bonds in crystalline state have attracted our attention. Some relative estimations and the features of electronic properties can be quite useful for understanding the nature of charge-assisted tetrel bonds.

The binding energy, E_b, between the halide anion and cation in the considered complexes varies from −52.28 to −82.67 kcal/mol (Table 2). These values do not fall out of the range that is determined in similar studies [8–11]. The BSSE correction, ΔE_{BSSE}, is negligible and influences the energy values of the third decimal place of kcal/mol units.

Table 2. Calculated binding energy E_b (kcal/mol), bond lengths D_{calc} (Å), electron density $\varrho(r_{bcp})$ at bcp (a.u.) for Hal⁻···CH_3 minimum of electrostatic potential ESP_{min} (a.u.) and maximum of potential acting on an electron in molecule $PAEM_{max}$ (a.u.) along the line of tetrel bonds in complexes.

Refcode	Tetrel Bond	E_b	D_{calc}	$\varrho(r_{bcp})$	ESP_{min}	$PAEM_{max}$
GETQIF	Cl⁻···CH_3Y	−81.07	2.8262	0.019	−0.038	−0.4819
LONGEB	Cl⁻···CH_3Y	−71.05	2.8782	0.017	−0.063	−0.4454
TMHYZC	Cl⁻···CH_3Y	−82.67	2.8248	0.019	−0.034	−0.4874
VAPREJ	Cl⁻···CH_3Y	−81.80	2.8226	0.019	−0.034	−0.4870
ZENJAD	Cl⁻···CH_3Y	−52.28	2.9268	0.015	−0.107	−0.3883
FADXIR	Br⁻···CH_3Y	−81.09	2.9855	0.017	−0.015	−0.4741
LILLOH	Br⁻···CH_3Y	−78.67	2.9896	0.016	−0.020	−0.4664
POSTUM02	Br⁻···CH_3Y	−79.82	2.9802	0.017	−0.019	−0.4699
ZZZGVM01	Br⁻···CH_3Y	−77.97	2.9949	0.0165	−0.024	−0.4612
ZZZUQO03	Br⁻···CH_3Y	−76.63	3.0015	0.0162	−0.028	−0.4560

The considered tetrel bonds exhibit significantly shorter lengths in the models of complexes extracted from the crystalline environment. On average, the observed Hal⁻···C bond lengths in such complexes differ by ~17% from those in crystal structures. As a result, the different approaches for complexes and crystals calculations lead to the values of $\varrho(r_{bcp})$ that are almost twice higher in crystals. Obviously, the direct transfer of tetrel bond properties in isolated complexes to the crystals, neglecting the rest of interactions between a halide anion and crystalline environment, is not entirely correct. Comparing the properties of the CH_3–Y (Y=N, O) covalent bonds in complexes and isolated cations, we see that the participation of CH_3-group in tetrel bond with Hal⁻ weakens the CH_3–Y covalent bond. Therefore, the tetrel bonding elongates the covalent bond of a methyl group by 0.01–0.04 Å, and the values of $\varrho(r_{bcp})$ for the CH_3–Y bonds decrease by ~8%.

It is useful to understand how the electronic properties of Hal⁻···CH_3Y tetrel bonds in complexes are related to the strength of complexes. In our opinion, the tetrel bonds belong to electrostatically driven interactions. In addition, the binding energy between two oppositely charged ions is much higher in comparison with neutral molecules. For this reason, the electrostatic properties of tetrel bonds have been analyzed first of all.

We found that the properties of both ESP and PAEM for the Hal⁻···CH_3Y tetrel bonds are linearly correlated with the binding energy, E_b, in complexes, as shown in Figure 6. The correlation coefficient for the minima of the electrostatic potential, ESP_{min}, on the line between Hal⁻ and C atoms is 0.917. For the maximum of PAEM along this line, $PAEM_{max}$, or PAEM barrier, the correlation coefficient is

0.985. It is important to note that in the relationship "PAEM$_{max}$ vs E$_b$", PAEM$_{max}$ for the tetrel bonds formed by Br$^-$ and Cl$^-$ fits strictly on the common line. This is a rare case among the established relationships between local properties of non-covalent bonds and the binding energy for bound fragments. For example, the electronic potential and kinetic energy densities at bcp do not allow constructing a good common relationship for Br$^-$ and Cl$^-$ rows (Figure S1). This finding has been discussed by us earlier for the halogen bonds formed by different atoms or fragments that play the role of halogen acceptors [45]. This fact has recently been illustrated in detail in Reference [46], where the large series of non-covalent interactions with different halide anions have been studied.

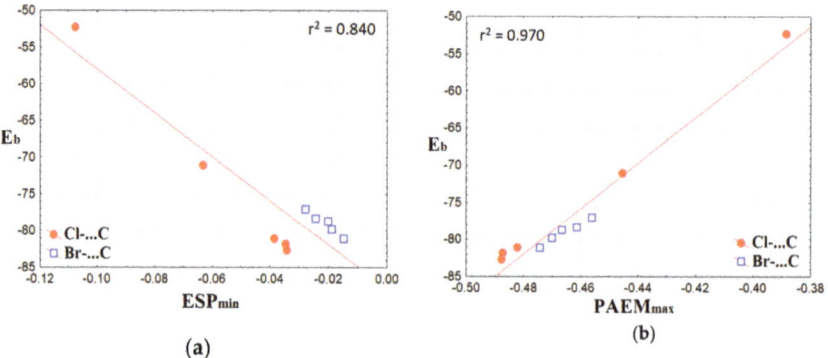

Figure 6. Binding energy in complexes vs minimum of electrostatic potential (**a**) and maximum of potential acting on an electron in a molecule, (**b**) along the line of tetrel bonds.

Note that the extreme values of ESP$_{min}$ and PAEM$_{max}$ slightly differ from their local values at the bond critical points of tetrel bonds (Table S4). Nevertheless, the PAEM(r$_{bcp}$) values correlate with the binding energy better than ESP(r$_{bcp}$) (Figure S2). This is probably due to the fact that in our series the maximum of PAEM is closer to the tetrel bond critical point than the minimum of ESP (Figure 7). The relative location of ESP$_{min}$ and PAEM$_{max}$ in the common projection is demonstrated by examples of the weakest Cl$^-$···CH$_3$ (ZENJAD) and the strongest Br$^-$···CH$_3$ (FADXIR) tetrel bonds in our set. Though the gap between the PAEM$_{max}$ and the minimum of electron density is larger for the Cl$^-$···CH$_3$ tetrel bond, and ESP has a lower negative minimum, it can be seen that the PAEM barrier is higher in absolute value. It means that the Cl$^-$···CH$_3$ tetrel bond is weaker, and this is confirmed by the linear correlation between PAEM$_{max}$ and E$_b$. Moreover, the relative positions of ESP$_{min}$ and PAEM$_{max}$ along the tetrel bond line allow us to distinguish, which atom is the acceptor of electrons. As it has been noted earlier [45], and as can be seen from the above, PAEM$_{max}$ position is located closer to the electrophilic site, while the position of ESP$_{min}$ is closer to the electron donor.

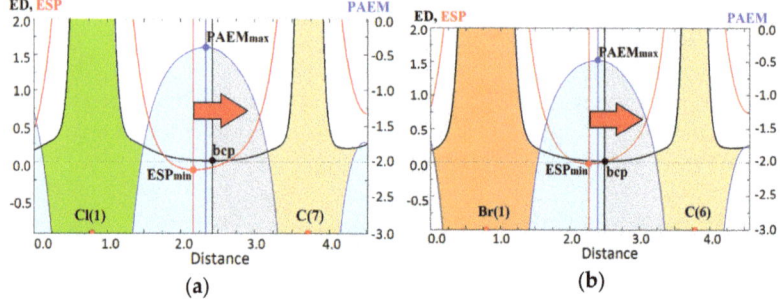

Figure 7. Potential acting on an electron in a molecule, a.u., (blue), electrostatic potential, a.u., (red) and electron density, a.u., (black) along the tetrel bond in (**a**) ZENJAD and (**b**) FADXIR complexes.

3. Materials and Methods

The structure optimization of molecular complexes consisting of organic cations and halide anions was carried out at M06-2X/aug-cc-pVDZ level [47–49] in GAMESS (v. 2017 R2, Mark Gordon's Quantum Theory Group, Ames Laboratory, Iowa State University, Ames, IA, USA [50]) with gradient convergence that equaled 0.00001. The optimized structures were tested for the absence of imaginary frequencies. The binding energy between cations and halide anions in electrically neutral complexes was estimated as $E_b = E_{com} - (E_{Hal} + E_{cat}) - \Delta E_{BSSE}$, where E_{com}, E_{cat}, E_{Hal} were the total energies of the optimized complex, relaxed isolated organic cation and halide anion. BSSE correction, ΔE_{BSSE}, was carried out taking into account the phantom orbitals in complexes calculated for compounds without energy relaxation, see Table S1 in Supporting Information.

All calculations with periodic boundary conditions were performed using CRYSTAL14 (v. 1.0.4, CRYSTAL Theoretical Chemistry Group, Chemistry Department, University of Turin, Turin, Italy [51]) at the B3LYP/6-31G** level for C, N, O, H atoms and DZVP basis set for halogen atoms [52,53] with Grimme dispersion correction D2 [54]. The structure relaxation was carried out with the atomic coordinate optimization only, with the fixed unit cell parameters for the purpose of maximum conformity to experimental data. The following convergence parameters have been used for all calculation: TOLDEG (root-mean-square on gradient) is less than 0.0001 a.u., TOLDEX (root-mean-square on estimated displacements) is less than 0.0003 a.u., TOLDEE (energy change between optimization steps threshold) is less than 10^{-10} a.u., TOLINTEG (truncation criteria for bielectronic integrals: overlap threshold for Coulomb integrals; penetration threshold for Coulomb integrals; overlap threshold for HF exchange integrals; pseudo-overlap for g and n HF exchange series) are 10, 10, 10, 10 and 16, respectively. The number of k-points in the Pack–Monkhorst net (in the irreducible part of Brillouin zone) was 125 or 170 depending on crystals; the number of k-points in the Gilat net was 729 or 1170, that corresponded to the set SHRINK 8 16 values. All calculations for isolated cations were performed using CRYSTAL17 (v. 1.0.2, CRYSTAL Theoretical Chemistry Group, Chemistry Department, University of Turin, Turin, Italy [55]) at the B3LYP/6-31G** level with the Grimme dispersion correction D2 and DOPING option to account for the cation positive charge.

The QTAIM analyses of electron density and electrostatic potential were carried out in TOPOND [56] in crystals and in AIMAll software package [57] for the complexes. PAEM and ESP distributions were computed using Multiwfn [58] program (Beijing Kein Research Center for Natural Sciences, Beijing, China).

The reported calculations were performed using the supercomputer resources of the South Ural State University [59].

4. Conclusions

In this computational study, the charge-assisted tetrel bonds in the crystals formed between halide anions and the methyl groups of organic cations such as $Hal^- \cdots CH_3Y$ (Hal^- = Cl, Br; Y = N, O) have been considered. The bond paths between the Hal^- and C atoms confirm the existence of these uncommon bonds in both the crystal structures and gas phase complexes. To define the type of $Hal^- \cdots CH_3Y$ bonding in crystals more precisely, we have suggested using the order of one-dimensional minima of electron density and electrostatic potential along the interatomic lines between the carbon atom of CH_3-group and the halide anion. This allowed us to apply a simple criterion which reveals that the carbon atom provides its electrophilic site for a typical tetrel bond formation.

The strong correlation between the binding energy in complexes and the extreme values of potential acting on an electron in a molecule calculated along the lines between the Hal^- and C atoms has been obtained. Therefore, PAEM extends and enforces the electronic criterion for revealing electrophilic sites and sheds some light on the nature of tetrel bonds. We may speculate that its application will be useful for the other electrostatically driven non-covalent interactions as well.

Supplementary Materials: The following are available online, Figure S1: Binding energy (kcal/mol) in complexes vs the potential (a) and kinetic (b) energy density (a.u.) at the bond critical point of tetrel bonds, Figure S2: Binding energy (kcal/mol) in complexes vs the electrostatic potential (a.u.) (a) and (b), potential acting on an electron in molecule (a.u.) at the bond critical point of tetrel bonds, Figure S3: ESP in the trimethylammonium chloride on the isosurface of electron density of 0.02 a.u, Table S1: The energy characteristics of Hal$^-$···CH$_3$–YR (Hal$^-$ = Cl, Br) complexes taken from crystal structures with listed refcodes, Table S2. Experimental and calculated tetrel and C–Y bond lengths D (Å), angles Hal$^-$···C–Y and electron density (a.u.) at bond critical points for considered crystal and cation structures calculated in CRYSTAL code. Table S3, Bond lengths D(Å), the characteristics of electron density, potential and kinetic energy densities (a.u.), electrostatic potential (a.u.), potential acting on an electron in molecule PAEM at bcp (a.u.) for Hal$^-$···CH$_3$ and Y–C bonds in complexes and cations calculated in GAMESS code.

Author Contributions: Conceptualization, V.T. and E.B.; investigation, Y.M.; formal analysis and project administration E.B.

Funding: This research was funded by the Russian Foundation for Basic Research, grant No. 17-03-00406, Ministry of Science and Higher Education of the Russian Federation (grant No. 4.1157.2017/4.6) and by the Government of the Russian Federation, Act 211, contract No. 02.A03.21.0011.

Conflicts of Interest: The authors declare no conflicts of interest.

References

1. Legon, A.C. Tetrel, Pnictogen and Chalcogen Bonds Identified in the Gas Phase before they had Names: A Systematic Look at Non-covalent Interactions. *Phys. Chem. Chem. Phys.* **2017**, *19*, 14884–14896. [CrossRef] [PubMed]
2. Cavallo, G.; Metrangolo, P.; Pilati, T.; Resnati, G.; Terraneo, G. Naming Interactions from the Electrophilic Site. *Cryst. Growth Des.* **2014**, *14*, 2697–2702. [CrossRef]
3. Terraneo, G.; Resnati, G. Bonding Matters. *Cryst. Growth Des.* **2017**, *17*, 1439–1440. [CrossRef]
4. Desiraju, G.R.; Parthasarathy, R. The Nature of Halogen···Halogen Interactions: Are Short Halogen Contacts Due to Specific Attractive Forces or Due to Close Packing of Nonspherical Atoms? *J. Am. Chem. Soc.* **1989**, *111*, 8725–8726. [CrossRef]
5. Desiraju, G.R.; Ho, P.S.; Kloo, L.; Legon, A.C.; Marquardt, R.; Metrangolo, P.; Politzer, P.; Resnati, G.; Rissanen, K. Definition of the Halogen Bond (IUPAC Recommendations 2013). *Pure Appl. Chem.* **2013**, *85*, 1711–1713. [CrossRef]
6. Politzer, P.; Murray, J.S.; Clark, T. Halogen Bonding: An Electrostatically-Driven Highly Directional Noncovalent Interaction. *Phys. Chem. Chem. Phys.* **2010**, *12*, 7748–7757. [CrossRef]
7. Politzer, P.; Murray, J.S.; Clark, T.; Resnati, G. The σ-hole Revisited. *Phys. Chem. Chem. Phys.* **2017**, *19*, 32166–32178. [CrossRef]
8. Del Bene, J.E.; Alkorta, I.; Elguero, J. Anionic Complexes of F$^-$ and Cl$^-$ with Substituted Methanes: Hydrogen, Halogen, and Tetrel Bonds. *Chem. Phys. Lett.* **2016**, *655–656*, 115–119. [CrossRef]
9. Scheiner, S. Tetrel Bonding as a Vehicle for Strong and Selective Anion Binding. *Molecules* **2018**, *23*, 1147. [CrossRef]
10. Esrafili, M.D.; Asadollahi, S.; Mousavian, P. Anionic Tetrel Bonds: An ab Initio Study. *Chem. Phys. Lett.* **2018**, *691*, 394–400. [CrossRef]
11. Esrafili, M.D.; Mousavian, P. Strong Tetrel Bonds: Theoretical Aspects and Experimental Evidence. *Molecules* **2018**, *23*, 2642. [CrossRef]
12. Grabowski, S.J. Tetrel bond–σ-hole bond as a preliminary stage of the S$_N$2 reaction. *Phys. Chem. Chem. Phys.* **2014**, *16*, 1824–1834. [CrossRef]
13. Mani, D.; Arunan, E. The X–C···Y (X = O/F, Y = O/S/F/Cl/Br/N/P) 'carbon bond' and hydrophobic interactions. *Phys. Chem. Chem. Phys.* **2013**, *15*, 14377–14383. [CrossRef]
14. Bauza, A.; Frontera, A. RCH$_3$···O Interactions in Biological Systems: Are They Trifurcated H-Bonds or Noncovalent Carbon Bonds? *Crystals* **2016**, *6*, 26. [CrossRef]
15. Garcia-LLinas, X.; Bauza, A.; Seth, S.K.; Frontera, A. Importance of R−CF$_3$···O Tetrel Bonding Interactions in Biological Systems. *J. Phys. Chem. A* **2017**, *121*, 5371–5376. [CrossRef] [PubMed]
16. Pal, P.; Konar, S.; Lama, P.; Das, K.; Bauza, A.; Frontera, A.; Mukhopadhyay, S. On the Importance of Noncovalent Carbon Bonding Interactions in the Stabilization of a 1D Co(II) Polymeric Chain as Precursor of a Novel 2D Coordination Polymer. *J. Phys. Chem. B* **2016**, *120*, 6803–6811. [CrossRef]

17. Tsirelson, V.G.; Stash, A.I.; Potemkin, V.A.; Rykounov, A.A.; Shutalev, A.D.; Zhurova, E.A.; Zhurov, V.V.; Pinkerton, A.A.; Gurskaya, G.V.; Zavodnik, V.E. Molecular and Crystal Properties of Ethyl-4,6-dimethyl-2-thioxo-1,2,3,4-tetrahydropyrimidine-5-carboxylate from Experimental and Theoretical Electron Densities. *Acta Cryst. B* **2006**, *B62*, 676–688. [CrossRef]
18. Thomas, S.P.; Pavan, M.S.; Row, T.N.G. Experimental Evidence for 'Carbon Bonding' in the Solid State from Charge Density Analysis. *Chem. Commun.* **2014**, *50*, 49–51. [CrossRef]
19. Scheiner, S. Steric Crowding in Tetrel Bonds. *J. Phys. Chem. A* **2018**, *122*, 2550–2562. [CrossRef]
20. Zierkiewicz, W.; Michalczyk1, M.; Scheiner, S. Comparison between Tetrel Bonded Complexes Stabilized by σ and π Hole Interactions. *Molecules* **2018**, *23*, 1416. [CrossRef]
21. Scheiner, S. Systematic Elucidation of Factors That Influence the Strength of Tetrel Bonds. *J. Phys. Chem. A* **2017**, *121*, 5561–5568. [CrossRef] [PubMed]
22. Liu, M.; Li, Q.; Scheiner, S. Comparison of Tetrel Bonds in Neutral and Protonated Complexes of PyridineTF$_3$ and FuranTF$_3$ (T = C, Si, and Ge) with NH$_3$. *Phys. Chem. Chem. Phys.* **2017**, *19*, 5550–5559. [CrossRef] [PubMed]
23. Laconsay, C.J.; Galbraith, J.M. A Valence Bond Theory Treatment of Tetrel Bonding Interactions. *Comp. Theor. Chem.* **2017**, *1116*, 202–206. [CrossRef]
24. Grabowski, S.J. Tetrel Bonds with π-Electrons Acting as Lewis Bases—Theoretical Results and Experimental Evidences. *Molecules* **2018**, *23*, 1183. [CrossRef] [PubMed]
25. Zierkiewicz, W.; Michalczyk1, M.; Scheiner, S. Implications of Monomer Deformation for Tetrel and Pnicogen Bonds. *Phys. Chem. Chem. Phys.* **2018**, *20*, 8832–8841. [CrossRef] [PubMed]
26. Stash, A.I.; Chen, Y.S.; Kovalchukova, O.V.; Tsirelson, V.G. Electron Density, Electrostatic Potential, and Spatial Organization of Ammonium Hydrooxalate Oxalic Acid Dihydrate Heteromolecular Crystal from Data of Diffraction Experiment at 15 K Using Synchrotron Radiation and Theoretical Calculations. *Russ. Chem. Bull.* **2013**, *62*, 1752–1763. [CrossRef]
27. Bader, R.F.W. *Atoms in Molecules: A Quantum Theory*; Clarendon Press: Oxford, UK, 1990; pp. 1–438.
28. Bader, R.F.W.; Carroll, M.T.; Cheeseman, J.R.; Chang, C. Properties of Atoms in Molecules: Atomic Volumes. *J. Am. Chem. Soc.* **1987**, *109*, 7968–7979. [CrossRef]
29. Tsirelson, V.G.; Avilov, A.S.; Lepeshov, G.G.; Kulygin, A.K.; Stahn, J.; Pietsch, U.; Spence, J.C.H. Quantitative Analysis of the Electrostatic Potential in Rock-Salt Crystals Using Accurate Electron Diffraction Data. *J. Phys. Chem.* **2001**, *B105*, 5068–5074. [CrossRef]
30. Tsirelson, V.G.; Shishkina, A.V.; Stash, A.I.; Parsons, S. The Experimental and Theoretical QTAIMC Study of the Atomic and Molecular Interactions in Dinitrogen Tetroxide. *Acta Crystallogr. Sect. B Struct. Sci. Cryst. Eng. Mater.* **2009**, *B65*, 647–658. [CrossRef] [PubMed]
31. Mata, I.; Molins, E.; Alkorta, I.; Espinosa, E. Topological Properties of the Electrostatic Potential in Weak and Moderate N···H Hydrogen Bonds. *J. Phys. Chem.* **2007**, *A111*, 6425–6433. [CrossRef]
32. Bartashevich, E.V.; Yushina, I.D.; Kropotina, K.K.; Muhitdinova, S.E.; Tsirelson, V.G. Testing the Tools for Revealing and Characterizing the Iodine-Iodine Halogen Bond in Crystals. *Acta Crystallogr. Sect. B Struct. Sci. Cryst. Eng. Mater.* **2017**, *B73*, 217–226. [CrossRef] [PubMed]
33. Pathak, R.K.; Gadre, S.R. Maximal and Minimal Characteristics of Molecular Electrostatic Potentials. *J. Chem. Phys.* **1990**, *93*, 1770–1773. [CrossRef]
34. Bartashevich, E.V.; Yushina, I.D.; Stash, A.I.; Tsirelson, V.G. Halogen Bonding and Other Iodine Interactions in Crystals of Dihydrothiazolo(oxazino)quinolinium Oligoiodides from the Electron-Density Viewpoint. *Cryst. Growth Des.* **2014**, *14*, 5674–5684. [CrossRef]
35. Bartashevich, E.V.; Yushina, I.D.; Muhitdinova, S.E.; Tsirelson, V.G. Electronic Criterion for Categorizing the Chalcogen and Halogen Bonds: Sulfur—Iodine Interactions in Crystals. *Acta Crystallogr. Sect. B Struct. Sci. Cryst. Eng. Mater.* **2019**, in press. [CrossRef]
36. Zhao, D.-X.; Gong, L.-D.; Yang, Z.-Z. The Relations of Bond Length and Force Constant with the Potential Acting on an Electron in a Molecule. *J. Phys. Chem. A* **2005**, *109*, 10121–10128. [CrossRef] [PubMed]
37. Bartashevich, E.V.; Tsirelson, V.G. A Comparative View on the Potential Acting on an Electron in a Molecule and the Electrostatic Potential through the Typical Halogen Bonds. *J. Comput. Chem.* **2018**, *39*, 573–580. [CrossRef] [PubMed]

38. Bartashevich, E.V.; Mukhitdinova, S.E.; Tsirelson, V.G. Characterizing the Halogen and Chalcogen Bonds in Crystals: PAEM vs ESP. In *Book of Abstracts of International Union of Crystallography (IUCr)'s Sagamore XIX Conference on Quantum Crystallography, Halifax, Canada*; Mount Saint Vincent University's Printshop: Halifax, NS, Canada, 2018; pp. 93–95.
39. Groom, C.R.; Bruno, I.J.; Lightfoot, M.P.; Ward, S.C. The Cambridge Structural Database. *Acta Crystallogr. Sect. B Struct. Sci. Cryst. Eng. Mater.* **2016**, *72*, 171–179. [CrossRef] [PubMed]
40. Bondi, A. Van der Waals Volumes and Radii. *J. Phys. Chem.* **1964**, *68*, 441–451. [CrossRef]
41. Tiritiris, I.; Kantlehner, W. Crystal Structure of N,N,N',N'-tetramethylchloroformamidinium chloride, $[C_5H_{12}N_2Cl]Cl$. *Z. Kristallogr. New Cryst. Struct.* **2008**, *223*, 345–346. [CrossRef]
42. Burg, A.B. Restudy of the Action of Sulfur Dioxide on Dry Trimethylamine Oxide: Iodine Oxidation and Lewis Acid Chemistry of the Most Reactive Product, $(CH_3)_2(H)NCH_2SO_3$. *Inorg. Chem.* **1989**, *28*, 1295–1300. [CrossRef]
43. Clark, G.R.; Shaw, G.L.; Surman, P.W.J.; Taylor, M.J.; Steele, D. Preparation, Structure and Vibrational Spectrum of the Dimethylmethyleniminium Ion, Including the Role of Cationic Polymers in its Formation. *J. Chem. Soc. Faraday Trans.* **1994**, *90*, 3139–3144. [CrossRef]
44. Silvi, B.; Savin, A. Classification of Chemical Bonds Based on Topological Analysis of Electron Localization Functions. *Nature* **1994**, *371*, 683–686. [CrossRef]
45. Bartashevich, E.V.; Tsirelson, V.G. Interplay between Non-covalent Interactions in Complexes and Crystals with Halogen Bonds. *Russ. Chem. Rev.* **2014**, *83*, 1181–1203. [CrossRef]
46. Kuznetsov, M.L. Can Halogen Bond Energy be Reliably Estimated from Electron Density Properties at Bond Critical Point? The Case of the $(A)nZ–Y\cdots X–$ (X, Y = F, Cl, Br) Interactions. *Int. J. Quantum Chem.* **2018**. [CrossRef]
47. Dunning, T.H., Jr. Gaussian Basis Sets for Use in Correlated Molecular Calculations. I. The Atoms Boron through Neon and Hydrogen. *J. Chem. Phys.* **1989**, *90*, 1007–1023. [CrossRef]
48. Wilson, A.K.; Woon, D.E.; Peterson, K.A.; Dunning, T.H., Jr. Gaussian Basis Sets for Use in Correlated Molecular Calculations. IX. The Atoms Gallium through Krypton. *J. Chem. Phys.* **1999**, *110*, 7667–7676. [CrossRef]
49. Zhao, Y.; Truhlar, D.G. The M06 Suite of Density Functionals for Main Group Thermochemistry, Thermochemical Kinetics, Noncovalent Interactions, Excited States, and Transition Elements: Two New Functionals and Systematic Testing of Four M06-class Functionals and 12 Other Functionals. *Theoret. Chem. Acc.* **2008**, *120*, 215–241. [CrossRef]
50. Schmidt, M.W.; Baldridge, K.K.; Boatz, J.A.; Elbert, S.T.; Gordon, M.S.; Jensen, J.H.; Koseki, S.; Matsunaga, N.; Nguyen, K.A.; Su, S.; et al. General Atomic and Molecular Electronic Structure System. *J. Comput. Chem.* **1993**, *14*, 1347–1363. [CrossRef]
51. Dovesi, R.; Saunders, V.R.; Roetti, C.; Orlando, R.; Zicovich-Wilson, C.M.; Pascale, F.; Civalleri, B.; Doll, K.; Harrison, N.M.; Bush, I.J.; et al. *CRYSTAL14 User's Manual*; University of Torino: Torino, Italy, 2014.
52. Becke, A.D. Density-Functional Thermochemistry. III. The Role of Exact Exchange. *J. Chem. Phys.* **1993**, *98*, 5648–5652. [CrossRef]
53. Lee, C.; Yang, W.; Parr, R.G. Development of the Colle-Salvetti Correlation-Energy Formula into a Functional of the Electron Density. *Phys. Rev. B Condens. Matter Mater. Phys.* **1988**, *37*, 785–789. [CrossRef]
54. Grimme, S. Semi-empirical GGA-Type Density Functional Constructed with a Long-Range Dispersion Correction. *J. Comput. Chem.* **2006**, *27*, 1787–1799. [CrossRef] [PubMed]
55. Dovesi, R.; Saunders, V.R.; Roetti, C.; Orlando, R.; Zicovich-Wilson, C.M.; Pascale, F.; Civalleri, B.; Doll, K.; Harrison, N.M.; Bush, I.J.; et al. *CRYSTAL17 User's Manual*; University of Torino: Torino, Italy, 2017.
56. Gatti, C.; Casassa, S. *Topond14 User's Manual*; University of Torino: Torino, Italy, 2014.
57. Keith, T.A. AIMALL, Version 12.06.03, 2012 Professional. Available online: http://aim.tkgristmill.com (accessed on 20 February 2019).

58. Lu, T.; Chen, F. Multiwfn: A multifunctional wavefunction analyzer. *J. Comput. Chem.* **2012**, *33*, 580–592. [CrossRef]
59. Kostenetskiy, P.; Semenikhina, P. SUSU Supercomputer Resources for Industry and Fundamental Science. In Proceedings of the Global Smart Industry Conference (GloSIC), Chelyabinsk, Russia, 13–15 November 2018; pp. 1–7. [CrossRef]

© 2019 by the authors. Licensee MDPI, Basel, Switzerland. This article is an open access article distributed under the terms and conditions of the Creative Commons Attribution (CC BY) license (http://creativecommons.org/licenses/by/4.0/).

Article

Crystallographic and Computational Characterization of Methyl Tetrel Bonding in S-Adenosylmethionine-Dependent Methyltransferases

Raymond C. Trievel [1,*] and Steve Scheiner [2]

1. Department of Biological Chemistry, University of Michigan, Ann Arbor, MI 48109, USA
2. Department of Chemistry and Biochemistry, Utah State University, Logan, UT 84322, USA; steve.scheiner@usu.edu
* Correspondence: rtrievel@umich.edu; Tel.: +1-734-647-0889; Fax: +1-734-763-4581

Received: 5 October 2018; Accepted: 20 October 2018; Published: 13 November 2018

Abstract: Tetrel bonds represent a category of non-bonding interaction wherein an electronegative atom donates a lone pair of electrons into the sigma antibonding orbital of an atom in the carbon group of the periodic table. Prior computational studies have implicated tetrel bonding in the stabilization of a preliminary state that precedes the transition state in S_N2 reactions, including methyl transfer. Notably, the angles between the tetrel bond donor and acceptor atoms coincide with the prerequisite geometry for the S_N2 reaction. Prompted by these findings, we surveyed crystal structures of methyltransferases in the Protein Data Bank and discovered multiple instances of carbon tetrel bonding between the methyl group of the substrate S-adenosylmethionine (AdoMet) and electronegative atoms of small molecule inhibitors, ions, and solvent molecules. The majority of these interactions involve oxygen atoms as the Lewis base, with the exception of one structure in which a chlorine atom of an inhibitor functions as the electron donor. Quantum mechanical analyses of a representative subset of the methyltransferase structures from the survey revealed that the calculated interaction energies and spectral properties are consistent with the values for bona fide carbon tetrel bonds. The discovery of methyl tetrel bonding offers new insights into the mechanism underlying the S_N2 reaction catalyzed by AdoMet-dependent methyltransferases. These findings highlight the potential of exploiting these interactions in developing new methyltransferase inhibitors.

Keywords: noncovalent bond; sigma-hole; charge transfer; molecular electrostatic potential; tetrel bond; methylation; methyltransferase; methyl transfer; S-adenosylmethionine; AdoMet; SAM; S_N2 reaction

---

1. Introduction

Methyltransferases represent a ubiquitous class of enzymes that methylate a vast array of small molecules and macromolecules and participate in numerous biological processes, including metabolism, signal transduction, and gene expression [1–3]. The majority of these enzymes utilize the methyl donor S-adenosylmethionine (AdoMet) whose methyl group is rendered highly reactive through its bonding to a sulfonium cation in the substrate. AdoMet-dependent methyltransferases catalyze an S_N2 reaction wherein a nucleophilic atom, such as oxygen, nitrogen, or sulfur, attacks the electrophilic methyl carbon atom of AdoMet, with the sulfur atom displaced as the leaving group [4]. The reaction mechanism of these enzymes has been a subject of intense study for over 40 years [5] and has led to the proposal of several different models for catalysis. These models include (1) compression or compaction of nucleophile, electrophile, and leaving groups along the reaction coordinate [6–9], (2) formation of near attack conformers (NACs) that align the nucleophile and methyl group in a productive geometry for the S_N2 reaction [10–13], (3) electrostatic pre-organization within the active site that promotes methyl transfer [14,15], and (4) cratic effects involving the free energy of association of the substrates in a catalytically favorable alignment within the active site [16,17]. Despite these models, the methyltransferase mechanism remains a topic of active debate.

Recent structure–function studies of methyltransferases have explored the interactions between their active sites and the AdoMet sulfonium cation. A survey of high-resolution crystal structures of methyltransferases in the Protein Databank (PDB) identified unconventional carbon–oxygen (CH···O) hydrogen bonds between the AdoMet methyl group and oxygen atoms within the active sites of different classes of these enzymes [18]. Quantum mechanical (QM) calculations demonstrated that the AdoMet methyl group forms relatively strong CH···O hydrogen bonds due to its polarization by the neighboring sulfonium cation [18–20]. Correlatively, structural and biochemical characterization of the protein lysine N-methyltransferase (KMT) SET7/9 and the reactivation domain of methionine synthase demonstrated that these hydrogen bonds promote high affinity binding to AdoMet compared to the methyl transfer product S-adenosylhomocysteine (AdoHcy), thus mitigating product inhibition [18,21]. Moreover, CH···O and CH···N interactions with the AdoMet methyl group have been proposed to contribute to transition state stabilization in several methyltransferases, including SET7/9, SET8, NSD2, and glycine N-methyltransferase [18,22–24].

In addition to unconventional hydrogen bonding, chalcogen bonding between the AdoMet sulfur cation and the active sites of methyltransferases has also been observed [25]. A chalcogen bond is defined as a non-bonded interaction wherein a Lewis base donates a lone pair of electrons into the sigma antibonding (σ^*) orbital of an atom from the Group VI elements (oxygen group) of the periodic table [26]. Structural and functional characterization of an S···O chalcogen bond between AdoMet and an asparagine residue in the active site of SET7/9 demonstrated that this interaction enhances the binding affinity for the substrate relative to AdoHcy and modestly augments the rate of methyl transfer [25]. Together, these results illustrate that carbon hydrogen bonding and sulfur chalcogen bonding between the AdoMet sulfonium cation and residues in the methyltransferase active site can enhance the enzyme's binding affinity for the substrate and promote the methyl transfer reaction.

Beyond hydrogen bonding and chalcogen bonding, there is a third unconventional interaction that can occur with sulfonium cations involving a σ^* orbital of a carbon atom [27]. This interaction is termed a tetrel bond and occurs when an atom from the Group IV elements (carbon group) of the periodic table accepts a lone pair of electrons from an electronegative atom [28,29]. In the case of AdoMet, this interaction can occur with the σ^* orbital of the methyl carbon atom that corresponds to the sulfur–carbon (S–CH$_3$) bond.

Although aliphatic carbon atoms typically form weak tetrel bonds compared to other Group IV elements, QM calculations have demonstrated that a methyl carbon atom bonded to a sulfonium ion can form relatively strong tetrel interactions due to polarization by the adjacent cation [27]. Notably, the geometry of the tetrel bond, in which the interaction angle between the Lewis base (X) and S–CH$_3$ bond is approximately linear, precludes strong methyl CH···X hydrogen bonding due to acute hydrogen bond angles [27,30]. Experimental evidence for carbon tetrel interactions first emerged from a survey of the Cambridge Structural Database, which identified over 700 small molecule crystal structures displaying C···O tetrel bonds, including multiple interactions involving methyl groups [31]. In addition, recent studies by Frontera and colleagues have revealed crystallographic evidence of methyl and trifluoromethyl C···O tetrel bonding between proteins and various ligands [30,32]. Pertinent to methyltransferases, a computational analysis by Grabowski directly implicated tetrel bonding between an electrophilic tetrel atom and a nucleophile as a preliminary state that precedes the transition state in S$_N$2 reactions, including methyl transfer [33]. Collectively, these findings prompted us to examine structures of AdoMet-dependent methyltransferases to ascertain whether methyl tetrel bonding occurs in these enzymes. The results of our structural survey coupled with corroborative QM calculations demonstrate the existence of the tetrel bonding in methyltransferases, furnishing insights into the potential roles of these interactions in ligand binding and S$_N$2 catalysis.

2. Material and Methods

2.1. PDB Survey

Crystal structures of methyltransferase/AdoMet complexes with a resolution of ≤2.50 Å were downloaded from the PDB and visually examined for the presence of carbon tetrel bonding to the AdoMet methyl group. Tetrel bonds between the AdoMet methyl group and an electronegative atom (X) of a small molecule inhibitor, solvent molecule, or ion were defined as exhibiting: (1) an θ(S–C···X) interaction angle between 160° and 180° (where S and C are the sulfur and methyl carbon atoms of AdoMet, respectively) and (2) a C···X interaction distance less than or equal to sum of the van der Waals radii of the carbon and electronegative atoms, specifically R(C···O) ≤3.25 Å and R(C···Cl) ≤3.5 Å (carbon, oxygen, and chlorine van der Waals radii were defined as 1.75 Å, 1.5 Å, and 1.75 Å, respectively) [34]. These geometric parameters are consistent with the formal definition of halogen bonding, a related category of interactions that are considered an archetype for σ-hole bonding [35]. For crystal structures displaying potential carbon tetrel bonds, the electron density corresponding to AdoMet and the electron donor were visually inspected using the program Coot to confirm the integrity of the model [36,37]. Structures that displayed ambiguous electron density for the ligands were omitted from the survey. For the structure of the DhpI phosphonate O-methyltransferase (accession code 3OU6.pdb), the AdoMet molecules were remodeled in the electron density maps using the real space refinement and geometry tools in Coot. The remodeled AdoMet coordinates were then used to measure the tetrel bond geometries (Table 1). Finally, in cases where two or more structures of a given methyltransferase possess the same tetrel bond donor, such as interactions involving water molecules and the COMT/AdoMet/DNC/Mg^{2+} complexes, only the highest resolution structure of the wild type enzyme is reported in Table 1.

Table 1. Crystallographic survey of methyl tetrel bonding in AdoMet-dependent methyltransferases.

Enzyme	PDB Code	Resolution (Å)	Ligand	Electron Donor (X)	R(C···X) Length (Å) [A]	θ(S–C···X) Angle (°) [A]
ASH1L	4YNM	2.19	H$_2$O	O	2.99 (B)	163 (B)
Bud23	4QTU	2.12	Ethylene glycol	O	3.04 (B)	173 (B)
COMT	2CL5	1.6	BIA 8-176 [B]	O	2.70 (A), 2.69 (B)	173 (A), 172 (B)
COMT	3S68	1.85	Tolcapone [C]	O	2.50	166
COMT	4XUC	1.8	Compound 18 [D]	O	2.64	175
COMT	4XUD	2.4	Compound 32 [E]	O	2.73	166
COMT	5LSA	1.5	3,5-Dinitrocatechol	O	2.71	173
DhpI	3OU6	2.3	Sulfate	O	3.00 (A), 3.09 (B), 2.97 (C)	175 (A), 175 (B), 176 (C)
G9A	5VSC	1.4	H$_2$O	O	3.14 (A), 3.17 (B)	166 (A), 168 (B)
GLP	5TTG	1.66	H$_2$O	O	3.15 (A), 3.24 (B)	168 (A), 169 (B)
MMSET	5LSU	2.14	H$_2$O	O	3.13 (B)	160 (B)
PrmA	2NXE	1.75	H$_2$O	O	3.08 (B)	171 (B)
PRMT5	5EML	2.39	H$_2$O	O	3.09 (A)	163 (A)
RsmF	3M6V	1.82	H$_2$O	O	3.20 (A), 3.23 (B)	164 (A), 162 (B)
SMYD2	3S7B	2.42	AZ505 [F]	O	2.77	169
SMYD2	3TG4	2.0	Glycerol	O	3.23	176
SMYD2	5ARG	1.99	SGC Probe BAY-598 [G]	Cl	3.43	175

Table 1. Cont.

Enzyme	PDB Code	Resolution (Å)	Ligand	Electron Donor (X)	R(C⋯X) Length (Å) [A]	θ(S–C⋯X) Angle (°) [A]
SMYD3	3QWP	1.53	Glycerol	O	3.01	163
SMYD3	5CCL	1.5	Oxindole compound [H]	O	2.89	164
SMYD3	5CCM	2.3	EPZ030456 [I]	O	2.78	168

Note: [A]: A, B, and C denote the protein chains in the asymmetric unit of the crystal structure; [B]: (3,4- dihydroxy-2-nitrophenyl)(phenyl)methanone; [C]: (3,4-dihydroxy-5-nitrophenyl)(4- methylphenyl)methanone; [D]: 1-(biphenyl-3-yl)-3-hydroxypyridin-4(1H)-one; [E]: [1-(biphenyl-3-yl)-5-hydroxy-4-oxo-1,4-dihydropyridin-3-yl]boronic acid; [F]: N-cyclohexyl-N~3~-[2-(3,4-dichlorophenyl)ethyl]-N-(2-{[2-(5-hydroxy-3-oxo-3,4-dihydro-2H-1,4-benzoxazin-8-yl)ethyl]amino}ethyl)-beta-alaninamide; [G]: N-[1-(N'-cyano-N-[3-(difluoromethoxy)phenyl]carbamimidoyl)-3-(3,4-dichlorophenyl)-4,5-dihydro-1H-pyrazol-4-yl]-N-ethyl-2-hydroxyacetamide; [H]: 2-oxanylidene-N-piperidin-4-yl-1,3-dihydroindole-5-carboxamide; [I]: 6-chloranyl-2-oxanylidene-N-[(1S,5R)-8-[4-[(phenylmethyl)amino]piperidin-1-yl]sulfonyl-8-azabicyclo[3.2.1]octan-3-yl]-1,3-dihydroindole-5-carboxamide.

2.2. QM Calculations

Quantum calculations were carried out within the framework of the Gaussian-09 program. Active site models for SMYD2 (5ARG.pdb), SMYD3 (5CCL.pdb), COMT (5LSA.pdb), and G9A (5VSC.pdb) were generated from their respective crystallographic coordinates, with the heavy atom (non-hydrogen) positions fixed. Hydrogen atom positions were not derived from the enzymes' X-ray structures but were added to the models followed by optimization of their positions at the M06-2X/6-31 + G** level. Energetics and NBO analyses [38,39] were performed at the M06-2X level with a larger aug-cc-pVDZ basis set. Interaction energies were evaluated as the difference in energy between the full system on one hand, and the sum of its components, as defined in the text, on the other. These quantities were corrected for the basis set superposition error with the counterpoise method. Spectral properties were computed at the M06-2X/6-31 + G** level. Interactions of each system with a polarizable medium were estimated via the CPCM method [40]. NMR data were computed with the GIAO approximation [41,42].

3. Results

3.1. Methyltransferase Structural Survey

A comprehensive survey of methyltransferase crystal structures (≤2.5 Å resolution) in the Protein Data Bank (PDB) was conducted to determine whether these enzymes exhibit evidence of methyl tetrel bonding. The survey comprised 269 structures and identified 20 nonredundant structures that display interaction geometries consistent with tetrel bonding between the AdoMet methyl group and ligands, ions, or solvent molecules within the active site (Table 1). Notably, no methyl tetrel bonding was observed between AdoMet and residues in the methyltransferases because the active sites of these enzymes preferentially orient the methyl group for nucleophilic attack by the methyl acceptor substrate. This finding explains the preponderance of AdoMet methyl tetrel bonding with ligands, ions, and solvent occupying the acceptor substrate binding cleft and thus the overall low percentage of methyltransferase structures displaying tetrel interactions. In contrast, CH⋯O hydrogen bonding between the AdoMet methyl group and active site residues was observed in a high proportion of methyltransferase structures, as these interactions mediate substrate recognition by the enzymes and promote the alignment of the methyl group during the S_N2 reaction [18,43].

The majority of the interactions observed in the PDB survey represent methyl C⋯O tetrel bonds, with the exception of a single structure displaying a C⋯Cl tetrel interaction between AdoMet and a small molecule inhibitor. The enzymes exhibiting tetrel bonding belong to either the (1) canonical class I methyltransferases (also known as the Rossmann fold-like or seven β-stranded methyltransferases) or (2) the Suppressor of variegation, Enhancer of Zeste, and Trithorax (SET) domain class of KMTs. This finding is not unexpected, given that these two classes are among the most abundant methyltransferases [2,3]. Furthermore, several members of the class I methyltransferases and SET domain KMTs are drug targets [44–48], resulting in the determination of multiple structures of these enzymes bound to various ligands.

Among the class I methyltransferases, several inhibitor-bound structures of catechol O-methyltransferase (COMT) display interactions indicative of C···O tetrel bonding (Table 1). COMT catalyzes the methylation of the hydroxyl groups of catechol substrates, such as norepinephrine, epinephrine, and dopamine, representing an initial step in their degradation [49]. Given its role in catechol catabolism, COMT represents an important drug target for treating neurological disorders such as Parkinson's Disease and schizophrenia [45,49]. The COMT inhibitors identified in the survey represent catechol or catechol-like substrate analogs that mimic the binding of the substrate in the active site [46,50–52], as illustrated by the ternary complex of the enzyme bound to AdoMet and 3,5-dinitrocatechol (DNC) (Figure 1a). The interaction distances between the AdoMet methyl group and the oxygen atoms in the catechol analog inhibitors (R(C···O) = 2.5–2.8 Å) are considerably shorter than the sum of the carbon and oxygen van der Waals radii (3.25 Å), indicative of strong tetrel bonding. Correlatively, Vidgren et al. noted a 2.6 Å C···O interaction between the AdoMet methyl group and oxygen atom of DNC in the first crystal structure of COMT [53]. These short interaction distances are presumably a consequence of the protonation state of the catechol hydroxyl group participating in the tetrel bond. Catechol substrates and analog inhibitors have been posited to bind to COMT as a deprotonated catecholate due to stabilization of the phenoxide anion through resonance with the aromatic ring and its substituents, as well as by coordination to the Mg^{2+} cation in the active site [9]. The effect of the catecholate charge on AdoMet methyl C···O tetrel bonding is investigated computationally in Section 3.2.

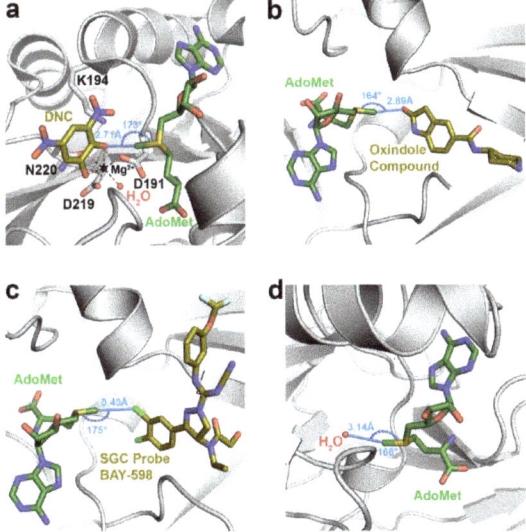

Figure 1. Representative examples of methyl tetrel bonding in crystal structures of AdoMet-dependent methyltransferases. AdoMet and small molecule inhibitors are depicted with green and yellow carbon atoms, respectively. Methyl tetrel bonding distances R(C···X) and angles θ(S–C···X) are denoted in blue. (**a**) COMT bound to AdoMet, DNC, and an Mg^{2+} ion (PDB accession code 5LSA). Key active site residues are illustrated, including the Mg^{2+}-coordinating residues and the catalytic Lys194. (**b**) The SET domain KMT SMYD3 bound to AdoMet and an oxindole-containing inhibitor (5CCL). (**c**) SMYD2/AdoMet/SGC Probe BAY-598 ternary complex (5ARG). (**d**) SET domain KMT G9A bound to AdoMet and Inhibitor 13 (not shown) (5VSC). A water molecule in the substrate lysine binding channel of the enzyme engages in a methyl C···O tetrel bond with AdoMet.

The SET domain KMTs represent the second methyltransferase class exhibiting evidence of AdoMet methyl tetrel bonding. These KMTs comprise several sub-classes, including the SET and MYND (Myeloid-Nervy-DEAF1) Domain-containing (SMYD) methyltransferases [54]. The human genome encodes five SMYD homologs, several of which have been implicated in cancer and cardiovascular disease, rendering them targets for drug design [54–56]. In particular, multiple structures of SMYD2 and SMYD3 in complex with various small molecule inhibitors have been determined [57–59]. Several of these structures display methyl C···O tetrel bonding (Table 1), as illustrated in the ternary complex of SMYD3, AdoMet, and an oxindole-containing compound (Figure 1b). Notably, the SMYD2 and SMYD3 inhibitors that form methyl C···O tetrel bonds with AdoMet are structurally dissimilar, unlike the catechol-based inhibitors of COMT. Unique among the structures in the survey, the SMYD3 inhibitor SGC Probe Bay-598 engages in an unusual C···Cl tetrel bond with the methyl group of AdoMet (Figure 1c). The length of this tetrel interaction (3.43 Å) is longer than that observed in C···O tetrel bonding due to the larger van der Waals radius of chlorine (Table 1). In summary, the inhibitor-bound structures of SMYD2 and SMYD3 illustrate that structurally diverse molecules can engage in AdoMet methyl tetrel bonding and that the electron donor is not limited to oxygen atoms, as halogens and potentially other Lewis bases can engage in tetrel interactions with the substrate.

In addition to interactions with small molecule inhibitors, the PDB survey also uncovered evidence of tetrel bonding between the AdoMet methyl group and solvent molecules as well as ions bound within methyltransferase active sites (Table 1). There are several structures that display C···O tetrel bonding between AdoMet and water molecules, including the class I methyltransferases PrmA, PRMT5, and RsmF, as well as the SET domain KMTs ASH1L, MMSET, GLP, and G9A (Figure 1d). Similarly, the hydroxyl groups of ethylene glycol and glycerol engage in methyl C···O tetrel bonding with AdoMet, as observed in the structures of SMYD2, SMYD3, and the class I enzyme Bud23 (Table 1). In addition to solvent molecules, methyl C···O tetrel bonding is observed between AdoMet and a sulfate anion in the structure of the phosphonate O-methyltransferase DhpI. The sulfate anion bound in the enzyme's active site has been proposed to mimic the phosphonate group of the methyl acceptor substrate [60], suggesting that the sulfate may function as a non-reactive substrate analog, similar to the catechol-based inhibitors of COMT. Consistent with this observation, the C···O tetrel bonds between AdoMet and sulfate are generally shorter and closer to linearity than the tetrel interactions involving solvent molecules (Table 1). Thus, these interactions may potentially represent a Michaelis complex-like state in DhpI and mimic the reaction coordinate for phosphonate methylation. Finally, the finding that AdoMet methyl tetrel bonds involving solvent molecules tend to be longer (R(C···O) = 3.0–3.25 Å) than the interactions observed in the complexes with inhibitors and substrate analogs (Table 1) implies that the solvent interactions may be energetically weaker. This observation is examined in Section 3.2.

3.2. Computational Results

After completing the PBD survey, we selected four methyltransferase structures for computational analysis to investigate the theoretical energies and spectroscopic properties of the observed tetrel bonds with the AdoMet methyl group. These structures include the COMT/AdoMet/DNC/Mg^{2+}, SMYD3/AdoMet/oxindole, and SMYD2/AdoMet/SGC BAY-598 complexes, as well as the G9A structure exhibiting a C···O tetrel bond between AdoMet and a water molecule (Figure 1). For the purposes of the QM calculations, AdoMet was represented as the sulfonium cation MeS$^+$(Et)$_2$, as previously reported [21,25,61]. The ability of the methyl group of this moiety to engage in a tetrel bond was first examined by computing its molecular electrostatic potential (MEP), as illustrated in Figure 2. As a cation, the MEP is positive at all positions with the most positive regions highlighted in blue. There is a region of blue along the extension of the S–CH$_3$ bond, corresponding to the σ* orbital that is also referred to as a σ-hole. A maximum occurs on the isodensity surface (0.05 au) with a value of +120 kcal/mol. It is in this region of the surface that Coulombic attraction with a Lewis base may occur.

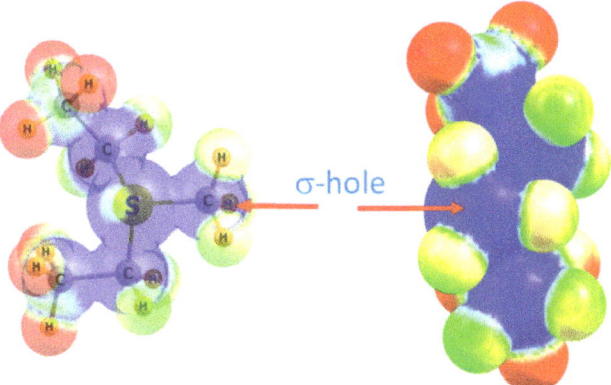

Figure 2. Two views of the molecular electrostatic potential surrounding the MeS$^+$(Et)$_2$ sulfonium cation. Right view looks directly down the H$_3$C–S axis. Blue and red colors correspond respectively to +0.40 and +0.30 au.

The geometries of the pair of relevant interacting groups in the structures of SMYD2, SMYD3, G9A, and COMT are depicted in Figures 3 and 4. The Lewis base in SMYD2 is represented by the o-dichlorobenzene group of the SGC BAY-598 compound, whereas the oxygen electron donor in the SMYD3 complex is represented by the oxindole moiety of the inhibitor. In G9A, a water molecule serves as the Lewis base. In the COMT model, the oxygen electron donor was modeled as a DNC phenoxide anion. The pKa value for the methyl-interacting hydroxyl group is estimated to be ~3.3 and thus has been predicted to bind to the enzyme in a deprotonated state [9]. Together, these four systems cover a range of attributes of potential tetrel bonds. The SMYD2 complex involves a Cl atom as the Lewis base, whereas the more typical oxygen atom assumes this role in the SMYD3, G9A, and COMT complexes. While the first three systems pair the MeS$^+$(Et)$_2$ cation with a neutral partner, the COMT model contains a formal negative charge from the phenoxide anion of DNC.

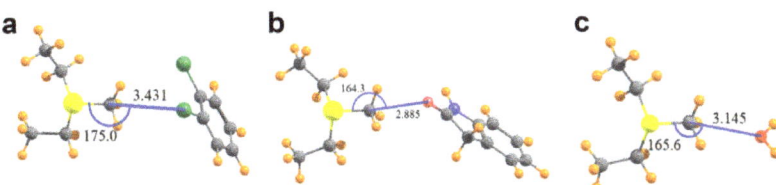

Figure 3. Molecular structures of the models used to computationally analyze AdoMet methyl tetrel bonding in SET domain KMTs. (**a**) Model of MeS$^+$(Et)$_2$ and dichlorobenzene in the SMYD2 structure; (**b**) MeS$^+$(Et)$_2$ and oxindole in the SMYD3 structure; (**c**) MeS$^+$(Et)$_2$ and a water molecule bound in the active site of G9A. Values for the tetrel bond distances and angles are reported in Angstrom and degrees, respectively.

As illustrated in Table 2, the putative tetrel bond is longest in SMYD2 with an intermolecular R(C···Cl) distance of 3.43 Å, and shortest in COMT with R(C···O) = 2.71 Å. All are reasonably close to linearity, the least of which is the θ(S–C···O) angle of 164° in SMYD3/AdoMet/oxindole complex. Table 2 also reports the interaction energies between the two monomers as E$_{int}$, where a negative quantity indicates an attractive interaction. As a frame of reference, an O–H···O hydrogen bond in a water dimer has an interaction energy of −5.8 kcal/mol when calculated at this level of QM theory [18]. The SMYD2/AdoMet/SGC Probe BAY-598 complex containing the longest of the tetrel bonds, with R(C···Cl) = 3.43 Å, is bound by −5.2 kcal/mol. The shorter bond of 2.89 Å in SMYD3 is associated with nearly twice the interaction energy, despite the 10° loss of linearity. Even a C···O length exceeding 3.14 Å in the G9A/AdoMet/H$_2$O complex is associated with a substantial bond energy of −7.0 kcal/mol. A much larger increment, raising the bonding energy to −65.7 kcal/mol, occurs when the partner subunit is negatively charged.

Table 2. Properties of tetrel bond in indicated systems, where X indicates nature of electron donor atom. Energetics are reported in kcal/mol.

Structure	PDB	X	R(C···X) (Å)	θ(S–C···X) (°)	E_{int}	E^T	E^H
SMYD2	5ARG	Cl	3.431	175.0	−5.2	0.63	0.10
SMYD3	5CCL	O	2.885	164.3	−9.0	0.62	0.38
G9A	5VSC	O	3.145	165.6	−7.0	0.46	0.16
COMT	5LSA	O$^-$	2.712	172.7	−65.7	1.33	0.16

E^T: $X_{lp} \rightarrow \sigma^*(SC)$ E^H: $X_{lp} \rightarrow \sigma^*(CH)$.

In order to probe the nature of the interaction, the wave function was analyzed by the NBO procedure which considers charge transfers from one molecular orbital to another. E^T represents the perturbation energy consequence of transfer from the Lewis base (X) lone pair to the σ*(C–S) antibonding orbital, indicative of tetrel bond formation. Because of the proximity of CH protons to the nucleophile, there is the alternate possibility of a CH···X hydrogen bond, which would manifest itself by a transfer into the σ*(H–C) antibonding orbital. Such a possibility is measured by E^H, which is reported in the last column of Table 2. A glance at the last two columns makes it clear that, while there may be a small amount of hydrogen bonding, particularly in the SMYD3 complex with the least linear θ(S–C···O) angle, the interaction is nonetheless dominated by E^T and tetrel bonding. Most importantly, this tetrel bonding is quite strong, as much as −9 kcal/mol for the neutral nucleophile, rising to more than −60 kcal/mol when the latter is an anion.

An additional means to establish the presence of a tetrel bond, which can also distinguish this sort of interaction from a CH···X hydrogen bond is by means of NMR and IR spectral data. A recent study [62] computed these quantities for a range of different complexes in which a methyl group is situated close to a nucleophile, in arrangements much like those considered here. In the case where a tetrel bond is unequivocally present, the chemical shielding of the methyl carbon nucleus is reduced by some 2–14 ppm, relative to the uncomplexed Lewis acid, depending upon the particular system. The methyl protons are deshielded as well, but by much smaller amounts, generally less than 1 ppm. This pattern effectively reverses in the case of a CH···X hydrogen bond where it is the methyl protons that are more strongly deshielded than the carbon nucleus. The vibrational frequencies of the methyl group can also be used to characterize the presence of a tetrel or hydrogen bond. Most diagnostic are the symmetric stretch and bend. The former undergoes a small blue shift in a tetrel bond, but a much larger red shift when it is a hydrogen bond that is present. The symmetric bend, or umbrella mode, is strongly red-shifted for a tetrel bond, but turns toward a blue shift for a hydrogen bond.

With these patterns in mind, Table 3 provides further confirmation of the tetrel bonds that are present in these systems. The methyl carbon atom is deshielded by between 2 and 6 ppm, an amount much larger in magnitude than the deshielding of the methyl protons, less than 1 ppm. The symmetric stretching frequency rises by a small amount, and the umbrella bend is very substantially red-shifted. All of these trends fit perfectly into the aforementioned spectroscopic fingerprint of a tetrel bond. Note also that the quantitative values of these changes follow the same order as do the interaction energies of the three systems listed in Table 2, with the largest changes associated with the COMT complex.

A glance at Figure 4a suggests the likelihood of a hydrogen bond connecting one of the three methyl CH protons with a nitro oxygen atom, since these two atoms lie only 2.16 Å apart. In addition, E for this secondary interaction amounts to −6.29 kcal/mol, larger than that for the tetrel bond itself (this quantity is not reported in Table 2 as it refers to a separate interaction that does not involve the methyl carbon atom directly). Moreover, the electron density at the AIM bond critical points, generally considered an accurate barometer of noncovalent bond strength, are 0.015 and 0.017 au for the C···O tetrel bond and CH···O hydrogen bond, respectively.

Table 3. Calculated changes in the NMR chemical shift (Δσ, ppm) and the symmetric stretching and bending frequencies (Δν, cm^{-1}) within the methyl group of MeS$^+$(Et)$_2$ caused by complexation.

Structure	PDB	Δσ$_C$	Δσ$_H$	Δν$_{str}$	Δν$_{bend}$
SMYD2	5ARG	−1.94	−0.14	0.8	−11.5
SMYD3	5CCL	−2.27	−0.28	5.7	−38.3
G9A	5VSC	−2.36	−0.29	3.9	−22.9
COMT	5LSA	−6.29	−0.95	4.8	−56.8

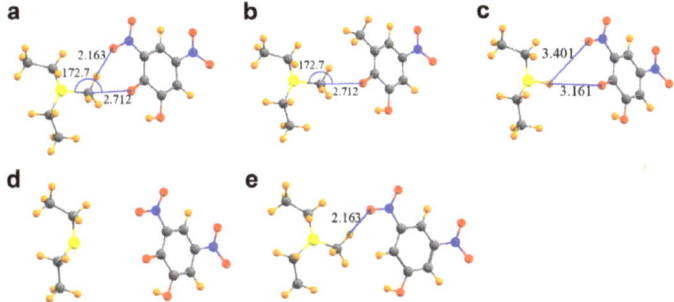

Figure 4. Molecular structures of COMT models used to probe C···O tetrel bonding between AdoMet and DNC. (a) MeS$^+$(Et)$_2$ and the phenoxide anion of DNC; (b) Model in which the 3-nitro moiety of DNC is replaced with a methyl group; (c) Complex wherein the methyl group of MeS$^+$(Et)$_2$ is replaced with a hydrogen atom, yielding HS$^+$(Et)$_2$; (d) Model of DNC and the thioether S(Et)$_2$ representing the product AdoHcy; (e) Complex in which the phenoxide oxygen atom of DNC is substituted by a hydrogen atom.

These two seemingly similar attractive forces need to be disentangled so as to better estimate the interaction energy of the tetrel bond itself. One way to evaluate the latter quantity is to remove the hydrogen bond entirely. Replacement of the nitro (NO$_2$) moiety by a simple methyl group deletes any possible hydrogen bond, while leaving the tetrel bond intact. The interaction energy of the resulting derivative dimer in Figure 4b is −62.1 kcal/mol, less attractive than the full complex by 3.6 kcal/mol, providing one estimate of the hydrogen bond energy. The replacement of the nitro moiety by a methyl group has secondary effects in that, for example, the removal of the electron-withdrawing nitro group would tend to make the phenoxide oxygen atom a bit more potent Lewis base, which would in turn amplify the interaction energy. One can effectively eliminate both the tetrel and hydrogen bonds by replacing the methyl group by a hydrogen atom. This hydrogen atom in Figure 4c is too far away from either the phenoxide oxygen (3.16 Å) or the nitro oxygen atom (3.40 Å) to engage in any substantive bond. The interaction energy in this case is reduced to −60.5 kcal/mol, 5.2 kcal/mol less than the full system, which includes both sorts of bonds. Much of the remaining attractive force resides in the simple cation···anion Coulombic ion pair interaction. If the methyl group is removed from MeS$^+$(Et)$_2$, leaving behind a neutral S(Et)$_2$ (representing the product AdoHcy) in Figure 4d, while also deleting both the tetrel and hydrogen bond, the interaction energy is reduced to 0.

Still another scheme to dissect the total interaction into its component segments arises if the phenoxide oxygen atom of DNC is replaced by a hydrogen atom, which would eliminate the tetrel bond, while retaining the CH···O hydrogen bond in Figure 4e. If this exchange occurs while retaining the negative charge of the DNC (a doublet), the interaction energy is −58.4 kcal/mol. This quantity is some 7.3 kcal/mol less attractive than that with the phenoxide oxygen, providing an estimate of the tetrel bond energy. On the other hand, if the modified DNC is made electrically neutral (a singlet), the interaction energy is reduced to zero. In other words, in the absence of the strong Coulombic ion pair interaction, any CH···O hydrogen bond in this system is quite weak despite the close R(H···O) distance of 2.16 Å.

We next consider the effects of other residues on the foregoing analysis. Unlike the KMTs G9A, SMYD2, and SMYD3, COMT possesses an active site Mg^{2+} ion that promotes the deprotonation of the catechol substrate's reactive hydroxyl group, forming the phenoxide anion through metal ion catalysis [49]. Notably, the phenoxide anion of DNC directly coordinates to this metal ion (Figure 5a). Thus, additional models were generated to examine the effect of the Mg^{2+} ion on the methyl C···O tetrel bond between AdoMet and DNC. In the model that only adds the Mg^{2+} ion (Figure 5a), its divalent charge acts to repel the $MeS^+(Et)_2$ cation, such that the interaction of these two species, without the DNC, amounts to 111.2 kcal/mol. The interaction energy of the $MeS^+(Et)_2$ with the DNC···Mg^{2+} pair cannot overcome this strong repulsive force, so is +48.2 kcal/mol. When the pure $MeS^+(Et)_2$···Mg^{2+} repulsion is subtracted from this quantity, one finds that the interaction between $MeS^+(Et)_2$ and DNC is attractive, in the amount of −63.0 kcal/mol. This quantity differs from the −65.7 kcal/mol pure $MeS^+(Et)_2$···DNC interaction, in the complete absence of Mg^{2+} (see Table 2) by 2.7 kcal/mol. In other words, the presence of the divalent cation reduces the tetrel/hydrogen bond energy by only 4%.

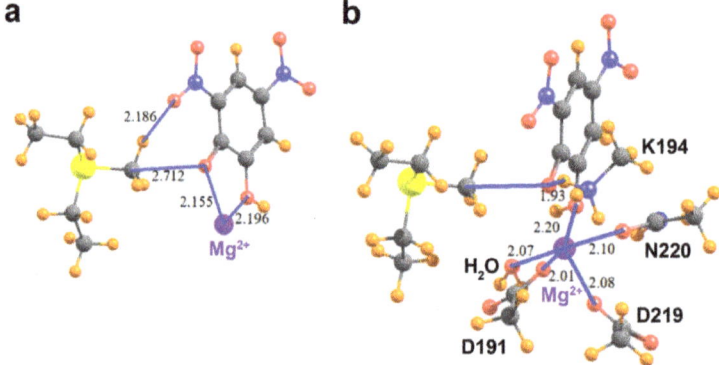

Figure 5. Molecular structures of COMT active site models. (**a**) $MeS^+(Et)_2$, the phenoxide anion of DNC, and the active site Mg^{2+} ion; (**b**) The model depicted in (**a**) that also includes the residues and water molecule that coordinate the Mg^{2+} ion and the catalytic residue Lys194.

As the Mg^{2+} ion is coordinated by several residues in COMT, it is important to assess how the metal coordination affects tetrel bonding between AdoMet and DNC. Within the enzyme's active site, the side chains of Asp191, Asp219, and Asn220 coordinate to Mg^{2+} with the last coordination site occupied by a water molecule. To represent these interactions, the aspartate and asparagine side chains were modeled as acetate and acetamide, respectively, and the water molecule was included to complete the metal's coordination sphere (Figure 5b). In addition, this model included the catalytic base, Lys194, that deprotonates the reactive hydroxyl group of the catechol substrate [49]. The lysine side chain was represented by a methyl ammonium cation and forms an NH···O hydrogen bond with the phenoxide anion of DNC, mimicking the deprotonation of the catechol hydroxyl group prior to methyl transfer. The presence of the Asp191 and Asp219 carboxylate anions tempers the effects of the Mg^{2+} to some degree. The interaction energy of the $MeS^+(Et)_2$ with the new system elements, i.e., all exclusive of the DNC, is +47.4 kcal/mol. The total interaction when the DNC is added to the entire system is −16.3 kcal/mol. After subtraction of the repulsion between the Lewis acid and the extraneous elements, the interaction energy of the $MeS^+(Et)_2$ with DNC is −63.7 kcal/mol. This quantity represents a very small decrease relative to the naked $MeS^+(Et)_2$/DNC dimer, in the amount of 3%. In summary, the interaction energy computed for a pair of species, interacting directly via a combined tetrel and hydrogen bond, is scarcely affected by the addition of surrounding groups, provided one takes proper account of the newly introduced pairwise interactions.

It is worth noting that the calculations above placed each system into an in vacuo situation, removed from the electrostatic or polarizing influence of neighboring groups. It is anticipated that these external effects will ameliorate some of the strongest Coulombic forces between the two groups, at least one of which carries a full charge. In an effort to estimate how much the surrounding protein environment might weaken these charge-assisted interactions [63], the various systems listed in Table 2 were placed within a polarizable continuum of dielectric constant $\varepsilon = 4$, a value that is commonly taken as the average value within a protein interior and that has been used several times [64–66] to good effect. Indeed, a weakening was observed, reducing the interaction energies by a factor between two and three. Nonetheless, the interactions remain strong, as large as -19.2 kcal/mol for the COMT system.

4. Discussion

As in the case of halogen, chalcogen, and pnictogen bonds, tetrel bonds are much stronger for elements in lower rows of the periodic table, e.g., Si and Ge. For this reason, the majority of computational work [67–77] has been dedicated to these stronger interactions. However, it is to the nominally weaker carbon tetrel bonds that this work is devoted, due in part to their prevalence in biological systems. While this survey and calculations have focused specifically on AdoMet-dependent methyltransferases, there is ample evidence that carbon tetrel bonds occur in a host of other systems. Thorough reviews of a variety of structures yield numerous interactions, on the order of thousands, where a powerful Lewis base is poised in the precise position, near to the extension of the R–C bond axis, that is consistent with a tetrel bond [32,78,79]. In the specific case of methyl groups, Guru Row and coworkers [31] identified more than 700 structures in the CSD where the interaction fits the geometrical requirements, and provided confirmation based on analyses of experimental charge density. In general, methyl groups are capable of forming only weak tetrel bonds [80] without the amplification that arises from either the presence of electron-withdrawing substituents, a strong base, charge assistance in the form of either a cationic Lewis acid or anionic base, or cooperative effects [33,74,81–91]. For example, the tetrel bond between $S(CH_3)_3^+$ and N-methylacetamide amounts to -13.7 kcal/mol, but to only -1.9 kcal/mol for the uncharged analogue $S(CH_3)_2$ [27].

Our finding here that the interaction can be better described as a tetrel bond than as a trifurcated CH···O hydrogen bond to a methyl group is consistent with earlier calculations [30]. This group has also performed calculations that confirm the presence of tetrel bonds in selected structures from the PDB [32], although that work was limited to the highly substituted CF_3 rather than pure methyl groups. Nonetheless, clear evidence was presented for the presence of tetrel bonds to this sp^3-hybridized carbon atom.

The discovery of carbon tetrel bonding in AdoMet dependent methyltransferases has important ramifications with respect to our understanding of the catalytic mechanism of these enzymes. As illustrated in prior computational studies, tetrel bonding between the electrophile and nucleophile represents a preliminary state that precedes the transition state in methyl transfer and other S_N2 reactions [33]. Much of our present knowledge of S_N2 catalysis in methyltransferases is derived from decades of paradigmatic studies of COMT and has resulted in several models for the reaction mechanism of these enzymes. Serendipitously, crystal structures of COMT bound to AdoMet and substrate analog inhibitors, such as DNC, have provided strong evidence for C···O tetrel bonding, given the close interaction distances of the AdoMet methyl group and phenoxide anion of DNC, as well as the S–C···O interaction angle that approaches linearity (Table 1 and Figure 1a). Thus, the tetrel interaction not only establishes the prerequisite geometry for the S_N2 reaction, but also aligns the lone pair of electrons of the nucleophile with the σ^* orbital of the AdoMet methyl carbon atom. This orbital alignment promotes the formation of the bond between the methyl carbon and the nucleophile in the transition state.

The identification of carbon tetrel bonding also furnishes experimental explanations for certain models of the methyltransferase catalytic mechanism that are based on studies of COMT and other enzymes. The compression/compaction model postulates that the methyltransferase active site essentially squeezes the distances along the S···CH$_3$···X reaction coordinate in the transition state (where X is the nucleophile) [6,7]. As observed in the COMT/AdoMet/DNC/Mg^{2+} complex (Figure 1a), the C···O tetrel bond between the AdoMet methyl group and phenoxide anion of DNC, representing a catechol substrate, is 2.71 Å, 0.54 Å shorter than the carbon–oxygen van der Waals contact distance of 3.25 Å. This short interaction distance is illustrative of the close approach of the two substrates postulated in the compression model. Importantly, the C···O tetrel interaction is attractive in nature (Table 2) and would result in an electron charge transfer from the nucleophile to the methyl carbon atom. In turn, this transfer would polarize and weaken the S–CH$_3$ bond during the S$_N$2 reaction [33]. The presence of methyl tetrel bonds observed in COMT also concurs with the NAC model for the catalytic mechanism of COMT and other methyltransferases. In molecular dynamic simulations, the NACs that transiently formed between AdoMet and a catechol bound to COMT were defined as having a C···O interaction distance of <3.2 Å and an S–C···O interaction angle ≥165° [12,92]. This geometry mirrors the interaction distances and angles observed for tetrel bonding (Figure 1 and Table 1). It is conceivable that the favorable energy of C···O tetrel bonding would increase the frequency in which AdoMet and the catechol substrate are brought into a catalytically productive alignment that is conducive to methyl transfer.

Knowledge of carbon tetrel bonding can also be leveraged in the design of methyltransferase inhibitors. Indeed, the PDB survey illustrates several examples of inhibitors of COMT and SET domain KMTs that form tetrel bonds with the AdoMet methyl group (Table 1 and Figure 1). In the case of catechol analog inhibitors of COMT, the addition of nitro groups or other electron withdrawing moieties to the catechol ring (such as in DNC), or substitution of the catechol phenyl ring with pyridine or other six-membered heterocycles diminishes the nucleophilicity of the reactive hydroxyl group, abrogating methyl transfer with AdoMet. It has been proposed that these inhibitors bind to the COMT and AdoMet as a deprotonated phenoxide anion [9], which can engage in strong methyl C···O tetrel bonding with AdoMet. The calculated interaction energy of the C···O tetrel interaction between the DNC phenoxide anion and the AdoMet methyl group is substantially stronger than tetrel interactions involving neutral oxygen atoms in carbonyl and hydroxyl groups (Table 2). In agreement with these findings, nitrocatechol inhibitors of COMT, such as DNC, tolcapone, and entacapone, display equilibrium inhibitory constants (K_I) that are in the low nanomolar to picomolar range [93,94]. The tight binding of these inhibitors to the enzyme may be mediated in part by the strong C···O tetrel bonding between the phenoxide anion and AdoMet methyl group.

In addition to COMT, inhibitors of the SMYD KMTs display methyl tetrel bonding with AdoMet (Table 1). With respect to SMYD2, structure–activity relationship (SAR) analysis of the BAY-598 inhibitor revealed a preference for a chlorine atom in the 3-position of the phenyl moiety [58]. This chlorine atom corresponds to the Lewis base that forms the C···Cl tetrel bond with AdoMet (Figure 1c). Kinetic analysis of BAY-598 demonstrated uncompetitive inhibition toward AdoMet, indicating the inhibitor binds exclusively to the SMYD2/AdoMet binary complex and thus does not recognize the free enzyme. It is conceivable that the C···Cl tetrel bond between AdoMet and SGC Probe Bay-598 in SMYD2 may contribute to this uncompetitive inhibition by promoting recognition of the inhibitor by the enzyme/substrate complex. Through compound screening and SAR analysis, Mitchell and coworkers identified an oxindole class of inhibitors that selectively inhibits SMYD3 [57]. Crystallographic characterization of the initial oxindole hit (Figure 1b) and the SAR-optimized inhibitor EPZ030456 bound to SMYD3 and AdoMet revealed methyl C···O tetrel bonding with the oxygen atom of the oxindole moiety (Table 1). The structural conservation of the methyl C···O tetrel bond formed by the oxindole inhibitors suggests that this interaction may be important for recognition and selectivity for SMYD3.

5. Conclusions

The survey of AdoMet-bound methyltransferase structures in the PDB uncovered a number of geometries that strongly indicate the presence of methyl tetrel bonding to electronegative atoms in small molecule inhibitors, solvent molecules, and ions. The C···X distances between the AdoMet methyl carbon atom and the Lewis base vary over a wide range, but all of the C···O interactions reported from this survey are less than 3.25 Å, with some distances as short as 2.5 Å. Further, the Lewis base is located close to the extension of the S–CH$_3$ bond of AdoMet, with θ(S–C···O) angles within 20° of linearity. QM analysis of a selected set of these interactions revealed unequivocal evidence of methyl tetrel bonds, rather than what might naively be considered a trifurcated CH···O hydrogen bond. The interaction energies of these selected tetrel bonds varied between −5 and −9 kcal/mol, comparable to or stronger than the paradigmatic hydrogen bond between a pair of water molecules (−5.8 kcal/mol). In the case of a Lewis base bearing a full negative charge, the tetrel bond energy was amplified to more than −60 kcal/mol. It is thus clear that our understanding of the forces present within biological systems must include tetrel bonding on the same footing as the venerable hydrogen bond. Finally, the discovery of AdoMet methyl tetrel bonding in methyltransferases illustrates that this interaction may have a fundamental role in the catalytic mechanism of these enzymes and thus merits further investigation. An understanding of this novel interaction can be applied in structure-guided design of potent inhibitors of methyltransferases implicated in disease.

Author Contributions: R.C.T. conducted the PDB survey for methyl tetrel bonding, and S.S. performed the QM calculations and analysis. R.C.T. and S.S. wrote the paper.

Funding: These studies were supported by NSF grant CHE-1508492 to R. Trievel.

Acknowledgments: We wish to thank S. Horowitz for his insightful comments and feedback on this work.

Conflicts of Interest: The authors declare no conflict of interest.

References

1. Schubert, H.L.; Blumenthal, R.M.; Cheng, X. Many paths to methyltransfer: A chronicle of convergence. *Trends Biochem. Sci.* **2003**, *28*, 329–335. [CrossRef]
2. Petrossian, T.C.; Clarke, S.G. Uncovering the Human Methyltransferasome. *Mol. Cell. Proteom.* **2011**, *10*, M110.000976. [CrossRef] [PubMed]
3. Petrossian, T.; Clarke, S. Bioinformatic Identification of Novel Methyltransferases. *Epigenomics* **2009**, *1*, 163–175. [CrossRef] [PubMed]
4. Woodard, R.W.; Tsai, M.D.; Floss, H.G.; Crooks, P.A.; Coward, J.K. Stereochemical course of the transmethylation catalyzed by catechol O-methyltransferase. *J. Biol. Chem.* **1980**, *255*, 9124–9127. [PubMed]
5. Hegazi, M.F.; Borchard, R.T.; Schowen, R.L. Letter: SN2-like transition state for methyl transfer catalyzed by catechol-O-methyl-transferase. *J. Am. Chem. Soc.* **1976**, *98*, 3048–3049. [CrossRef] [PubMed]
6. Gray, C.H.; Coward, J.K.; Schowen, K.B.; Schowen, R.L. Alpha-Deuterium and C-13 Isotope Effects for a Simple, Inter-Molecular Sulfur-to-Oxygen Methyl-Transfer Reaction–Transition-State Structures and Isotope Effects in Transmethylation and Transalkylation. *J. Am. Chem. Soc.* **1979**, *101*, 4351–4358. [CrossRef]
7. Mihel, I.; Knipe, J.O.; Coward, J.K.; Schowen, R.L. Alpha-Deuterium Isotope Effects and Transition-State Structure in an Intra-Molecular Model System for Methyl-Transfer Enzymes. *J. Am. Chem. Soc.* **1979**, *101*, 4349–4351. [CrossRef]
8. Zhang, J.; Klinman, J.P. Enzymatic methyl transfer: Role of an active site residue in generating active site compaction that correlates with catalytic efficiency. *J. Am. Chem. Soc.* **2011**, *133*, 17134–17137. [CrossRef] [PubMed]
9. Zhang, J.; Kulik, H.J.; Martinez, T.J.; Klinman, J.P. Mediation of donor-acceptor distance in an enzymatic methyl transfer reaction. *Proc. Natl. Acad. Sci. USA* **2015**, *112*, 7954–7959. [CrossRef] [PubMed]
10. Lau, E.Y.; Kahn, K.; Bash, P.A.; Bruice, T.C. The importance of reactant positioning in enzyme catalysis: A hybrid quantum mechanics/molecular mechanics study of a haloalkane dehalogenase. *Proc. Natl. Acad. Sci. USA* **2000**, *97*, 9937–9942. [CrossRef] [PubMed]

11. Kahn, K.; Bruice, T.C. Transition-state and ground-state structures and their interaction with the active-site residues in catechol-O-methyl transferase. *J. Am. Chem. Soc.* **2000**, *122*, 46–51. [CrossRef]
12. Lau, E.Y.; Bruice, T.C. Importance of correlated motions in forming highly reactive near attack conformations in catechol O-methyltransferase. *J. Am. Chem. Soc.* **1998**, *120*, 12387–12394. [CrossRef]
13. Zheng, Y.J.; Bruice, T.C. A theoretical examination of the factors controlling the catalytic efficiency of a transmethylation enzyme: Catechol O-methyltransferase. *J. Am. Chem. Soc.* **1997**, *119*, 8137–8145. [CrossRef]
14. Lameira, J.; Bora, R.P.; Chu, Z.T.; Warshel, A. Methyltransferases do not work by compression, cratic, or desolvation effects, but by electrostatic preorganization. *Proteins* **2015**, *83*, 18–30. [CrossRef] [PubMed]
15. Roca, M.; Marti, S.; Andres, J.; Moliner, V.; Tunon, I.; Bertran, J.; Williams, I.H. Theoretical modeling of enzyme catalytic power: Analysis of "cratic" and electrostatic factors in catechol O-methyltransferase. *J. Am. Chem. Soc.* **2003**, *125*, 7726–7737. [CrossRef] [PubMed]
16. Kollman, P.A.; Kuhn, B.; Donini, O.; Perakyla, M.; Stanton, R.; Bakowies, D. Elucidating the nature of enzyme catalysis utilizing a new twist on an old methodology: Quantum mechanical-free energy calculations on chemical reactions in enzymes and in aqueous solution. *Acc. Chem. Res.* **2001**, *34*, 72–79. [CrossRef] [PubMed]
17. Kuhn, B.; Kollman, P.A. QM-FE and molecular dynamics calculations on catechol O-methyltransferase: Free energy of activation in the enzyme and in aqueous solution and regioselectivity of the enzyme-catalyzed reaction. *J. Am. Chem. Soc.* **2000**, *122*, 2586–2596. [CrossRef]
18. Horowitz, S.; Dirk, L.M.; Yesselman, J.D.; Nimtz, J.S.; Adhikari, U.; Mehl, R.A.; Scheiner, S.; Houtz, R.L.; Al-Hashimi, H.M.; Trievel, R.C. Conservation and functional importance of carbon–oxygen hydrogen bonding in AdoMet-dependent methyltransferases. *J. Am. Chem. Soc.* **2013**, *135*, 15536–15548. [CrossRef] [PubMed]
19. Adhikari, U.; Scheiner, S. Magnitude and mechanism of charge enhancement of CH··O hydrogen bonds. *J. Phys. Chem. A* **2013**, *117*, 10551–10562. [CrossRef] [PubMed]
20. Horowitz, S.; Yesselman, J.D.; Al-Hashimi, H.M.; Trievel, R.C. Direct evidence for methyl group coordination by carbon–oxygen hydrogen bonds in the lysine methyltransferase SET7/9. *J. Biol. Chem.* **2011**, *286*, 18658–18663. [CrossRef] [PubMed]
21. Horowitz, S.; Adhikari, U.; Dirk, L.M.; Del Rizzo, P.A.; Mehl, R.A.; Houtz, R.L.; Al-Hashimi, H.M.; Scheiner, S.; Trievel, R.C. Manipulating unconventional CH-based hydrogen bonding in a methyltransferase via noncanonical amino acid mutagenesis. *ACS Chem. Biol.* **2014**, *9*, 1692–1697. [CrossRef] [PubMed]
22. Poulin, M.B.; Schneck, J.L.; Matico, R.E.; McDevitt, P.J.; Huddleston, M.J.; Hou, W.; Johnson, N.W.; Thrall, S.H.; Meek, T.D.; Schramm, V.L. Transition state for the NSD2-catalyzed methylation of histone H3 lysine 36. *Proc. Natl. Acad. Sci. USA* **2016**, *113*, 1197–1201. [CrossRef] [PubMed]
23. Linscott, J.A.; Kapilashrami, K.; Wang, Z.; Senevirathne, C.; Bothwell, I.R.; Blum, G.; Luo, M. Kinetic isotope effects reveal early transition state of protein lysine methyltransferase SET8. *Proc. Natl. Acad. Sci. USA* **2016**, *113*, E8369–E8378. [CrossRef] [PubMed]
24. Swiderek, K.; Tunon, I.; Williams, I.H.; Moliner, V. Insights on the Origin of Catalysis on Glycine N-Methyltransferase from Computational Modeling. *J. Am. Chem. Soc.* **2018**, *140*, 4327–4334. [CrossRef] [PubMed]
25. Fick, R.J.; Kroner, G.M.; Nepal, B.; Magnani, R.; Horowitz, S.; Houtz, R.L.; Scheiner, S.; Trievel, R.C. Sulfur-Oxygen Chalcogen Bonding Mediates AdoMet Recognition in the Lysine Methyltransferase SET7/9. *ACS Chem. Biol.* **2016**, *11*, 748–754. [CrossRef] [PubMed]
26. Wang, W.Z.; Ji, B.M.; Zhang, Y. Chalcogen Bond: A Sister Noncovalent Bond to Halogen Bond. *J. Phys. Chem. A* **2009**, *113*, 8132–8135. [CrossRef] [PubMed]
27. Scheiner, S. Comparison of CH···O, SH···O Chalcogen, and Tetrel Bonds Formed by Neutral and Cationic Sulfur-Containing Compounds. *J. Phys. Chem. A* **2015**, *119*, 9189–9199. [CrossRef] [PubMed]
28. Bauza, A.; Mooibroek, T.J.; Frontera, A. Tetrel-bonding interaction: Rediscovered supramolecular force? *Angew. Chem.* **2013**, *52*, 12317–12321. [CrossRef] [PubMed]
29. Murray, J.S.; Lane, P.; Politzer, P. Expansion of the sigma-hole concept. *J. Mol. Model.* **2009**, *15*, 723–729. [CrossRef] [PubMed]
30. Bauza, A.; Frontera, A. RCH3···O Interactions in Biological Systems: Are They Trifurcated H-Bonds or Noncovalent Carbon Bonds? *Crystals* **2016**, *6*, 26. [CrossRef]
31. Thomas, S.P.; Pavan, M.S.; Guru Row, T.N. Experimental evidence for 'carbon bonding' in the solid state from charge density analysis. *Chem. Commun.* **2014**, *50*, 49–51. [CrossRef] [PubMed]

32. Garcia, L.X.; Bauza, A.; Seth, S.K.; Frontera, A. Importance of R–CF3···O Tetrel Bonding Interactions in Biological Systems. *J. Phys. Chem. A* **2017**, *121*, 5371–5376. [CrossRef] [PubMed]
33. Grabowski, S.J. Tetrel bond-sigma-hole bond as a preliminary stage of the SN2 reaction. *Phys. Chem. Chem. Phys.* **2014**, *16*, 1824–1834. [CrossRef] [PubMed]
34. Taylor, R.; Kennard, O. Crystallographic Evidence for the Existence of C-H···O, C-H···N, and C-H···C1 Hydrogen-Bonds. *J. Am. Chem. Soc.* **1982**, *104*, 5063–5070. [CrossRef]
35. Desiraju, G.R.; Ho, P.S.; Kloo, L.; Legon, A.C.; Marquardt, R.; Metrangolo, P.; Politzer, P.; Resnati, G.; Rissanen, K. Definition of the halogen bond (IUPAC Recommendations 2013). *Pure Appl. Chem.* **2013**, *85*, 1711–1713. [CrossRef]
36. Emsley, P.; Cowtan, K. Coot: Model-building tools for molecular graphics. *Acta Crystallogr. D* **2004**, *60*, 2126–2132. [CrossRef] [PubMed]
37. Emsley, P.; Lohkamp, B.; Scott, W.G.; Cowtan, K. Features and development of Coot. *Acta Crystallogr. D* **2010**, *66*, 486–501. [CrossRef] [PubMed]
38. Reed, A.E.; Curtiss, L.A.; Weinhold, F. Intermolecular Interactions from a Natural Bond Orbital, Donor-Acceptor Viewpoint. *Chem. Rev.* **1988**, *88*, 899–926. [CrossRef]
39. Reed, A.E.; Weinhold, F.; Curtiss, L.A.; Pochatko, D.J. Natural Bond Orbital Analysis of Molecular-Interactions–Theoretical-Studies of Binary Complexes of HF, H_2O, NH_3, N_2, O_2, F_2, CO, and CO_2 with HF, H_2O, and NH_3. *J. Chem. Phys.* **1986**, *84*, 5687–5705. [CrossRef]
40. Barone, V.; Cossi, M. Quantum calculation of molecular energies and energy gradients in solution by a conductor solvent model. *J. Phys. Chem. A* **1998**, *102*, 1995–2001. [CrossRef]
41. Ditchfield, R. Self-Consistent Perturbation-Theory of Diamagnetism 1. Gauge-Invariant Lcao Method for Nmr Chemical-Shifts. *Mol. Phys.* **1974**, *27*, 789–807. [CrossRef]
42. Cheeseman, J.R.; Trucks, G.W.; Keith, T.A.; Frisch, M.J. A comparison of models for calculating nuclear magnetic resonance shielding tensors. *J. Chem. Phys.* **1996**, *104*, 5497–5509. [CrossRef]
43. Couture, J.F.; Hauk, G.; Thompson, M.J.; Blackburn, G.M.; Trievel, R.C. Catalytic roles for carbon–oxygen hydrogen bonding in SET domain lysine methyltransferases. *J. Biol. Chem.* **2006**, *281*, 19280–19287. [CrossRef] [PubMed]
44. Kaniskan, H.U.; Martini, M.L.; Jin, J. Inhibitors of Protein Methyltransferases and Demethylases. *Chem. Rev.* **2018**, *118*, 989–1068. [CrossRef] [PubMed]
45. Ma, Z.; Liu, H.; Wu, B. Structure-based drug design of catechol-*O*-methyltransferase inhibitors for CNS disorders. *Br. J. Clin. Pharmacol.* **2014**, *77*, 410–420. [CrossRef] [PubMed]
46. Kiss, L.E.; Soares-da-Silva, P. Medicinal chemistry of catechol *O*-methyltransferase (COMT) inhibitors and their therapeutic utility. *J. Med. Chem.* **2014**, *57*, 8692–8717. [CrossRef] [PubMed]
47. Gnyszka, A.; Jastrzebski, Z.; Flis, S. DNA methyltransferase inhibitors and their emerging role in epigenetic therapy of cancer. *Anticancer. Res.* **2013**, *33*, 2989–2996. [PubMed]
48. Morera, L.; Lubbert, M.; Jung, M. Targeting histone methyltransferases and demethylases in clinical trials for cancer therapy. *Clin. Epigenetics* **2016**, *8*, 57. [CrossRef] [PubMed]
49. Bonifacio, M.J.; Palma, P.N.; Almeida, L.; Soares-da-Silva, P. Catechol-*O*-methyltransferase and its inhibitors in Parkinson's disease. *CNS Drug Rev.* **2007**, *13*, 352–379. [CrossRef] [PubMed]
50. Ellermann, M.; Lerner, C.; Burgy, G.; Ehler, A.; Bissantz, C.; Jakob-Roetne, R.; Paulini, R.; Allemann, O.; Tissot, H.; Grunstein, D.; et al. Catechol-*O*-methyltransferase in complex with substituted 3'-deoxyribose bisubstrate inhibitors. *Acta Crystallogr. D Biol. Crystallogr.* **2012**, *68*, 253–260. [CrossRef] [PubMed]
51. Harrison, S.T.; Poslusney, M.S.; Mulhearn, J.J.; Zhao, Z.; Kett, N.R.; Schubert, J.W.; Melamed, J.Y.; Allison, T.J.; Patel, S.B.; Sanders, J.M.; et al. Synthesis and Evaluation of Heterocyclic Catechol Mimics as Inhibitors of Catechol-*O*-methyltransferase (COMT). *ACS Med. Chem. Lett.* **2015**, *6*, 318–323. [CrossRef] [PubMed]
52. Palma, P.N.; Rodrigues, M.L.; Archer, M.; Bonifacio, M.J.; Loureiro, A.I.; Learmonth, D.A.; Carrondo, M.A.; Soares-da-Silva, P. Comparative study of ortho- and meta-nitrated inhibitors of catechol-*O*-methyltransferase: Interactions with the active site and regioselectivity of *O*-methylation. *Mol. Pharmacol.* **2006**, *70*, 143–153. [CrossRef] [PubMed]
53. Vidgren, J.; Svensson, L.A.; Liljas, A. Crystal structure of catechol *O*-methyltransferase. *Nature* **1994**, *368*, 354–358. [CrossRef] [PubMed]
54. Spellmon, N.; Holcomb, J.; Trescott, L.; Sirinupong, N.; Yang, Z. Structure and function of SET and MYND domain-containing proteins. *Int. J. Mol. Sci.* **2015**, *16*, 1406–1428. [CrossRef] [PubMed]

55. Kudithipudi, S.; Jeltsch, A. Role of somatic cancer mutations in human protein lysine methyltransferases. *BBA-Rev. Cancer* **2014**, *1846*, 366–379. [CrossRef] [PubMed]
56. Tracy, C.; Warren, J.S.; Szulik, M.; Wang, L.; Garcia, J.; Makaju, A.; Russell, K.; Miller, M.; Franklin, S. The Smyd Family of Methyltransferases: Role in Cardiac and Skeletal Muscle Physiology and Pathology. *Curr. Opin. Physiol.* **2018**, *1*, 140–152. [CrossRef] [PubMed]
57. Mitchell, L.H.; Boriack-Sjodin, P.A.; Smith, S.; Thomenius, M.; Rioux, N.; Munchhof, M.; Mills, J.E.; Klaus, C.; Totman, J.; Riera, T.V.; et al. Novel Oxindole Sulfonamides and Sulfamides: EPZ031686, the First Orally Bioavailable Small Molecule SMYD3 Inhibitor. *ACS Med. Chem. Lett.* **2016**, *7*, 134–138. [CrossRef] [PubMed]
58. Eggert, E.; Hillig, R.C.; Koehr, S.; Stockigt, D.; Weiske, J.; Barak, N.; Mowat, J.; Brumby, T.; Christ, C.D.; Ter Laak, A.; et al. Discovery and Characterization of a Highly Potent and Selective Aminopyrazoline-Based in Vivo Probe (BAY-598) for the Protein Lysine Methyltransferase SMYD2. *J. Med. Chem.* **2016**, *59*, 4578–4600. [CrossRef] [PubMed]
59. Ferguson, A.D.; Larsen, N.A.; Howard, T.; Pollard, H.; Green, I.; Grande, C.; Cheung, T.; Garcia-Arenas, R.; Cowen, S.; Wu, J.; et al. Structural basis of substrate methylation and inhibition of SMYD2. *Structure* **2011**, *19*, 1262–1273. [CrossRef] [PubMed]
60. Lee, J.H.; Bae, B.; Kuemin, M.; Circello, B.T.; Metcalf, W.W.; Nair, S.K.; van der Donk, W.A. Characterization and structure of DhpI, a phosphonate O-methyltransferase involved in dehydrophos biosynthesis. *Proc. Natl. Acad Sci. USA* **2010**, *107*, 17557–17562. [CrossRef] [PubMed]
61. Fick, R.J.; Clay, M.C.; Vander Lee, L.; Scheiner, S.; Al-Hashimi, H.; Trievel, R.C. Water-Mediated Carbon–oxygen Hydrogen Bonding Facilitates S-Adenosylmethionine Recognition in the Reactivation Domain of Cobalamin-Dependent Methionine Synthase. *Biochemistry* **2018**, *57*, 3733–3740. [CrossRef] [PubMed]
62. Scheiner, S. Ability of IR and NMR Spectral Data to Distinguish between a Tetrel Bond and a Hydrogen Bond. *J. Phys. Chem. A* **2018**, in press. [CrossRef] [PubMed]
63. Lee, K.-M.; Chen, J.C.C.; Chen, H.-Y.; Lin, I.J.B. A triple helical structure supported solely by C–H···O hydrogen bonding. *Chem. Commun.* **2012**, *48*, 1242–1244. [CrossRef] [PubMed]
64. Simonson, T.; Perahia, D. Internal and Interfacial Dielectric-Properties of Cytochrome-C from Molecular-Dynamics in Aqueous-Solution. *Proc. Natl. Acad. Sci. USA* **1995**, *92*, 1082–1086. [CrossRef] [PubMed]
65. Dwyer, J.J.; Gittis, A.G.; Karp, D.A.; Lattman, E.E.; Spencer, D.S.; Stites, W.E.; Garcia-Moreno, B. High apparent dielectric constants in the interior of a protein reflect water penetration. *Biophys. J.* **2000**, *79*, 1610–1620. [CrossRef]
66. Smith, P.E.; Brunne, R.M.; Mark, A.E.; Vangunsteren, W.F. Dielectric-Properties of Trypsin-Inhibitor and Lysozyme Calculated from Molecular-Dynamics Simulations. *J. Phys. Chem.* **1993**, *97*, 2009–2014. [CrossRef]
67. Roy, S.; Drew, M.G.B.; Bauza, A.; Frontera, A.; Chattopadhyay, S. Non-covalent tetrel bonding interactions in hemidirectional lead(ii) complexes with nickel(ii)-salen type metalloligands. *New J. Chem.* **2018**, *42*, 6062–6076. [CrossRef]
68. Dong, W.; Li, Q.; Scheiner, S. Comparative Strengths of Tetrel, Pnicogen, Chalcogen, and Halogen Bonds and Contributing Factors. *Molecules* **2018**, *23*, 1681. [CrossRef] [PubMed]
69. Shen, S.; Zeng, Y.; Li, X.; Meng, L.; Zhang, X. Insight into the π-hole···π-electrons tetrel bonds between F_2ZO (Z = C, Si, Ge) and unsaturated hydrocarbons. *Int. J. Quantum Chem.* **2018**, *118*, e25521. [CrossRef]
70. Zierkiewicz, W.; Michalczyk, M.; Scheiner, S. Comparison between Tetrel Bonded Complexes Stabilized by σ and π Hole Interactions. *Molecules* **2018**, *23*, 1416. [CrossRef] [PubMed]
71. Zierkiewicz, W.; Michalczyk, M.; Scheiner, S. Implications of monomer deformation for tetrel and pnicogen bonds. *Phys. Chem. Chem. Phys.* **2018**, *20*, 8832–8841. [CrossRef] [PubMed]
72. Grabowski, S.J. Hydrogen bonds, and s-hole and p-hole bonds–mechanisms protecting doublet and octet electron structures. *Phys. Chem. Chem. Phys.* **2017**, *19*, 29742–29759. [CrossRef] [PubMed]
73. Grabowski, S.J.; Sokalski, W.A. Are Various σ-Hole Bonds Steered by the Same Mechanisms? *ChemPhysChem.* **2017**, *18*, 1569–1577. [CrossRef] [PubMed]
74. Liu, M.; Li, Q.; Scheiner, S. Comparison of tetrel bonds in neutral and protonated complexes of pyridineTF$_3$ and furanTF$_3$ (T = C, Si, and Ge) with NH$_3$. *Phys. Chem. Chem. Phys.* **2017**, *19*, 5550–5559. [CrossRef] [PubMed]
75. Scheiner, S. Steric Crowding in Tetrel Bonds. *J. Phys. Chem. A* **2018**, *122*, 2550–2562. [CrossRef] [PubMed]

76. Scheiner, S. Assembly of Effective Halide Receptors from Components. Comparing Hydrogen, Halogen, and Tetrel Bonds. *J. Phys. Chem. A* **2017**, *121*, 3606–3615. [CrossRef] [PubMed]
77. Wei, Y.-X.; Li, H.-B.; Cheng, J.-B.; Li, W.-Z.; Li, Q.-Z. Prominent enhancing effects of substituents on the strength of π···σ-hole tetrel bond. *Int. J. Quantum Chem.* **2017**, *117*, e25448. [CrossRef]
78. Bauza, A.; Mooibroek, T.J.; Frontera, A. Tetrel Bonding Interactions. *Chem. Rec.* **2016**, *16*, 473–487. [CrossRef] [PubMed]
79. Southern, S.A.; Bryce, D.L. NMR Investigations of Noncovalent Carbon Tetrel Bonds. Computational Assessment and Initial Experimental Observation. *J. Phys. Chem. A* **2015**, *119*, 11891–11899. [CrossRef] [PubMed]
80. Scheiner, S. Systematic Elucidation of Factors That Influence the Strength of Tetrel Bonds. *J. Phys. Chem. A* **2017**, *121*, 5561–5568. [CrossRef] [PubMed]
81. Mani, D.; Arunan, E. The X–CY (X = O/F, Y = O/S/F/Cl/Br/N/P) 'carbon bond' and hydrophobic interactions. *Phys. Chem. Chem. Phys.* **2013**, *15*, 14377–14383. [CrossRef] [PubMed]
82. Esrafili, M.D.; Kiani, H.; Mohammadian-Sabet, F. Tuning of carbon bonds by substituent effects: An ab initio study. *Mol. Phys.* **2016**, *114*, 3658–3668. [CrossRef]
83. Del Bene, J.E.; Alkorta, I.; Elguero, J. Anionic complexes of F^- and Cl^- with substituted methanes: Hydrogen, halogen, and tetrel bonds. *Chem. Phys. Lett.* **2016**, *655*, 115–119. [CrossRef]
84. Liu, M.; Li, Q.; Cheng, J.; Li, W.; Li, H.-B. Tetrel bond of pseudohalide anions with XH_3F (X = C, Si, Ge, and Sn) and its role in S_N2 reaction. *J. Chem. Phys.* **2016**, *145*, 224310. [CrossRef] [PubMed]
85. Marín-Luna, M.; Alkorta, I.; Elguero, J. Cooperativity in Tetrel Bonds. *J. Phys. Chem. A* **2016**, *120*, 648–656. [CrossRef] [PubMed]
86. Martín-Fernández, C.; Montero-Campillo, M.M.; Alkorta, I.; Elguero, J. Weak interactions and cooperativity effects on disiloxane: A look at the building block of silicones. *Mol. Phys.* **2018**, *116*, 1539–1550. [CrossRef]
87. Esrafili, M.D.; Asadollahi, S.; Mousavian, P. Anionic tetrel bonds: An ab initio study. *Chem. Phys. Lett.* **2018**, *691*, 394–400. [CrossRef]
88. Esrafili, M.D.; Nurazar, R.; Mohammadian-Sabet, F. Cooperative effects between tetrel bond and other σ–hole bond interactions: A comparative investigation. *Mol. Phys.* **2015**, *113*, 3703–3711. [CrossRef]
89. Wei, Y.; Li, Q. Comparison for σ-hole and π-hole tetrel-bonded complexes involving cyanoacetaldehyde. *Mol. Phys.* **2018**, *116*, 222–230. [CrossRef]
90. Ghara, M.; Pan, S.; Kumar, A.; Merino, G.; Chattaraj, P.K. Structure, stability, and nature of bonding in carbon monoxide bound EX_3^+ complexes (E = group 14 element; X = H, F, Cl, Br, I). *J. Comput. Chem.* **2016**, *37*, 2202–2211. [CrossRef] [PubMed]
91. Scheiner, S. Comparison of halide receptors based on H, halogen, chalcogen, pnicogen, and tetrel bonds. *Faraday Discuss.* **2017**, *203*, 213–226. [CrossRef] [PubMed]
92. Lau, E.Y.; Bruice, T.C. Comparison of the dynamics for ground-state and transition-state structures in the active site of catechol *O*-methyltransferase. *J. Am. Chem. Soc.* **2000**, *122*, 7165–7171. [CrossRef]
93. Lotta, T.; Vidgren, J.; Tilgmann, C.; Ulmanen, I.; Melen, K.; Julkunen, I.; Taskinen, J. Kinetics of human soluble and membrane-bound catechol *O*-methyltransferase: A revised mechanism and description of the thermolabile variant of the enzyme. *Biochemistry* **1995**, *34*, 4202–4210. [CrossRef] [PubMed]
94. Byrne, J.M.; Tipton, K.F. Nitrocatechol derivatives as inhibitors of catechol-*O*-methyltransferase. *Biochem. Soc. Trans.* **1996**, *24*, 64S. [CrossRef] [PubMed]

© 2018 by the authors. Licensee MDPI, Basel, Switzerland. This article is an open access article distributed under the terms and conditions of the Creative Commons Attribution (CC BY) license (http://creativecommons.org/licenses/by/4.0/).

Article

Strong Tetrel Bonds: Theoretical Aspects and Experimental Evidence

Mehdi D. Esrafili * and Parisasadat Mousavian

Laboratory of Theoretical Chemistry, Department of Chemistry, University of Maragheh, Maragheh 5513864596, Iran; p.mosavian1327@gmail.com
* Correspondence: esrafili@maragheh.ac.ir; Tel.: +98-42-1223-7955; Fax: +98-42-1227-6060

Received: 25 September 2018; Accepted: 5 October 2018; Published: 15 October 2018

Abstract: In recent years, noncovalent interactions involving group-14 elements of the periodic table acting as a Lewis acid center (or tetrel-bonding interactions) have attracted considerable attention due to their potential applications in supramolecular chemistry, material science and so on. The aim of the present study is to characterize the geometry, strength and bonding properties of strong tetrel-bond interactions in some charge-assisted tetrel-bonded complexes. Ab initio calculations are performed, and the results are supported by the quantum theory of atoms in molecules (QTAIM) and natural bond orbital (NBO) approaches. The interaction energies of the anionic tetrel-bonded complexes formed between XF_3M molecule (X=F, CN; M=Si, Ge and Sn) and A^- anions ($A^-=F^-$, Cl^-, Br^-, CN^-, NC^- and N_3^-) vary between −16.35 and −96.30 kcal/mol. The M atom in these complexes is generally characterized by pentavalency, i.e., is hypervalent. Moreover, the QTAIM analysis confirms that the anionic tetrel-bonding interaction in these systems could be classified as a strong interaction with some covalent character. On the other hand, it is found that the tetrel-bond interactions in cationic tetrel-bonded $[p-NH_3(C_6H_4)MH_3]^+ \cdots Z$ and $[p-NH_3(C_6F_4)MH_3]^+ \cdots Z$ complexes (M=Si, Ge, Sn and Z=NH_3, NH_2CH_3, NH_2OH and NH_2NH_2) are characterized by a strong orbital interaction between the filled lone-pair orbital of the Lewis base and empty BD^*_{M-C} orbital of the Lewis base. The substitution of the F atoms in the benzene ring provides a strong orbital interaction, and hence improved tetrel-bond interaction. For all charge-assisted tetrel-bonded complexes, it is seen that the formation of tetrel-bond interaction is accompanied by significant electron density redistribution over the interacting subunits. Finally, we provide some experimental evidence for the existence of such charge-assisted tetrel-bond interactions in crystalline phase.

Keywords: noncovalent interaction; tetrel-bond; σ-hole; electrostatic potential; ab initio

1. Introduction

Over the past decades, there has been an increasing awareness of the importance of noncovalent interactions owing to their critical roles in various fields of chemistry and biochemistry, such as protein folding, molecular recognition, drug design and crystal packing [1–3]. Of the various noncovalent interactions, hydrogen-bonding (H-bonding) has emerged as the most extensively studied case [4–8]. It is typically formulated as an attractive Lewis acid-Lewis base interaction, D-H$\cdots$A, between the electron-deficient hydrogen atom of one molecule (D-H), acting as a bridge to an electron-rich site on the other molecule (A). However, much attention has been recently devoted to other types of noncovalent interactions like σ-hole bonding due to their useful applications in supramolecular chemistry, crystal engineering, and biochemistry [9–17]. A σ-hole bond [18–24] is a noncovalent interaction analogous to the H-bonding, in which a covalently bonded atom of groups 14–18 of the periodic table, rather than an H atom, serves a similar function as a bridge between two molecules. For example, the possibility of noncovalent interaction between some halocarbons and potential

Lewis bases has been known for some time [25,26] and has continued to be studied at a rapidly increasing rate in recent years [27–30]. It has been found that the halogen atom in these molecules is able to develop a positive σ-hole region on the outermost portion of the halogen atom, along the C–X atoms (X=F, Cl, Br, I). The emergence of such a positive area, which may seem surprising due to the high electronegativity of halogen atoms, is responsible for the high directionality and formation of an electrostatically driven interaction with a negative region on the Lewis base. Note also that besides these electrostatic effects, there are also polarization effects and a substantial charge-transfer from the Lewis base into the BD*$_{C-X}$ antibonding orbital [31–33], precisely analogous to the case of a H-bond. The σ-hole interaction involving the halogen atoms is also known as halogen-bonding in view of the concept of H-bonding. Furthermore, it is not only the halogen atoms which can act as a Lewis acid center, but the elements of groups 14, 15, 16 and 18 of the periodic table as well, in which the resulting σ-hole interaction is called a tetrel-bonding [34–39], pnicogen-bonding [40–44], chalcogen-bonding [45–49] and aerogen-bonding [50–53], respectively.

Generally, σ-hole interactions share many common physical and chemical properties with the more traditional H-bonding. They offer a rich array of possibilities to design and fabricate new materials with desired properties, in areas ranging from pharmaceuticals to crystal growth [54–57]. Such diverse applications of σ-hole interactions mainly originate from their directional tunability. For example, the strength and properties of tetrel-bonds can be tuned not only by changing the tetrel atom (group 14 elements) itself, but also by changing the electron withdrawing/accepting ability of the reminder of the molecule [37,58–61]. As a result, a broad range of interaction energies may be spanned by changing the C atom in the tetrel-bond donor into a Si, Ge or Sn atom. Meanwhile, the Lewis base moiety in the tetrel-bond interaction could vary from anions like F$^-$ or Cl$^-$ [34,62,63], through lone-pair electrons on nitrogen or oxygen [64–66], to π electrons in unsaturated bonds [67,68]. The latter may offer a further opportunity to tune the strength of tetrel-bonds and therefore expand their application scope [69,70].

A series of systematic experimental and theoretical studies have produced detailed descriptions of tetrel-bonds in either crystalline state or gas phase. However, these studies have mostly focused on the neutral complexes. For example, Mitzel et al. [71] have found short Si$\cdots$N contacts in the crystalline structure of Si(ONMe$_2$)$_4$ and related compounds. Thomas and coworkers have provided an experimental evidence for the tetrel-bonding in crystalline structures based on charge density analysis. Alkorta and coworkers have investigated tetrel-bonding interactions between SiXY$_3$ (X and Y=H, F, and Cl) and some electron-rich groups (NH$_3$, NCH, CNH, OH$_2$, and FH) [72]. A detailed computational study by Mani and Arunan [73,74] has also found unusual tetrel-bond interactions called "carbon bonding" in the complexes of methanol as the tetrelbond donor with different Lewis bases. The formation of the latter interactions has also been proposed as a preliminary stage of the SN2 reaction by Grabowski [75]. These studies clearly showed that tetrel-bonding is moderately strong and could act as a possible molecular linker in crystal engineering and supramolecular chemistry, similar to H-bonding.

Recently, Scheiner has reported [76] a detailed study on the ability of hydrogen, halogen, chalcogen, pnicogen, and tetrel-bonds as a potential halide (F$^-$, Cl$^-$, Br$^-$) receptor. It was found that the tetrel-bonding exhibits a quite larger tendency to bind to halides than other σ-hole interactions. In another study [77], the author has also shown that the addition of a -SnF$_3$ group to either an imidazolium or triazolium ion provides a strong halide receptor. Interestingly, the tetrel-bonding receptors bind far more strongly to each anion than an equivalent number of K$^+$ counterions. In the present study, we perform a systematic study on the strength and characteristic of charge-assisted tetrel-bond interactions in some model complexes (Schemes 1 and 2). The nature of anionic as well as cationic tetrel-bonds is analyzed by means of molecular electrostatic potential (MEP), quantum theory of atoms in molecules (QTAIM), natural bond orbital (NBO) and electron density difference (EDD) methods. The influence of different substituents on either Lewis acid or Lewis base is also studied in detail. Moreover, the characteristics of these charge-assisted complexes are compared with those of available neutral ones. Finally, we provide some experimental evidence for the existence of such charge-assisted tetrel-bonds in crystalline structures and supramolecular assemblies.

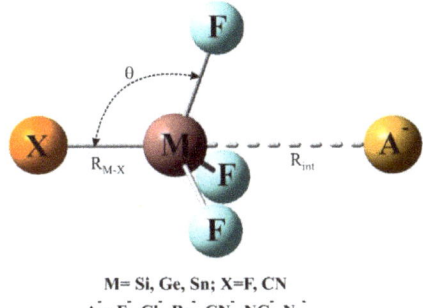

Scheme 1. Representative geometrical structure of anionic tetrel-bonded complexes.

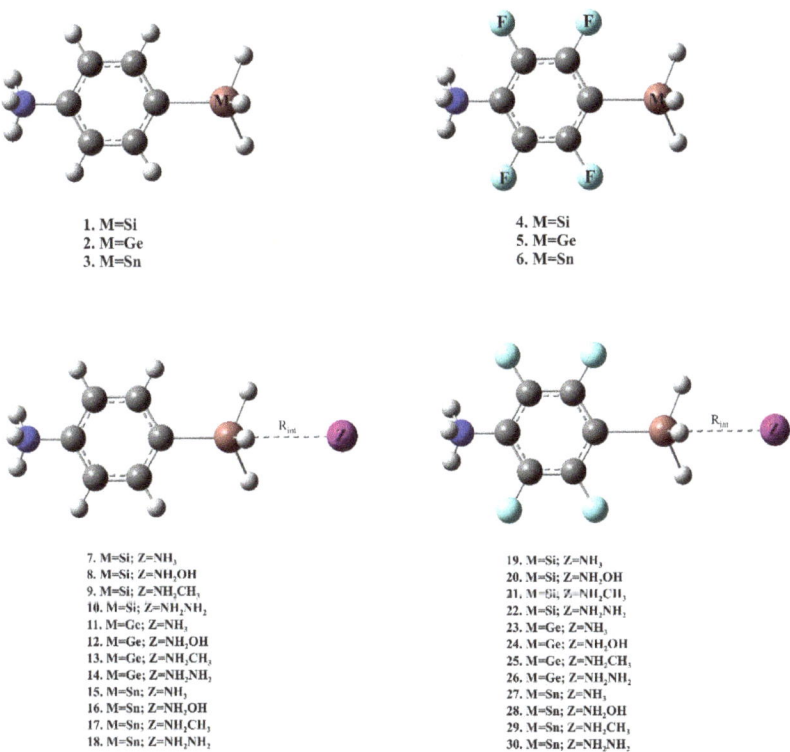

Scheme 2. Structure of monomers **1–6** and cationic tetrel-bonded complexes **7–30**.

2. Systems and Methods

In this work, we report the results of ab initio calculations to study charge-assisted tetrel-bond interactions for two different sets of model systems. In the first model, XF_3M molecule (X=F, CN; M=Si, Ge and Sn) interacts with A^- anions (A^- =F^-, Cl^-, Br^-, CN^-, NC^- and N_3^-). This allows us to check the possibility of anionic tetrel-bonding interaction in the mentioned complexes. The second model studied here includes the cationic tetrel-bonded $[p\text{-}NH_3(C_6H_4)MH_3]^+\cdots Z$ complexes, in which M=Si, Ge, Sn and Z=NH_3, NH_2CH_3, NH_2OH and N_2H_4. The H atoms of the benzene ring are additionally substituted by F atoms in order to study substituent effects.

All ab initio calculations were performed using the Gaussian 09 package [78]. The MP2 method was used, along with the aug-cc-pVTZ basis set to optimize geometries of the anionic $XF_3M:A^-$ tetrel-bonded complexes. Frequency calculations were performed at the same computational level to ensure that the optimized structures correspond a true minimum on the potential energy surface. In the case of the cationic tetrel-bonded systems, the geometry optimizations and the corresponding frequency calculations were performed at the MP2/aug-cc-pVDZ level. Single-point calculations with a larger aug-cc-pVTZ basis set were then performed using the aug-cc-pVDZ optimized geometries. The interaction energies for both sets of the complexes were computed at the MP2/aug-cc-pVTZ level, as the difference between the energy of the complex and the energy sum of the isolated monomers, and corrected for the basis set superposition error (BSSE) by using the Boys–Bernardi counterpoise method [79].

To evaluate the possible orbital interactions between the interacting monomers, the NBO analysis was performed with the NBO 5.0 program (Theoretical Chemistry Institute, University of Wisconsin, Madison, Wisconsin, United States) [80]. The most positive ($V_{S,max}$) and most negative ($V_{S,min}$) electrostatic potentials of the isolated monomers were obtained using the Wave Function Analysis-Surface Analysis Suite (WFA-SAS) [81]. The QTAIM analysis was performed by means of the AIM2000 program [82] with the MP2/aug-cc-pVTZ generated wave functions. To see the amount of electron density shift due to the complex formation, the EDD isosurfaces were computed with the help of MultiWFN [83]. These were obtained by subtracting the electron density of the complex with the sum of the electron densities of the interacting monomers with the geometries in the optimized complex.

3. Results and Discussion

3.1. Anionic Tetrel-Bonds

Scheme 1 indicates the general structure of anionic tetrel-bonded complexes $XF_3M:A^-$, in which the M atom of XF_3M acts as the Lewis acid site to interact with the excess electron density over the anions A^-. It should be mentioned that although there might be many minima on the potential energy surface of these complexes, we are interested here in the interaction involving the linear $X\text{-}M\cdots A^-$ arrangement. The corresponding optimized geometries at the MP2/aug-cc-pVTZ level are summarized in Figure S1 of Supporting Information. All these complexes are found to have a favorable $X-M\cdots A^-$ linear arrangement. The binding distances and interaction energies of these complexes are listed in Table 1. According to the previous studies [37,62,65,67,84,85], the formation of such anionic tetrel-bonding interactions can be largely attributed to the localization of a positive electrostatic potential over the M atom, in the extension of the M-X bond. Indeed, the MEP analysis of XF_3M monomers in Figure 1 reveals that the maximum positive electrostatic ($V_{S,max}$) of the Sn atom (96.5 kcal/mol) in SnF_4 is greater than that of Ge (70.2 kcal/mol) and Si (57.3 kcal/mol) in GeF_4 and SiF_4, respectively. In the case of MF_3CN monomers, it is seen that the σ-hole potential associated with the M atom becomes more positive as the size of the M atom increases. Consequently, it is expected that XF_3M molecules can participate in a σ-hole interaction with the σ-hole acting as a Lewis acid center, and the strongest acidic properties are predicted for the Sn atom of SnF_4 and SnF_3CN based on the MEP analysis.

Table 1. Binding distances (R_{int}, Å), X-M-F angles (θ, °), M-X bond lengths (R_{M-X}, Å) and their changes with respect to those of isolated MF$_3$X monomers (ΔR_{M-X}, Å), interaction energies (E_{int}, kcal/mol) of the anionic tetrel-bonded complexes, and the calculated local MEP minimum values ($V_{S,min}$, kcal/mol) of the anions.

Lewis Acid	Anion	R_{int}	θ	R_{M-X}	ΔR_{M-X}	E_{int}	$V_{S,min}$
SiF$_4$	F$^-$	1.679	90.0	1.679	0.105	−70.11	−175.0
	Cl$^-$	2.102	91.7	1.665	0.091	−25.00	−141.2
	Br$^-$	2.226	92.8	1.657	0.083	−16.35	−132.9
	NC$^-$	1.888	92.0	1.661	0.087	−35.03	−141.9
	CN$^-$	1.969	91.4	1.663	0.089	−38.86	−135.9
	N$_3^-$	1.904	91.7	1.661	0.087	−30.08	−133.5
SiF$_3$CN	F$^-$	1.662	91.4	1.970	0.145	−84.37	−175.0
	Cl$^-$	2.112	89.9	1.951	0.126	−37.29	−141.2
	Br$^-$	2.166	90.6	1.942	0.117	−27.60	−132.9
	NC$^-$	1.860	90.5	1.943	0.118	−46.59	−141.9
	CN$^-$	1.946	90.0	1.946	0.121	−50.72	−135.9
	N$_3^-$	1.868	89.9	1.945	0.12	−41.87	−133.5
GeF$_4$	F$^-$	1.773	90.0	1.773	0.085	−79.17	−175.0
	Cl$^-$	2.263	90.3	1.764	0.076	−41.26	−141.2
	Br$^-$	2.436	90.7	1.760	0.072	−32.99	−132.9
	NC$^-$	1.932	91.3	1.762	0.074	−46.62	−141.9
	CN$^-$	1.994	90.5	1.762	0.074	−53.57	−135.9
	N$_3^-$	1.940	90.7	1.762	0.074	−43.30	−133.5
GeF$_3$CN	F$^-$	1.762	90.5	1.994	0.118	−84.92	−175.0
	Cl$^-$	2.239	89.7	1.983	0.107	−46.32	−141.2
	Br$^-$	2.407	90.0	1.977	0.101	−37.73	−132.9
	NC$^-$	1.915	90.8	1.981	0.105	−50.96	−141.9
	CN$^-$	1.979	90.0	1.979	0.103	−58.35	−135.9
	N$_3^-$	1.921	90.0	1.981	0.105	−47.94	−133.5
SnF$_4$	F$^-$	1.930	90.0	1.930	0.048	−93.58	−175.0
	Cl$^-$	2.371	89.5	1.936	0.054	−61.84	−141.2
	Br$^-$	2.523	90.4	1.934	0.052	−54.44	−132.9
	NC$^-$	2.078	90.1	1.933	0.051	−62.50	−141.9
	CN$^-$	2.144	90.1	1.932	0.05	−70.26	−135.9
	N$_3^-$	2.083	90.3	1.932	0.05	−60.07	−133.5
SnF$_3$CN	F$^-$	2.144	90.1	2.144	0.091	−96.30	−175.0
	Cl$^-$	2.357	90.8	2.141	0.088	−64.11	−141.2
	Br$^-$	2.508	89.2	2.139	0.086	−56.59	−132.9
	NC$^-$	2.067	90.9	2.134	0.081	−64.21	−141.9
	CN$^-$	2.132	90.0	2.132	0.079	−72.26	−135.9
	N$_3^-$	2.073	90.1	2.133	0.080	−62.09	−133.5

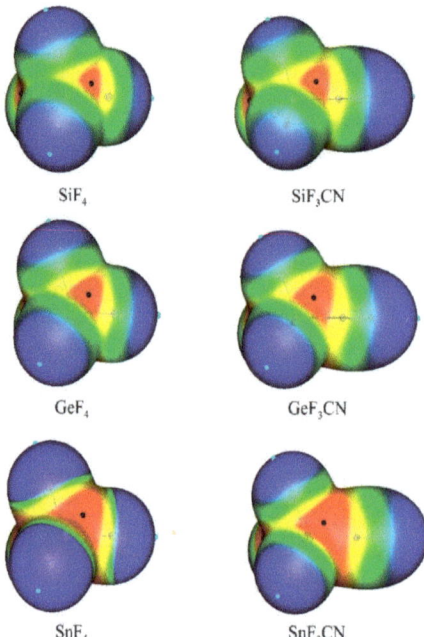

Figure 1. MEPs of isolated monomers. The color code, in kcal/mol, is: red > 40; 40 > yellow > 20; 20 > green > 0 and blue < 0. The small black and blue circles indicate surface maxima and minima, respectively.

As Table 1 indicates, the M···A$^-$ binding distances of XF$_3$M:A$^-$ complexes are in the range of 1.662–2.523 Å, which are much shorter than the sum of the van der Waals (vdW) radii of the interacting atoms [86]. This clearly shows the existence of a strong interaction between the XF$_3$M and A$^-$ moieties. In many cases, the M atom is characterized by pentavalency, i.e., is hypervalent. This is similar to the one described for the transition state structure of SN2 reaction between a tetrel atom center and anion species [75,87]. In fact, most of the M···A$^-$ binding distances are short enough to be considered covalent bonds which have lost some degree of covalency. The binding distances for a given anion increase in the order SiF$_3$CN < SiF$_4$ < GeF$_3$CN < GeF$_4$ < SnF$_3$CN < SnF$_4$. Note also that M···A$^-$ distances become longer in the order of Si < Ge < Sn when the anion is the same, which is similar to the order of vdW or the covalent radius of these atoms (Si < Ge < Sn). The interaction between the anion and M atom is also able to induce a large distortion in the XF$_3$M molecule, as evidenced by the calculated X-M-F angles (Table 1). For each set of the complexes, one can see that the X-M-F angles are close to 90°, which may provide further evidence for the strong interaction between the XF$_3$M and A$^-$ moieties.

From Table 1, one can see that the interaction energies of XF$_3$M:A$^-$ are very large and negative, indicating a strong interaction between XF$_3$M and A$^-$ subunits. These results are in agreement with recent reports that indicate that tetrel-bonding can be used as a vehicle for strong and selective anion binding [77,88]. Moreover, the interaction energies obtained here are in good agreement with those of other related studies [62,63,76,77,88]. Comparing interaction energies clearly indicates that for a given XF$_3$M, the value of interaction energy for F$^-$ is systematically larger than other anions. In fact, such large negative interaction energies together with the corresponding very short binding distances indicate that the M···F$^-$ interactions are mainly covalent in nature. Interacting with the same anion, SnF$_4$ tends to form stronger tetrel-bond interaction than other molecules, as characterized by a larger interaction energies in the corresponding complexes. This has been found for other tetrel-bonds, previously [37,58,75,84,89]; the increase of the interaction energy for the analogue complexes if the atomic number of the tetrel atom increases. This is connected with the electrostatic nature of these interactions due to the presence of a large positive electrostatic potential on the central atom of XF_3M molecules (Figure 1). Moreover, it is natural that the more negative electrostatic potential ($V_{S,min}$) associated with the anion forms a more stable M···A$^-$ interaction. However, as Figure 2 indicates, we found almost a poor linear correlation between the interaction energies of these complexes and $V_{S,min}$ values associated with the anions. Note that such a poor linear relationship between the E_{int} and $V_{S,min}$ values has already been described in the literature [90–92]. This is mainly related to the different nature of the A$^-$ moiety in these complexes, which provides a distinct contribution of other energy terms such as polarization or charge-transfer in these systems.

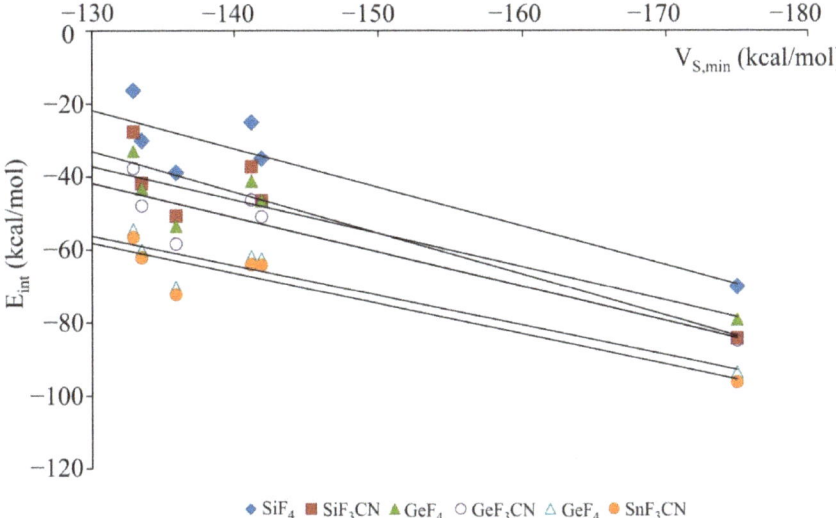

Figure 2. Correlation between the interaction energies and $V_{S,min}$ values associated with the A$^-$ anions in the binary XF_3M:A$^-$ complexes. The squared correlation coefficient (R^2) value is 0.833, 0.831, 0.846, 0.839, 0.868 and 0.872 for SiF_4, SiF_3CN, GeF_4, GeF_3CN, SnF_4 and SnF_3CN complexes, respectively.

The results of Table 1 also indicate that due to the formation of XF_3M:A$^-$ complexes, the M-X bonds are elongated. The magnitude of this bond elongation is in the range of 0.083–0.118 Å, 0.048–0.145 Å and 0.072–0.091 Å in the Si, Ge and Sn complexes, respectively. Note that the strongest M···F$^-$ interaction in these systems is characterized by a large elongation of M-X bond, which is much larger than the corresponding values in the Cl$^-$ or Br$^-$ complexes. Also, paired with the same anion, MF_4 complexes show a relatively smaller variation in the M–X bond distances than MF_3X analogues, which is based on the fact that the F is a poor leaving group than the CN. This result is consistent with the variation of interaction energy of these complexes, and suggests that Sn-X bond displays a larger red shift in the corresponding M-X stretching frequency than the Ge-X and Si-X ones. We will come back to this conclusion further on in our discussion when NBO analysis is illustrated.

To have a deeper understanding of the nature of anionic tetrel-bond interactions, we have performed the topological analysis of the electron density of XF_3M:A$^-$ complexes (Table 2). It is found that for each system considered, there exists a bond critical point (BCP) associated with the M···A$^-$ interaction. As seen, the strong tetrel-bond interactions in XF_3M:A$^-$ complexes are characterized by a large electron density value at the corresponding bond critical points (BCPs), which are much larger than those of at the neutral tetrel-bonded systems [38,59,87,93,94]. For a given M or X, the F$^-$ complexes exhibit the largest ρ_{BCP} value, while the smallest one corresponds to the Br$^-$ complexes.

Note also that, as predicted by the ρ_{BCP} values, the tetrel-bond interactions of the Sn complexes are stronger than those of Ge or Si ones. This is in line with the other related studies [65,72,75,95], where it was found that the ρ_{BCP} value is a good descriptor of the strength of interaction. Besides, almost a good exponential correlation was found between the binding distances and electron density values at the corresponding BCPs of XF$_3$M:A$^-$ complexes (Figure 3). Moreover, the Laplacian values of ρ_{BCP} are found to be positive and in the range of 0.053–0.936 au, which is indicative of closed-shell nature of these interactions [96]. Meanwhile, the negative values of total electron energy density at M$\cdots$A$^-$ BCPs, H$_{BCP}$, for all these complexes clearly confirm that the anionic tetrel-bond interactions could be classified as the strong interactions with some covalent character [97].

Table 2. Electron density (ρ_{BCP}, au), its Laplacian ($\nabla^2\rho_{BCP}$, au) and total electron energy density (H$_{BCP}$, au) at the M$\cdots$A$^-$ BCPs, and NBO stabilization energy (E$^{(2)}$, kcal/mol), atomic charge on the M atom (q$_M$, e), net charge-transfer (q$_{CT}$, e) and Wiberg bond index (WBI) values of the anionic tetrel-bonded complexes.

Lewis Acid	Anion	ρ_{BCP}	$\nabla^2\rho_{BCP}$	H$_{BCP}$	E$^{(2)}$	q$_M$	q$_{CT}$	WBI
SiF$_4$	F$^-$	0.112	0.879	-0.025	53.63	2.64	0.26	0.48
	Cl$^-$	0.065	0.152	-0.030	38.72	2.54	0.33	0.51
	Br$^-$	0.054	0.060	-0.028	32.06	2.55	0.31	0.49
	NC$^-$	0.085	0.421	-0.029	69.31	2.63	0.21	0.37
	CN$^-$	0.081	0.280	-0.045	47.92	2.50	0.34	0.56
	N$_3{}^-$	0.086	0.368	-0.034	45.64	2.42	0.59	0.41
SiF$_3$CN	F$^-$	0.117	0.936	-0.028	65.86	2.50	0.27	0.56
	Cl$^-$	0.070	0.183	-0.035	56.59	2.34	0.39	0.60
	Br$^-$	0.064	0.080	-0.034	49.40	2.32	0.40	0.60
	NC$^-$	0.093	0.301	-0.049	89.12	2.45	0.25	0.60
	CN$^-$	0.086	0.299	-0.049	61.17	2.30	0.39	0.60
	N$_3{}^-$	0.094	0.415	-0.040	60.75	2.42	0.58	0.46
GeF$_4$	F$^-$	0.130	0.774	-0.050	54.69	2.68	0.27	0.42
	Cl$^-$	0.084	0.130	-0.038	51.64	2.52	0.38	0.57
	Br$^-$	0.073	0.055	-0.032	46.01	2.50	0.40	0.58
	NC$^-$	0.107	0.364	-0.048	78.64	2.66	0.22	0.38
	CN$^-$	0.112	0.191	-0.058	55.20	2.51	0.38	0.58
	N$_3{}^-$	0.109	0.308	-0.053	53.60	2.60	0.58	0.44
GeF$_3$CN	F$^-$	0.134	0.800	-0.053	58.67	2.51	0.27	0.44
	Cl$^-$	0.089	0.135	-0.042	60.69	2.31	0.42	0.63
	Br$^-$	0.078	0.053	-0.036	55.20	2.28	0.46	0.65
	NC$^-$	0.112	0.381	-0.052	91.00	2.45	0.25	0.42
	CN$^-$	0.116	0.194	-0.062	61.82	2.29	0.42	0.63
	N$_3{}^-$	0.115	0.323	-0.058	59.72	2.40	0.58	0.48
SnF$_4$	F$^-$	0.113	0.687	-0.029	34.37	2.93	0.21	0.36
	Cl$^-$	0.081	0.195	-0.027	42.18	2.75	0.36	0.37
	Br$^-$	0.072	0.117	-0.024	40.60	2.71	0.40	0.60
	NC$^-$	0.098	0.378	-0.031	53.92	2.90	0.19	0.34
	CN$^-$	0.094	0.229	-0.038	37.04	2.77	0.33	0.53
	N$_3{}^-$	0.099	0.384	-0.066	34.80	2.84	0.27	0.40
SnF$_3$CN	F$^-$	0.101	0.702	-0.031	33.81	2.77	0.22	0.38
	Cl$^-$	0.084	0.199	-0.029	44.15	2.55	0.40	0.61
	Br$^-$	0.075	0.118	-0.026	42.58	2.50	0.45	0.66
	NC$^-$	0.101	0.385	-0.034	53.18	2.71	0.22	0.37
	CN$^-$	0.098	0.230	-0.040	34.38	2.55	0.37	0.58
	N$_3{}^-$	0.098	0.341	-0.037	31.10	2.65	0.54	0.44

As noted earlier, charge-transfer from the electron donor into the empty orbital of the electron acceptor also plays an important role in the formation and stabilization of tetrel-bonded complexes [61–63,85,87,98]. For the anionic tetrel-bonded complexes studied here, it is expected that there exists a stabilizing orbital-orbital interaction between the lone-pair orbital of the anion, LP (A$^-$), and empty anti-bonding M-X orbital of XF$_3$M molecule (BD*$_{M-X}$). The latter orbital interaction should be responsible for the elongation of M-X bonds and their red-shift upon the complexation. To confirm this, we performed NBO analysis on the XF$_3$M:A$^-$ complexes. Table 2 summarizes the calculated stabilization energy E$^{(2)}$ values due to the LP (A$^-$)→BD*$_{M-X}$ orbital interaction. As is evident, these E$^{(2)}$ values are quite large, especially for the A$^-$=CN$^-$ and F$^-$ complexes, which demonstrates the significant role of the mentioned orbital interaction in these systems. It is also found that for all complexes analyzed here, the formation of tetrel bonds results in an increase in the positive charge of the M atom due to its polarization in the presence of the negative charge of the anion (Table 2). For each set of the complexes, such polarization is largest in the F$^-$ complexes, which is consistent with the stronger tetrel-bond interaction in these systems. However, due to the variety of Lewis bases, it is not possible to find any regularity in the changes of the atomic charges here. The data in Table 2 also reveal that the net charge-transfer values (q$_{CT}$) for the XF$_3$M:A$^-$ complexes are very large, with values ranging from 0.21 to 0.58 e. Moreover, for a given M or X, the N$_3^-$ and Br$^-$ complexes are identified by a larger q$_{CT}$ values compared to other ones, which is most likely due to the large polarizability of these moieties. Hence, the larger elongation of the M-X bond in the latter complexes can be attributed to the more favorable charge-transfer, which results in the partial population of the antibonding BD*$_{M-X}$ orbital of XF$_3$M and its redshift. As also shown in Table 2, the Wiberg bond index (WBI) of the anionic tetrel-bonds is large, which verifies the formation of covalent M···A$^-$ interactions in these systems.

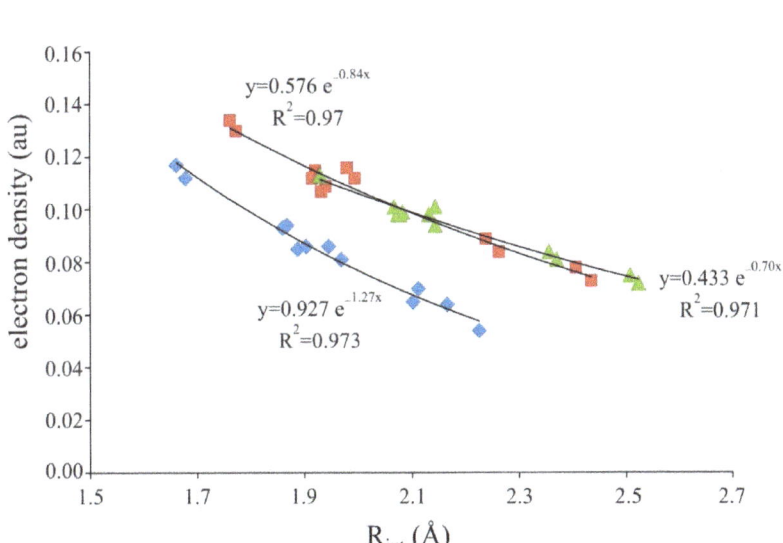

Figure 3. Exponential relationship between the M···A$^-$ binding distances and electron density at the corresponding BCPs of the XF$_3$M:A$^-$ complexes.

When the XF$_3$M molecule is paired with the A$^-$ anion, there is a mutual polarization between the two moieties, which can be verified using the EDD analysis. Figure 4 shows the EDD isosurfaces for some representative complexes of XF$_3$M:A$^-$, which were obtained by subtracting the electron density of the complex with the sum of the electron densities of the interacting monomers with the geometries in the optimized complex. Here, violet regions show a decreased electron density, while green areas refer to an increased electron density. As can be seen, the formation of these complexes leads to the appearance of a large electron density loss region over the M atom, facing the anion. The size of this electron density loss region becomes larger as the size of the M atom increases. Meanwhile, a large electron density accumulation is found between the M and A^-, which confirms the formation of a covalent $M \cdots A^-$ interaction in these systems. Moreover, the formation of anionic tetrel-bond interaction in these complexes tends to induce an accumulation of electron density on the F atom of XF_3M. One can also see the localization of a large electron density loss region over the anion, which is related to the polarization of these moieties in the presence of positive σ-hole on the M atom. Clearly, such electron density shift is larger for the Sn complexes than Ge and Si ones, due to more positive σ-hole potential associated with the former systems.

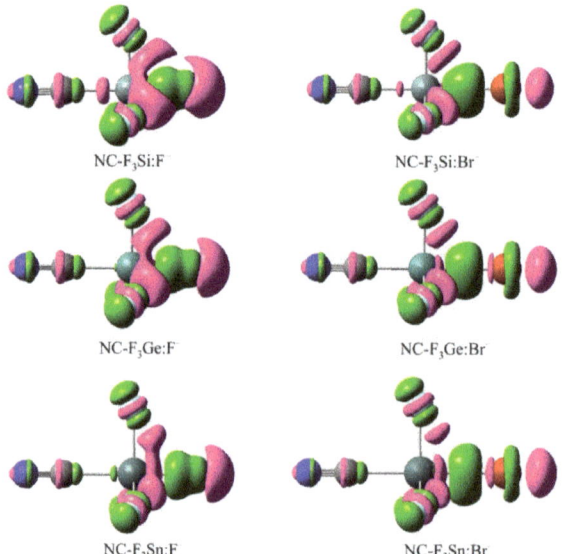

Figure 4. EDD isosurfaces (± 0.005 au) of some representative $XF_3M:A^-$ complexes. The violet and green regions indicate regions of decreased and increased electron densities, respectively.

3.2. Cationic Tetrel-Bonds

Scheme 2 depicts the general representation of cationic tetrel-bonded complexes **7–30** studied here. The corresponding optimized geometries are summarized in Figure S2 of the Supporting Information. Their intermolecular $M \cdots N$ distances and interaction energies are reported in Table 3. From Figure S2, one can see that all these complexes are characterized by a linear $C-M \cdots N$ interaction, in which nitrogen atom of the Lewis base is pointed towards the M atom of the Lewis acid. The binding distances are in the range of 2.175–2.567, 2.352–2.790 and 2.479–2.735 for the M=Si, Ge and Sn, respectively. All these binding distances are smaller than the sum of vdW radii of the respective atoms [86], which implies that there is an attractive interaction between the interacting molecules. For a given M, NH_2NH_2 forms always the shortest tetrel-bond distance, while the longest corresponds to NH_3. Moreover, the substitution of F atoms in the benzene ring tends to decrease the binding distances, which can be attributed to the increase of positive electrostatic potential on the M atom due to presence of the F atoms. This indicates that the formation of cationic tetrel-bond in these systems is, at least partly, a consequence of the electrostatic attraction between the nitrogen atom of Lewis bases and the M atom. As also expected, the $M \cdots N$ binding distances for a fixed nitrogen base increase in the order of Ge > Sn > Si can be related to the combination result of the interaction energy and the atomic radius of these atoms. The results of Table 3 also show that the formation of cationic tetrel-bonds in the binary complexes **7–30** leads to a significant increase in the C-M-H angles (θ), as evidenced by θ values close to 90°. This is in line with previous studies, where it was found for the strong tetrel-bonded complexes that the intermolecular interaction should be a preliminary stage of the SN2 reaction [75,87].

Table 3. Binding distances (R_{int}, Å), C-M-H angles (θ, °) and interaction energies (E_{int}, kcal/mol) of the cationic tetrel-bonded complexes.

Complex		R_{int}	θ	E_{int}
7	1+NH_3	2.567	101.1	−7.69
8	1+NH_2OH	2.459	101.4	−9.25
9	1+NH_2CH_3	2.291	98.2	−13.58
10	1+NH_2NH_2	2.245	97.5	−15.05
11	2+NH_3	2.790	102.6	−9.82
12	2+NH_2OH	2.724	103.0	−11.86
13	2+NH_2CH_3	2.538	100.2	−17.44
14	2+NH_2NH_2	2.486	99.9	−20.8
15	3+NH_3	2.735	100.5	−13.36
16	3+NH_2OH	2.671	100.2	−16.65
17	3+NH_2CH_3	2.568	97.7	−19.11
18	3+NH_2NH_2	2.548	98.1	−22.46
19	4+NH_3	2.330	99.4	−10.89
20	4+NH_2OH	2.277	99.0	−13.47
21	4+NH_2CH_3	2.195	96.7	−17.42
22	4+NH_2NH_2	2.175	97.0	−20.86
23	5+NH_3	2.527	100.3	−14.25
24	5+NH_2OH	2.498	101.1	−18.48
25	5+NH_2CH_3	2.370	98.4	−22.55
26	5+NH_2NH_2	2.352	98.6	−25.5
27	6+NH_3	2.586	99.0	−17.02
28	6+NH_2OH	2.574	100.3	−20.08
29	6+NH_2CH_3	2.487	98.1	−23.66
30	6+NH_2NH_2	2.479	98.7	−26.98

Considering the interaction energies in Table 3, it is found that the most strongly bound complex **30** has an interaction energy of −26.98 kcal/mol, while the most weakly bound complex **7** has an interaction energy of only −7.69 kcal/mol. These interaction energies are larger than the reported values for similar tetrel-bond interactions in the related neutral complexes [65,93]. Meanwhile, the calculated interaction energies for the Si and Ge complexes are close to those of tetrel-bonding interactions in the prorogated complexes of pyridine-MF_3 or furan-MF_3 with NH_3 [66]. Note that, for the same electron acceptor moiety, NH_2NH_2 tends to form more stable tetrel-bond interaction than others, which is not consistent with the $V_{S,min}$ value (in kcal/mol) associated with the nitrogen atom in thee bases: NH_3(−42.7) > NH_2CH_3 (−38.6) > NH_2NH_2 (−36.5) > NH_2OH (−28.8). This may be attributed in part to secondary interactions between these Lewis bases and the H atoms of -MH_3 moiety in the Lewis acid. Moreover, this can be explained in the manner of the negative hyperconjugation effect on the side of Lewis bases as suggested by Zierkiewicz and Michalczyk [92]. This finding clearly reveals that the $V_{S,min}$ value, a property in a single special point of the Lewis base, cannot be regarded as a good indicator of the cationic tetrel-bond interactions, and in addition to electrostatic effects other factors such as the polarization should play an important role in the stability and formation of these interactions.

The topological analysis of the electron density of the complexes **7–30** exhibits the presence of a BCP between the M atom of Lewis acid and N atom of the nitrogen bases. The electron densities at the M···N BCPs are between 0.024 au in **7** and 0.060 au in **30** (Table 4). It is interesting to note that, like weak tetrel-bond interactions in neutral complexes [38,59,87,93,94], we found an exponential correlation between the electron density at the BCP and the interatomic distances of these systems (Figure 5).

Importantly, the calculated squared correlation confection values (R^2) for these cationic terel-bonded complexes are larger than those of anionic ones (Figure 3). Moreover, the positive $\nabla^2\rho_{BCP}$ values associated with these complexes demonstrate that the cationic tetrel-bond interactions are within the closed-shell interaction regime. Meanwhile, negative total energy densities (H_{BCP}) are obtained for all complexes studied here, which confirm that the cationic tetrel-bonds have some covalent character. Note that the for each set of the complexes, most negative values of H_{BCP} correspond to the stronger interactions and to the greater values of ρ_{BCP} (Table 4).

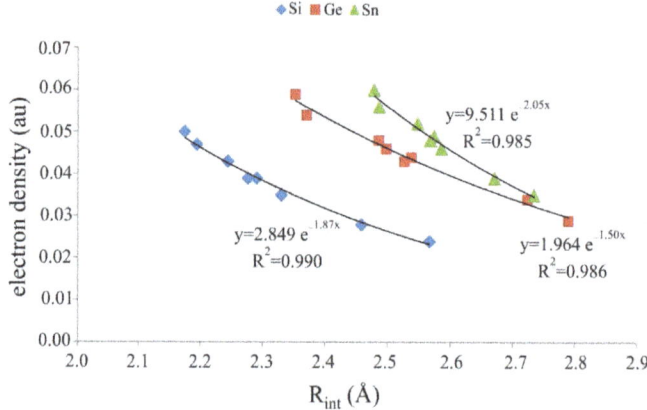

Figure 5. Exponential relationship between the M···N binding distances and electron density at the corresponding BCPs of the cationic tetrel-bonded complexes.

Table 4. Electron density (ρ_{BCP}, au), its Laplacian ($\nabla^2\rho_{BCP}$, au) and total electron energy density (H_{BCP}, au) at the M···Z BCPs, and NBO stabilization energies due to the LP(N) →BD*$_{M-C}$ orbital interaction ($E^{(2)}$, kcal/mol), net charge-transfer (q_{CT}, e) and Wiberg bond index (WBI) values of the cationic tetrel-bonded complexes.

Complex		ρ_{BCP}	$\nabla^2\rho_{BCP}$	H_{BCP}	$E^{(2)}$	q_{CT}	WBI
7	1+NH$_3$	0.024	0.049	−0.003	10.42	0.072	0.114
8	1+NH$_2$OH	0.028	0.066	−0.017	11.20	0.075	0.129
9	1+NH$_2$CH$_3$	0.039	0.084	−0.021	13.45	0.078	0.135
10	1+NH$_2$NH$_2$	0.043	0.111	−0.024	14.52	0.080	0.144
11	2+NH$_3$	0.029	0.055	−0.005	12.28	0.076	0.134
12	2+NH$_2$OH	0.034	0.062	−0.019	14.55	0.079	0.155
13	2+NH$_2$CH$_3$	0.044	0.088	−0.026	17.70	0.082	0.167
14	2+NH$_2$NH$_2$	0.048	0.099	−0.030	19.14	0.084	0.176
15	3+NH$_3$	0.035	0.090	−0.007	15.25	0.079	0.155
16	3+NH$_2$OH	0.039	0.072	−0.008	17.10	0.081	0.168
17	3+NH$_2$CH$_3$	0.048	0.083	−0.022	18.75	0.084	0.182
18	3+NH$_2$NH$_2$	0.052	0.111	−0.026	20.40	0.086	0.193
19	4+NH$_3$	0.035	0.053	−0.008	13.28	0.079	0.149
20	4+NH$_2$OH	0.039	0.090	−0.023	14.48	0.084	0.189
21	4+NH$_2$CH$_3$	0.047	0.136	−0.030	16.05	0.087	0.201
22	4+NH$_2$NH$_2$	0.050	0.141	−0.033	18.60	0.090	0.228
23	5+NH$_3$	0.043	0.113	−0.030	15.08	0.082	0.165
24	5+NH$_2$OH	0.046	0.097	−0.021	16.69	0.085	0.172
25	5+NH$_2$CH$_3$	0.054	0.121	−0.028	18.82	0.089	0.191
26	5+NH$_2$NH$_2$	0.059	0.123	−0.032	20.80	0.094	0.206
27	6+NH$_3$	0.046	0.089	−0.010	17.04	0.092	0.180
28	6+NH$_2$OH	0.049	0.093	−0.026	18.92	0.095	0.196
29	6+NH$_2$CH$_3$	0.056	0.131	−0.033	19.77	0.096	0.212
30	6+NH$_2$NH$_2$	0.060	0.141	−0.036	22.50	0.098	0.225

According to the NBO analysis, there is a noticeable charge-transfer interaction between the lone-pair bonding orbital LP(N) of nitrogen atom of the Lewis bases and the BD*$_{M-C}$ antibonding orbital of the Lewis acid. A similar interaction between the lone-pair of nitrogen and the BD*$_{M-C}$ anti-bonding orbital (M=C, Si, Ge) was analyzed recently for the prorogated complexes of pyridine-MF$_3$ with NH$_3$ [66]. This orbital interaction is responsible for a negligible elongation of the M-C bond in these complexes. The stabilization energies E$^{(2)}$ associated with the latter orbital interaction are in the range of 9.22–18.60, 12.28–20.20 and 15.25–22.50 kcal/mol for the Si, Ge and Sn complexes, respectively. There is an almost linear correlation between these E$^{(2)}$ values and interaction energies of these complexes (see Figure S3, Supporting Information), which demonstrates that the charge-transfer interaction also plays an important role in the stability of these systems. Note also that the F-substituted complexes **19–30** are characterized by quite a large E$^{(2)}$ value with respect to the **7–18**. We note that in addition to the LP(N) → BD*$_{M-C}$ orbital interaction, there is also some weak orbital interactions between the BD orbital of M-H to the BD*$_{O-H}$, BD*$_{C-H}$ or BD*$_{N-H}$ antibonding orbital of NH$_2$OH, NH$_2$CH$_3$ or NH$_2$NH$_2$ with the stabilization energies in the range of 1.20–6.85 kcal/mol. The calculated NBO charges also show a significant net charge transfer (q$_{CT}$) in these complexes with values ranging from 0.072 e in **7** and 0.098 e in **30**. As expected, relatively larger q$_{CT}$ values are found for the Sn complexes, which indicates that there exists a relationship between the size of transferred charge between the interacting monomers and interaction energy. This is evident in Figure S4 of the Supporting Information, where a linear correlation is seen between these quantities. Additionally, the data in Table 4 shows that the obtained WBI values for these complexes vary from 0.114 to 0.225, which suggests that these cationic tetrel-bonds have a considerable covalent character.

Finally, we would like to highlight the electron density redistribution within and between the monomers upon the formation of the cationic tetrel-bonded complexes. Figure 6 shows the EDD plots of some selected complexes. As is evident, the formation of tetrel-bond interaction in these complexes makes a large electron density accumulation region over the nitrogen atom of the Lewis base. The degree of the accumulation depends on the strength of the tetrel-bond, and increases in the order of **15** > **11** > **7** and **27** > **23** > **19**. In contrast, the lone-pair orbital of the nitrogen atom induces an electron density loss area on the M atom. Note that such mutual polarization between the interacting monomers is almost similar as that in other studied tetrel-bonded systems [38,90,93].

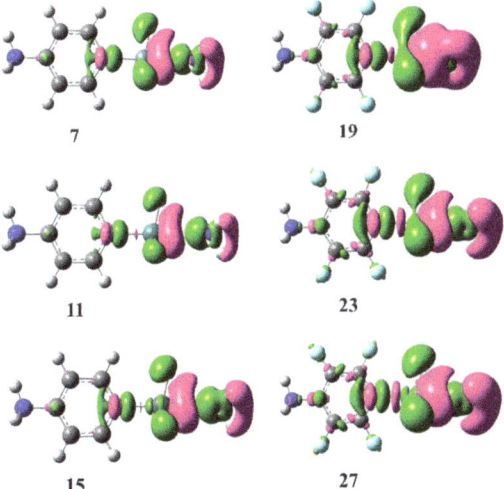

Figure 6. EDD isosurfaces (±0.001 au) of some representative cationic tetrel-bonded complexes. The violet and green regions indicate regions of decreased and increased electron densities, respectively.

3.3. Experimental Evidence for Charge-Assisted Tetrel-Bonds

It is noteworthy that the existence of charge-assisted tetrel-bonds described theoretically here may also be confirmed experimentally. To this end, the Cambridge Structure Database (CSD) [99] was examined to analyze whether anionic or cationic tetrel-bonding could be a generally occurring interaction within crystal structures (CSD, version 5.34, November 2012, including three updates). In Figure 7, we show some selected examples and the CSD reference codes of crystal structures in which the anionic tetrel-bonding interaction is observed. In all these complexes, the tetrel-bonding is highly directional, as evidenced by R-M···A$^-$ bonding angles close to 180° (R is the electron-withdrawing atom or group attached to the M atom). A quite interesting experimental finding that reports the existence of anionic tetrel-bonding interactions is the formation of crystalline spherosilicate structures, where a fluoride ion is perfectly centered within the octasilsesquioxane cage [100,101]. The X-ray structures are indicated in Figure 7 (WAVYEZ and WAVYAV), where tetrel-bond interactions are confirmed by relatively short contacts between the Si atom and F$^-$ anion, which are significantly shorter than the sum of the corresponding vdW radii. Likewise, there also exist attractive anionic tetrel-bonding interactions between the Si atoms and encapsulated Cl$^-$ anion in FOSDUO. In Figure 7, we also show the crystalline structures of three binuclear pentacoordinate silicon complexes of diketopiperazine, which gives evidence for the covalent-bonding between the Si atom and the F$^-$, Cl$^-$ and OSO$_2$BF$_3^-$ anions [102]. Note that the shorter Si···F$^-$ bond distances compared to the Si···Cl$^-$ confirm our earlier finding that the F$^-$ has a larger tendency to interact with the tetrel atom than the Cl$^-$.

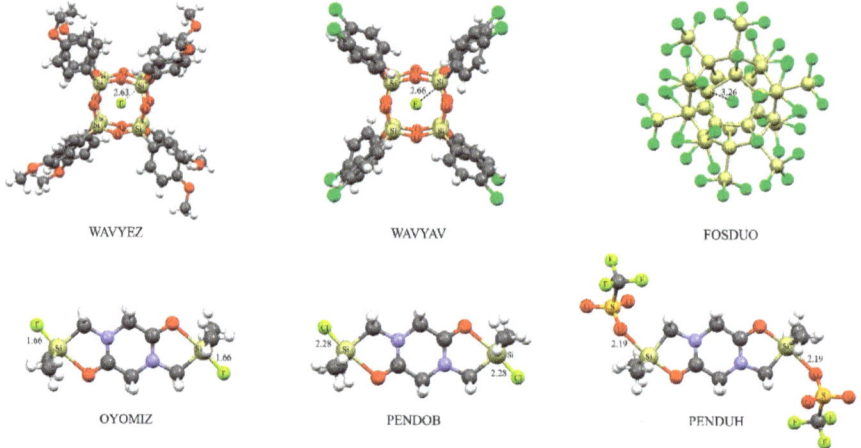

Figure 7. Crystalline structure of some selected anionic tetrel-bonded complexes.

The existence of short tetrel-bond interaction between the cationic tetrel atom and a potential nitrogen base has been already suggested for the complex [Sn(Me)$_3$(NH$_3$)$_2$][N(SO$_2$Me)$_2$] [103] (Figure 8). Here, the positively charged Sn(Me)$_3$NH$_3$ moiety forms a strong Sn···N interaction with the NH$_3$ molecule. Meanwhile, there is a short H-bonding interaction between the latter NH$_3$ molecule and the negatively charged N(SO$_2$Me)$_2$ moiety in this complex. The formation of this H-bonding interaction is able to greatly modulate the strength and properties of the tetrel-bonding, as evidenced by the previous theoretical study about the cooperativity effects between the tetrel-bonding and H-bonding interactions [59].

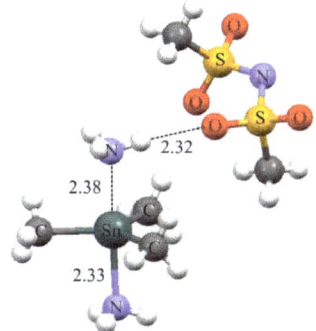

Figure 8. Crystalline structure of the [Sn(Me)$_3$(NH$_3$)$_2$][N(SO$_2$Me)$_2$] complex.

4. Conclusions

Using the ab initio calculations, the geometry, interaction energy and bonding properties of anionic and cationic tetrel-bonded complexes were investigated. Our results indicated that these interactions are highly directional due to the localization of a positive electrostatic potential on the tetrel atom and might serve as a molecular linker in supramolecular assemblies. The strength of these charge-assisted tetrel-bonds increases with the increase of the atomic number of the Lewis acid center (Si < Ge < Sn). The QTAIM and NBO approaches were used to deepen the understanding of the nature of the charge-assisted tetrel-bonds. The formation of the anionic and cationic tetrel-bonds results in a sizable electron density redistribution over the interacting subunits, and an increase of the polarization of M-X or M-C bond. In particular, the M atom in very strong tetrel-bonded complexes XF$_3$M:A$^-$ is characterized by pentavalency, i.e., is hypervalent. Moreover, the application of such charge-assisted tetrel-bonds in crystal materials were characterized and evidenced by a CSD search. The results of this study may provide some new insights into the role of tetrel-bonding interactions in crystalline structure and supramolecular chemistry.

Supplementary Materials: The following are available online. Figure S1: Optimized structure of the anionic tetrel-bonded complexes, Figure S2: Optimized structure of the cationic tetrel-bonded complexes **7–30**, Figure S3: Correlation between the stabilization energy, due to the LP(N) → BD*M-C orbital interaction, and interaction energies of cationic tetrel-bonded complexes **7–30**, Figure S4: Correlation between the net charge-transfer and interaction energies of cationic tetrel-bonded complexes **7–30**.

Author Contributions: Conceptualization, M.D.E.; Investigation, P.M., Writing-Original Draft, P.M.; Writing-Review & Editing, M.D.E.

Funding: This research received no external funding.

Acknowledgments: The authors would like to thank the "Computational Center of University of Maragheh" for its technical support of this work.

Conflicts of Interest: The authors declare they have no conflict of interest.

References

1. Müller-Dethlefs, K.; Hobza, P. Noncovalent interactions: A challenge for experiment and theory. *Chem. Rev.* **2000**, *100*, 143–168. [CrossRef] [PubMed]
2. Strekowski, L.; Wilson, B. Noncovalent interactions with DNA: An overview. *Mutat. Res.-Fund. Mol. M.* **2007**, *623*, 3–13. [CrossRef] [PubMed]
3. Riley, K.E.; Hobza, P. Noncovalent interactions in biochemistry. *WIREs Comput. Mol. Sci.* **2011**, *1*, 3–17. [CrossRef]
4. Alkorta, I.; Elguero, J. Carbenes and silylenes as hydrogen bond acceptors. *J. Phys. Chem.* **1996**, *100*, 19367–19370. [CrossRef]

5. Scheiner, S. *Hydrogen Bonding. A Theoretical Perspective*; Oxford University Press: New York, NY, USA, 1997.
6. Rozas, I.; Alkorta, I.; Elguero, J. Behavior of ylides containing N, O, and C atoms as hydrogen bond acceptors. *J. Am. Chem. Soc.* **2000**, *122*, 11154–11161. [CrossRef]
7. Dannenberg, J.J. The nature of the hydrogen bond: Outline of a comprehensive hydrogen bond theory. *J. Am. Chem. Soc.* **2010**, *132*, 3229–3230. [CrossRef]
8. Jing, B.; Li, Q.; Gong, B.; Li, R.; Liu, Z.; Li, W.; Cheng, J.; Sun, J. Hydrogen bond and σ-hole interaction in $M_2C=S\cdots HCN$ (M=H, F, Cl, Br, HO, H_3C, H_2N) complex: Dual roles of C=S group and substitution effect. *Int. J. Quantum Chem.* **2012**, *112*, 1491–1498. [CrossRef]
9. Metrangolo, P.; Resnati, G.; Pilati, T.; Biella, S. *Halogen Bonding in Crystal Engineering*; Springer: Berlin, Germany, 2008.
10. Eskandari, K.; Zariny, H. Halogen bonding: A lump–hole interaction. *Chem. Phys. Lett.* **2010**, *492*, 9–13. [CrossRef]
11. Ji, B.; Wang, W.; Deng, D.; Zhang, Y. Symmetrical bifurcated halogen bond: Design and synthesis. *Cryst. Growth Des.* **2011**, *11*, 3622–3628. [CrossRef]
12. Metrangolo, P.; Murray, J.S.; Pilati, T.; Politzer, P.; Resnati, G.; Terraneo, G. The fluorine atom as a halogen bond donor, viz. a positive site. *CrystEngComm* **2011**, *13*, 6593–6596. [CrossRef]
13. Politzer, P.; Riley, K.E.; Bulat, F.A.; Murray, J.S. Perspectives on halogen bonding and other σ-hole interactions: Lex parsimoniae (Occam's Razor). *Comput. Theor. Chem.* **2012**, *998*, 2–8. [CrossRef]
14. Stone, A.J. Are halogen bonded structures electrostatically driven? *J. Am. Chem. Soc.* **2013**, *135*, 7005–7009. [CrossRef] [PubMed]
15. Ho, P.S. Biomolecular halogen bonds. In *Halogen Bonding I*; Springer: Weinheim, Germany, 2014; pp. 241–276.
16. Lv, H.; Zhuo, H.-Y.; Li, Q.-Z.; Yang, X.; Li, W.-Z.; Cheng, J.-B. Halogen bonds with N-heterocyclic carbenes as halogen acceptors: A partially covalent character. *Mol. Phys.* **2014**, *112*, 3024–3032. [CrossRef]
17. Novák, M.; Foroutan-Nejad, C.; Marek, R. Asymmetric bifurcated halogen bonds. *Phys. Chem. Chem. Phys.* **2015**, *17*, 6440–6450. [CrossRef] [PubMed]
18. Clark, T.; Hennemann, M.; Murray, J.S.; Politzer, P. Halogen bonding: The σ-hole. *J. Mol. Model.* **2007**, *13*, 291–296. [CrossRef] [PubMed]
19. Murray, J.S.; Lane, P.; Clark, T.; Politzer, P. σ-hole bonding: Molecules containing group VI atoms. *J. Mol. Model.* **2007**, *13*, 1033–1038. [CrossRef] [PubMed]
20. Murray, J.S.; Concha, M.C.; Lane, P.; Hobza, P.; Politzer, P. Blue shifts vs. red shifts in σ-hole bonding. *J. Mol. Model.* **2008**, *14*, 699–704. [CrossRef] [PubMed]
21. Politzer, P.; Murray, J.S.; Concha, M.C. σ-hole bonding between like atoms; a fallacy of atomic charges. *J. Mol. Model.* **2008**, *14*, 659–665. [CrossRef] [PubMed]
22. Murray, J.S.; Lane, P.; Politzer, P. Expansion of the σ-hole concept. *J. Mol. Model.* **2009**, *15*, 723–729. [CrossRef] [PubMed]
23. Murray, J.S.; Lane, P.; Clark, T.; Riley, K.E.; Politzer, P. σ-Holes, π-holes and electrostatically-driven interactions. *J. Mol. Model.* **2012**, *18*, 541–548. [CrossRef] [PubMed]
24. Bundhun, A.; Ramasami, P.; Murray, J.; Politzer, P. Trends in σ-hole strengths and interactions of F3MX molecules (M=C, Si, Ge and X=F, Cl, Br, I). *J. Mol. Model.* **2013**, *19*, 2739–2746. [CrossRef] [PubMed]
25. Hassel, O.; Hvoslef, J. The structure of bromine 1, 4-dioxanate. *Acta Chem. Scand.* **1954**, *8*, 873. [CrossRef]
26. Hassel, O. Structural aspects of interatomic charge-transfer bonding. *Science* **1970**, *170*, 497–502. [CrossRef] [PubMed]
27. Abate, A.; Saliba, M.; Hollman, D.J.; Stranks, S.D.; Wojciechowski, K.; Avolio, R.; Grancini, G.; Petrozza, A.; Snaith, H.J. Supramolecular halogen bond passivation of organic–inorganic halide perovskite solar cells. *Nano Lett.* **2014**, *14*, 3247–3254. [CrossRef] [PubMed]
28. Saccone, M.; Dichiarante, V.; Forni, A.; Goulet-Hanssens, A.; Cavallo, G.; Vapaavuori, J.; Terraneo, G.; Barrett, C.J.; Resnati, G.; Metrangolo, P. Supramolecular hierarchy among halogen and hydrogen bond donors in light-induced surface patterning. *J. Mater. Chem. C* **2015**, *3*, 759–768. [CrossRef]
29. Jungbauer, S.H.; Huber, S.M. Cationic multidentate halogen-bond donors in halide abstraction organocatalysis: Catalyst optimization by preorganization. *J. Am. Chem. Soc.* **2015**, *137*, 12110–12120. [CrossRef] [PubMed]

30. Cavallo, G.; Metrangolo, P.; Milani, R.; Pilati, T.; Priimagi, A.; Resnati, G.; Terraneo, G. The halogen bond. *Chem. Rev.* **2016**, *116*, 2478–2601. [CrossRef] [PubMed]
31. Vasylyeva, V.; Catalano, L.; Nervi, C.; Gobetto, R.; Metrangolo, P.; Resnati, G. Characteristic redshift and intensity enhancement as far-IR fingerprints of the halogen bond involving aromatic donors. *CrystEngComm* **2016**, *18*, 2247–2250. [CrossRef]
32. Oliveira, V.; Kraka, E.; Cremer, D. The intrinsic strength of the halogen bond: Electrostatic and covalent contributions described by coupled cluster theory. *Phys. Chem. Chem. Phys.* **2016**, *18*, 33031–33046. [CrossRef] [PubMed]
33. Řezáč, J.; de la Lande, A. On the role of charge transfer in halogen bonding. *Phys. Chem. Chem. Phys.* **2017**, *19*, 791–803. [CrossRef] [PubMed]
34. Bauzá, A.; Mooibroek, T.J.; Frontera, A. Tetrel-Bonding Interaction: Rediscovered Supramolecular Force? *Angew. Chem. Int. Ed.* **2013**, *125*, 12543–12547. [CrossRef]
35. Bauzá, A.; Ramis, R.; Frontera, A. Computational study of anion recognition based on tetrel and hydrogen bonding interaction by calix [4] pyrrole derivatives. *Comput. Theor. Chem.* **2014**, *1038*, 67–70. [CrossRef]
36. Servati Gargari, M.; Vladimir, S.; Bauzá, A.; Frontera, A.; McArdle, P.; Van Derveer, D.; Weng Ng, S.; Mahmoudi, G. Design of lead (II) metal–organic frameworks based on covalent and tetrel-bonding. *Chem. Eur. J.* **2015**, *21*, 17951–17958. [CrossRef] [PubMed]
37. Li, Q.; Guo, X.; Yang, X.; Li, W.; Cheng, J.; Li, H.-B. A σ-hole interaction with radical species as electron donors: Does single-electron tetrel-bonding exist? *Phys. Chem. Chem. Phys.* **2014**, *16*, 11617–11625. [CrossRef] [PubMed]
38. Rezaei, Z.; Solimannejad, M.; Esrafili, M.D. Interplay between hydrogen bond and single-electron tetrel-bond: $H_3C\cdots COX_2\cdots HY$ and $H_3C\cdots CSX_2\cdots HY$ (X=F, Cl; Y=CN, NC) complexes as a working model. *Comput. Theor. Chem.* **2015**, *1074*, 101–106. [CrossRef]
39. Scheiner, S. Assembly of effective halide receptors from components. comparing hydrogen, halogen, and tetrel-bonds. *J. Phys. Chem. A* **2017**, *121*, 3606–3615. [CrossRef] [PubMed]
40. Alkorta, I.; Sánchez-Sanz, G.; Elguero, J.; Del Bene, J.E. Influence of hydrogen bonds on the P··· P pnicogen bond. *J. Chem. Theory Comput.* **2012**, *8*, 2320–2327. [CrossRef] [PubMed]
41. Del Bene, J.E.; Alkorta, I.; Sánchez-Sanz, G.; Elguero, J. Interplay of F–H . . . F hydrogen bonds and P . . . N pnicogen bonds. *J. Phys. Chem. A* **2012**, *116*, 9205–9213. [CrossRef] [PubMed]
42. Alkorta, I.; Elguero, J.; Del Bene, J.E. Pnicogen bonded complexes of PO_2X (X=F, Cl) with nitrogen bases. *J. Phys. Chem. A* **2013**, *117*, 10497–10503. [CrossRef] [PubMed]
43. Alkorta, I.; Elguero, J.; Solimannejad, M. Single electron pnicogen bonded complexes. *J. Phys. Chem. A* **2014**, *118*, 947–953. [CrossRef] [PubMed]
44. Alkorta, I.; Elguero, J.; Grabowski, S.J. Pnicogen and hydrogen bonds: Complexes between PH3X(+) and PH2X systems. *Phys. Chem. Chem. Phys.* **2015**, *17*, 3261–3272. [CrossRef] [PubMed]
45. Scheiner, S. Sensitivity of noncovalent bonds to intermolecular separation: Hydrogen, halogen, chalcogen, and pnicogen bonds. *CrystEngComm* **2013**, *15*, 3119–3124. [CrossRef]
46. Scheiner, S. Detailed comparison of the pnicogen bond with chalcogen, halogen, and hydrogen bonds. *Int. J. Quantum Chem.* **2013**, *113*, 1609–1620. [CrossRef]
47. Esrafili, M.D.; Mohammadian-Sabet, F. An ab initio study on chalcogen–chalcogen bond interactions in cyclic $(SHX)_3$ complexes (X=F, Cl, CN, NC, CCH, OH, OCH_3, NH_2). *Chem. Phys. Lett.* **2015**, *628*, 71–75. [CrossRef]
48. Esrafili, M.D.; Mohammadian-Sabet, F. Does single-electron chalcogen bond exist? Some theoretical insights. *J. Mol. Model.* **2015**, *21*, 65. [CrossRef] [PubMed]
49. Esrafili, M.D.; Mohammadian-Sabet, F.; Baneshi, M.M. An ab initio investigation of chalcogen–hydride interactions involving HXeH as a chalcogen bond acceptor. *Struct. Chem.* **2016**, *27*, 785–792. [CrossRef]
50. Bauzá, A.; Frontera, A. Aerogen bonding interaction: A new supramolecular force? *Angew. Chem. Int. Ed.* **2015**, *54*, 7340–7343. [CrossRef] [PubMed]
51. Bauzá, A.; Frontera, A. π-Hole aerogen bonding interactions. *Phys. Chem. Chem. Phys.* **2015**, *17*, 24748–24753. [CrossRef] [PubMed]
52. Esrafili, M.D.; Asadollahi, S.; Vakili, M. Investigation of substituent effects in aerogen-bonding interaction between $ZO3$ (Z=Kr, Xe) and nitrogen bases. *Int. J. Quantum Chem.* **2016**, *116*, 1254–1260. [CrossRef]

53. Frontera, A.; Bauzá, A. Concurrent aerogen bonding and lone pair/anion–π interactions in the stability of organoxenon derivatives: A combined CSD and ab initio study. *Phys. Chem. Chem. Phys.* **2017**, *19*, 30063–30068. [CrossRef] [PubMed]
54. Brezgunova, M.E.; Lieffrig, J.; Aubert, E.; Dahaoui, S.; Fertey, P.; Lebègue, S.b.; Ángyán, J.n.G.; Fourmigué, M.; Espinosa, E. Chalcogen bonding: Experimental and theoretical determinations from electron density analysis. Geometrical preferences driven by electrophilic–nucleophilic interactions. *Cryst. Growth Des.* **2013**, *13*, 3283–3289. [CrossRef]
55. Xu, Z.; Yang, Z.; Liu, Y.; Lu, Y.; Chen, K.; Zhu, W. Halogen bond: Its role beyond drug–target binding affinity for drug discovery and development. *J. Chem. Inf. Model.* **2014**, *54*, 69–78. [CrossRef] [PubMed]
56. Adhikari, U.; Scheiner, S. Effects of charge and substituent on the S···N chalcogen bond. *J. Phys. Chem. A* **2014**, *118*, 3183–3192. [CrossRef] [PubMed]
57. Gilday, L.C.; Robinson, S.W.; Barendt, T.A.; Langton, M.J.; Mullaney, B.R.; Beer, P.D. Halogen bonding in supramolecular chemistry. *Chem. Rev.* **2015**, *115*, 7118–7195. [CrossRef] [PubMed]
58. Bauzá, A.; Mooibroek, T.J.; Frontera, A. Tetrel-bonding interactions. *Chem. Rec.* **2016**, *16*, 473–487. [CrossRef] [PubMed]
59. Tang, Q.; Li, Q. Interplay between tetrel-bonding and hydrogen bonding interactions in complexes involving F_2XO (X=C and Si) and HCN. *Comput. Theor. Chem.* **2014**, *1050*, 51–57. [CrossRef]
60. Mahmoudi, G.; Bauzá, A.; Frontera, A. Concurrent agostic and tetrel-bonding interactions in lead (II) complexes with an isonicotinohydrazide based ligand and several anions. *Dalton Trans.* **2016**, *45*, 4965–4969. [CrossRef] [PubMed]
61. Nziko, P.V.; Scheiner, S. Comparison of π-hole tetrel-bonding with σ-hole halogen bonds in complexes of XCN (X=F, Cl, Br, I) and NH3. *Phys. Chem. Chem. Phys.* **2016**, *18*, 3581–3590. [CrossRef] [PubMed]
62. Del Bene, J.E.; Alkorta, I.; Elguerob, J. Anionic complexes of F^- and Cl^- with substituted methanes: Hydrogen, halogen, and tetrel-bonds. *Chem. Phys. Lett.* **2016**, *655–656*, 115–119. [CrossRef]
63. Esrafili, M.D.; Asadollahi, S.; Mousavian, P. Anionic tetrel-bonds: An ab initio study. *Chem. Phys. Lett.* **2018**, *691*, 394–400. [CrossRef]
64. Solimannejad, M.; Orojloo, M.; Amani, S. Effect of cooperativity in lithium bonding on the strength of halogen bonding and tetrel-bonding:$(LiCN)_n$···$ClYF_3$ and $(LiCN)_n$···YF_3Cl (Y=C, Si and $n = 1$–5) complexes as a working model. *J. Mol. Model.* **2015**, *21*, 183. [CrossRef] [PubMed]
65. Liu, M.; Li, Q.; Li, W.; Cheng, J. Tetrel-bonds between $PySiX_3$ and some nitrogenated bases: Hybridization, substitution, and cooperativity. *J. Mol. Graph. Model.* **2016**, *65*, 35–42. [CrossRef] [PubMed]
66. Liu, M.; Li, Q.; Scheiner, S. Comparison of tetrel-bonds in neutral and protonated complexes of pyridine TF_3 and furan TF_3 (T=C, Si, and Ge) with NH_3. *Phys. Chem. Chem. Phys.* **2017**, *19*, 5550–5559. [CrossRef] [PubMed]
67. Grabowski, S.J. Tetrel-bonds with π-electrons acting as Lewis bases-theoretical results and experimental evidences. *Molecules* **2018**, *23*, 1183. [CrossRef] [PubMed]
68. Shen, S.; Zeng, Y.; Li, X.; Meng, L.; Zhang, X. Insight into the π-hole···π-electrons tetrel-bonds between F_2ZO (Z=C, Si, Ge) and unsaturated hydrocarbons. *Int. J. Quantum Chem.* **2018**, *118*, e25521. [CrossRef]
69. García-LLinás, X.; Bauzá, A.; Seth, S.K.; Frontera, A. Importance of R–CF_3···O tetrel-bonding interactions in biological systems. *J. Phys. Chem. A* **2017**, *121*, 5371–5376. [CrossRef] [PubMed]
70. Bauzá, A.; Frontera, A. Tetrel-bonding interactions in perchlorinated cyclopenta-and cyclohexatetrelanes: Acombined DFT and CSD study. *Molecules* **2018**, *23*, 1770. [CrossRef] [PubMed]
71. Mitzel, N.W.; Losehand, U. β-donor bonds in compounds containing SiON fragments. *Angew. Chem. Int. Ed.* **1997**, *36*, 2807–2809. [CrossRef]
72. Alkorta, I.; Rozas, I.; Elguero, J. Molecular complexes between silicon derivatives and electron-rich groups. *J. Phys. Chem. A* **2001**, *105*, 743–749. [CrossRef]
73. Mani, D.; Arunan, E. The X–C···Y (X=O/F, Y=O/S/F/Cl/Br/N/P)'carbon bond'and hydrophobic interactions. *Phys. Chem. Chem. Phys.* **2013**, *15*, 14377–14383. [CrossRef] [PubMed]
74. Mani, D.; Arunan, E. The X–C···π (X=F, Cl, Br, Cn) Carbon Bond. *J. Phys. Chem. A* **2014**, *118*, 10081–10089. [CrossRef] [PubMed]

75. Grabowski, S.J. Tetrel-bond–σ-hole bond as a preliminary stage of the SN$_2$ reaction. *Phys. Chem. Chem. Phys.* **2014**, *16*, 1824–1834. [CrossRef] [PubMed]
76. Scheiner, S. Comparison of halide receptors based on H, halogen, chalcogen, pnicogen, and tetrel-bonds. *Faraday Discuss.* **2017**, *203*, 213–226. [CrossRef] [PubMed]
77. Scheiner, S. Tetrel-bonding as a vehicle for strong and selective anion binding. *Molecules* **2018**, *23*, 1147. [CrossRef] [PubMed]
78. Frisch, M.J.; Trucks, G.W.; Schlegel, H.B.; Scuseria, G.E.; Robb, M.A.; Cheeseman, J.R.; Scalmani, G.; Barone, V.; Mennucci, B.; Petersson, G.A.; et al. *Gaussian 09*; Gaussian, Inc.: Wallingford, CT, USA, 2009.
79. Boys, S.F.; Bernardi, F. The calculation of small molecular interactions by the differences of separate total energies. Some procedures with reduced errors. *Mol. Phys.* **1970**, *19*, 553–566. [CrossRef]
80. Glendening, E.; Badenhoop, J.; Reed, A.; Carpenter, J.; Bohmann, J.; Morales, C.; Weinhold, F. NBO 5.0. Theoretical Chemistry Institute, University of Wisconsin: Madison, WI, USA, 2001.
81. Bulat, F.; Toro-Labbé, A.; Brinck, T.; Murray, J.; Politzer, P. Quantitative analysis of molecular surfaces: Areas, volumes, electrostatic potentials and average local ionization energies. *J. Mol. Model.* **2010**, *16*, 1679–1691. [CrossRef] [PubMed]
82. Biegler-König, F.; Schönbohm, J.; Derdau, R.; Bayles, D. AIM2000. *J. Comput. Chem.* **2001**, *22*, 545–559.
83. Lu, T.; Chen, F. Multiwfn: A multifunctional wavefunction analyzer. *J. Comput. Chem.* **2012**, *33*, 580–592. [CrossRef] [PubMed]
84. Li, Q.-Z.; Zhuo, H.-Y.; Li, H.-B.; Liu, Z.-B.; Li, W.-Z.; Cheng, J.-B. Tetrel–hydride interaction between XH$_3$F (X=C, Si, Ge, Sn) and HM (M=Li, Na, BeH, MgH). *J. Phys. Chem. A* **2014**, *119*, 2217–2224. [CrossRef] [PubMed]
85. Del Bene, J.E.; Elguero, J.; Alkorta, I. Complexes of CO$_2$ with the Azoles: Tetrel-bonds, hydrogen bonds and other secondary interactions. *Molecules* **2018**, *23*, 906. [CrossRef] [PubMed]
86. Bondi, A. van der Waals volumes and radii. *J. Phys. Chem.* **1964**, *68*, 441–451. [CrossRef]
87. Liu, M.; Li, Q.; Cheng, J.; Li, W.; Li, H.-B. Tetrel-bond of pseudohalide anions with XH$_3$F (X=C, Si, Ge, and Sn) and its role in SN2 reaction. *J. Chem. Phys.* **2016**, *145*, 224310. [CrossRef] [PubMed]
88. Scheiner, S. Highly selective halide receptors based on chalcogen, pnicogen, and tetrel-bonds. *Chem. Eur. J.* **2016**, *22*, 18850–18858. [CrossRef] [PubMed]
89. Esrafili, M.D.; Mohammadian-Sabet, F. Exploring σ-hole bonding in XH$_3$Si···HMY (X=H, F, CN; M=Be, Mg; Y=H, F, CH$_3$) complexes: A "tetrel-hydride" interaction. *J. Mol. Model.* **2015**, *21*, 60. [CrossRef] [PubMed]
90. Esrafili, M.D.; Mohammadian-Sabet, F. Cooperativity of tetrel-bonds tuned by substituent effects. *Mol. Phys.* **2016**, *114*, 1528–1538. [CrossRef]
91. Esrafili, M.D.; Mohammadian-Sabet, F. σ-Hole bond tunability in YO$_2$X$_2$: NH$_3$ and YO$_2$X$_2$: H$_2$O complexes (X=F, Cl, Br; Y=S, Se): Trends and theoretical aspects. *Struct. Chem.* **2016**, *27*, 617–625. [CrossRef]
92. Zierkiewicz, W.; Michalczyk, M. On the opposite trends of correlations between interaction energies and electrostatic potentials of chlorinated and methylated amine complexes stabilized by halogen bond. *Theor. Chem. Acc.* **2017**, *136*, 125. [CrossRef]
93. Esrafili, M.D.; Mohammadian-Sabet, F. Tuning tetrel-bonds via cation–π interactions: An ab initio study on concerted interaction in M$^+$–C$_6$H$_5$XH$_3$–NCY complexes (M=Li, Na, K; X=Si, Ge; Y=H, F, OH). *Mol. Phys.* **2016**, *114*, 83–91. [CrossRef]
94. Guo, X.; Liu, Y.-W.; Li, Q.-Z.; Li, W.-Z.; Cheng, J.-B. Competition and cooperativity between tetrel-bond and chalcogen bond in complexes involving F$_2$CX (X=Se and Te). *Chem. Phys. Lett.* **2015**, *620*, 7–12. [CrossRef]
95. Esrafili, M.D.; Mohammadirad, N.; Solimannejad, M. Tetrel-bond cooperativity in open-chain (CH$_3$CN)$_n$ and (CH$_3$NC)$_n$ clusters (n = 2–7): An ab initio study. *Chem. Phys. Lett.* **2015**, *628*, 16–20. [CrossRef]
96. Bader, R.F. A bond path: A universal indicator of bonded interactions. *J. Phys. Chem. A* **1998**, *102*, 7314–7323. [CrossRef]
97. Bader, R.F.W. *Atoms in Molecules: A Quantum Theory*; Oxford University Press: New York, NY, USA, 1990.
98. Dong, W.; Li, Q.; Scheiner, S. Comparative strengths of tetrel, pnicogen, chalcogen, and halogen bonds and contributing factors. *Molecules* **2018**, *23*, 1681. [CrossRef] [PubMed]
99. Groom, C.R.; Bruno, I.J.; Lightfoot, M.P.; Ward, S.C. The Cambridge structural database. *Acta Crystallogr. B* **2016**, *72*, 171–179. [CrossRef] [PubMed]

100. Taylor, P.G.; Bassindale, A.R.; El Aziz, Y.; Pourny, M.; Stevenson, R.; Hursthouse, M.B.; Coles, S.J. Further studies of fluoride ion entrapment in octasilsesquioxane cages; X-ray crystal structure studies and factors that affect their formation. *Dalton Trans.* **2012**, *41*, 2048–2059. [CrossRef] [PubMed]
101. Bassindale, A.R.; Parker, D.J.; Pourny, M.; Taylor, P.G.; Horton, P.N.; Hursthouse, M.B. Fluoride ion entrapment in octasilsesquioxane cages as models for ion entrapment in zeolites. Further examples, X-ray crystal structure studies, and investigations into how and why they may be formed. *Organometallics* **2004**, *23*, 4400–4405. [CrossRef]
102. Muhammad, S.; Bassindale, A.R.; Taylor, P.G.; Male, L.; Coles, S.J.; Hursthouse, M.B. Study of binuclear silicon complexes of diketopiperazine at SN2 reaction profile. *Organometallics* **2011**, *30*, 564–571. [CrossRef]
103. Blaschette, A.; Hippel, I.; Krahl, J.; Wieland, E.; Jones, P.G.; Sebald, A. Polysulfonylamine: XXXV. Synthese, Röntgenstrukturanalysen und hochaufgelöste Festkörper-NMR-Spektren der ionischen Organozinn (IV)-dimesylamide [Me$_3$Sn(NH$_3$)$_2$][N(SO2Me)$_2$] und [Me$_2$Sn(DMSO)$_4$][N(SO$_2$Me)$_2$]$_2$. *J. Organomet. Chem.* **1992**, *437*, 279–297. [CrossRef]

 © 2018 by the authors. Licensee MDPI, Basel, Switzerland. This article is an open access article distributed under the terms and conditions of the Creative Commons Attribution (CC BY) license (http://creativecommons.org/licenses/by/4.0/).

Article

Tetrel Bonding as a Vehicle for Strong and Selective Anion Binding

Steve Scheiner

Department of Chemistry and Biochemistry, Utah State University, Logan, UT 84322-0300, USA; steve.scheiner@usu.edu; Tel.: +1-435-797-7419

Received: 24 April 2018; Accepted: 9 May 2018; Published: 11 May 2018

Abstract: Tetrel atoms T (T = Si, Ge, Sn, and Pb) can engage in very strong noncovalent interactions with nucleophiles, which are commonly referred to as tetrel bonds. The ability of such bonds to bind various anions is assessed with a goal of designing an optimal receptor. The Sn atom seems to form the strongest bonds within the tetrel family. It is most effective in the context of a -SnF_3 group and a further enhancement is observed when a positive charge is placed on the receptor. Connection of the -SnF_3 group to either an imidazolium or triazolium provides a strong halide receptor, which can be improved if its point of attachment is changed from the C to an N atom of either ring. Aromaticity of the ring offers no advantage nor is a cyclic system superior to a simple alkyl amine of any chain length. Placing a pair of -SnF_3 groups on a single molecule to form a bipodal dicationic receptor with two tetrel bonds enhances the binding, but falls short of a simple doubling. These two tetrel groups can be placed on opposite ends of an alkyl diamine chain of any length although $SnF_3{}^+NH_2(CH_2)_nNH_2SnF_3{}^+$ with n between 2 and 4 seems to offer the strongest halide binding. Of the various anions tested, OH^- binds most strongly: $OH^- > F^- > Cl^- > Br^- > I^-$. The binding energy of the larger $NO_3{}^-$ and $HCO_3{}^-$ anions is more dependent upon the charge of the receptor. This pattern translates into very strong selectivity of binding one anion over another. The tetrel-bonding receptors bind far more strongly to each anion than an equivalent number of K^+ counterions, which leads to equilibrium ratios in favor of the former of many orders of magnitude.

Keywords: bipodal; Gibbs free energy; imidazolium; triazolium; counterion; deformation energy

1. Introduction

The detection, extraction, and transport of anions is an important element in a wide range of biological and chemical processes [1]. Biological evolution has developed a score of anion binding proteins usually with high selectivity. The sulphate-binding protein of Salmonella typhimurium [2] is an example of one that binds this anion via a number of H-bonds. Another protein is responsible for the binding and transport of phosphate [3] with very high specificity. Still another protein, present in blue-green algae, is highly specific for the nitrate anion [4] and another binds specifically to bicarbonate [5]. While the evolutionary process has developed some very specific and selective anion binding agents, modern technology lags behind. Many receptors make use of general electrostatic interactions and sometimes of H-bonds [6–12]. The thiourea molecule, for example, is a widely used [13–15] anion binder that takes advantage of its H-bonding capability. The guanidinium cation and its derivatives [16,17] have also found use in this regard. However, the anion receptors that have been developed to date still suffer from certain disadvantages. Their selectivity is not optimal and they are unable to detect the presence of a particular anion below a given concentration threshold. Furthermore, at this point in time, the biggest need is the development of highly selective receptors that can function in an aqueous rather than organic or biological environment. "Examples of receptors that are neutral or of low charge and operate in organic–aqueous mixtures are uncommon and those that function in 100% water are rarer still" [1].

One major advancement in this field arose from the growing recognition of the phenomenon of halogen bonds (XBs) [18–24] where an attractive force occurs between a halogen atom and an electron donor such as the lone pair of an amine. It was not long before researchers applied this concept in order to develop receptors that are highly selective for one anion over another [25–32]. The Beer group [33] found that substitution of H by Br enabled the consequent halogen bond to more effectively bind chloride and that receptors of this type could recognize [34] both chloride and bromide ions purely by virtue of XBs [35,36]. Chudzinski et al. [37] obtained quantitative estimates of the contribution of halogen bonding to the binding of anions to bipodal receptors and later [38] applied halogen bonds to develop pre-organized multi-dentate receptors capable of high-affinity anion recognition. Halogen bonding exerts selectivity for bromide over chloride or other anions in a set of tripodal receptors [39]. Our own group [40–46] has applied quantum chemical calculations toward solving this issue, which shows that the replacement of H in a series of H-bonding bidentate receptors by halogen atoms can influence their binding to halides. The work detailed a remarkable enhancement of both binding and selectivity especially when the H atom is replaced by I.

A rapidly burgeoning group of studies has extended the basic concepts of the XB to other atoms in the periodic table. Depending upon the particular family to which the atom in question belongs, these bonds have come to be known as chalcogen and pnicogen bonds [47–60]. Given the similarities, it was not surprising that noncovalent bonds of this sort can be every bit as useful as XBs in the context of anion binding and transport, which is being demonstrated in recent work [61–64]. Tetrel atoms (C, Si, Ge, Sn, Pb) seem capable of engaging in very similar interactions as well, which is becoming increasingly clear [65–73]. Therefore, there is every reason to believe that tetrel bonds might find a place in this constellation of noncovalent bonds that can function as integral components in anion receptors.

It was just this idea that motivated our group to recently perform calculations to examine how the latter type of bonds might compare with chalcogens in this context. A set of bidentate receptors, modeled closely after those in a prior experimental study, was constructed [43], which varied in whether the atoms that engaged directly with the anion were chalcogen, pnicogen, or tetrel. The transition from chalcogen to pnicogen to tetrel yielded not only progressively stronger binding to anions but also greater selectivity. In a quantitative sense, the binding energy of halides to a Ge-bonding bidentate receptor was as high as 63 kcal/mol and preferentially bound F^- over other halides with a selectivity of 27 orders of magnitude. These quantities are especially impressive since the receptor was electrically neutral, which forgoes the positive charge on many other such candidates. A follow-up study [46] delved somewhat more deeply into this issue by adding halogen atoms to the mix as well and by using a different bidentate receptor structure. It was found that with respect to Cl^- and Br^-, the binding is insensitive to the nature of the binding atom, viz H, halogen, chalcogen, or tetrel. However, there is a great deal of differentiation with respect to F^- where the order varies as tetrel > H ~pnicogen > halogen > chalcogen. The replacement of the various binding atoms by their analogues in the next row of the periodic table enhances the fluoride binding energy by 22% to 56%. The strongest fluoride binding agents utilize the tetrel bonds of the Sn atom while it is I-halogen bonds that are preferred for Cl^- and Br^-.

At this point then, there are sound reasons to believe that tetrel bonding offers a unique opportunity in the design of effective and highly selective receptors. However, there are a number of very important issues that remain to be resolved. In the first place, most of the prior calculations have been centered in the gas phase while it is in solution, especially in water, there is more urgent need for these receptors. Particularly when one is dealing with charged species, the effects of hydration can be expected to be especially strong, so in vacuo trends cannot be simply transferred to water but must be assessed directly. For example, hydration would stabilize the receptor/anion complex but would more heavily stabilize the separate individually charged species. Therefore, gas-phase trends may be radically different in water. It is for this reason that the calculations reported in this study are conducted in an aqueous environment.

Within the realm of tetrel bonds, there is a question as to which tetrel (T) atom would be most effective. Past work has suggested that tetrel bonding is strengthened when the T atom is enlarged, but this phenomenon relates to the gas phase and has not been thoroughly tested in water. The same question pertains to the finding that tetrel bonding is enhanced by electron-withdrawing substituents. How might the strength of tetrel bonding be affected if the tetravalent -TH$_3$ group is perfluorinated in water or likewise if the group possesses a positive charge? In a similar vein, most of the bipodal receptors that have come under experimental scrutiny are dications so it is important to assess how a double positive charge affects the binding. Within the context of the construction of the full receptor, the binding group has typically been placed by experimentalists on an imidazolium or triazolium group. Calculations can be used to compare and contrast a wider range of different groups and consider whether the aromaticity of this group is important or whether it even needs to be cyclic. One can address specificity by comparing the binding energetics of a number of various anions to each candidate tetrel-binding receptor. Lastly, since the extraction of an anion from solution by any receptor must overcome the attraction of this anion to counter-ions, this competition must be considered as well.

2. Systems and Methods

In the first set of tests, tetrel T atoms examined included the full {Si, Ge, Sn, Pb} set. These were placed into both a -TH$_3$ setting and its perfluorinated -TF$_3$ counterpart. One of the most commonly used groups to which anion-binding agents have been attached in the past is the imidazole species [9,27,33,34,39,42,74–78] so it is this group that is considered in the pilot set of calculations. Both TH$_3$ and TF$_3$ were, therefore, affixed to an imidazole moiety and comparisons were made to the same system after protonation of the ring to an imidazolium group. The primary anion used to test binding was Cl$^-$, which is representative of the entire halide set without the complications noted earlier for the smaller F$^-$, which was prone to engage in asymmetric covalent bonding with the receptor. Another reason for selecting chloride as the prototype anion is the close correspondence observed recently [79] between its calculated binding energy with a series of Lewis acids and the experimental trends arising from NMR measurements. Since this first battery of tests pointed to Sn as the most effective tetrel-bonding atom, it was the focus of the next testbed of calculations, which evaluated a wide range of groups that might replace imidazolium and perhaps enhance the anion binding. These replacements included both aromatic and nonaromatic, cyclic and noncyclic, and both mono and dications. Having established one or two prime candidates, calculations then turned to comparisons between different anions of chemical and biochemical importance including all four halides, OH$^-$, NO$_3^-$, and HCO$_3^-$. Since the receptor must be capable of pulling the anion of choice out of solution where it is closely associated with positive counter-ions, the receptor/anion binding was compared to that with K$^+$ cations as model counter-ions.

Calculations were carried out with the Gaussian-09 [80] set of programs. The M06-2X DFT functional [81] was used along with the aug-cc-pVDZ basis set. For the heavy atoms I, Pb, and Sn. The aug-cc-pVDZ-PP pseudopotential was taken from the EMSL library [82,83] so as to incorporate relativistic effects. This level of theory is appropriate for this task as evident by previous work by others [84–89] as well as by ourselves in dealing with very similar sorts of systems [40–42,90]. The geometries of the receptors and complexes were fully optimized with no restriction, which was assured as minima by the absence of imaginary vibrational frequencies. The binding energy of each anion with its receptor was calculated as the difference between the energy of the complex and the sum of the energies of separately optimized monomers. It was then corrected for basis set superposition error by the counterpoise [91,92] procedure. Gibbs free energies of each complexation reaction are computed at 298 K. To account for solvent effects, the polarizable conductor calculation model (CPCM) was applied [93] with water as the solvent. This approach treats the surroundings as a polarizable continuum with dielectric constant of 78 but does not include explicit water molecules.

3. Results

3.1. Receptors Containing Imidazole

The binding energies obtained when the Cl$^-$ anion was allowed to interact with each of the various tetrel-containing species are reported in Table 1. Several trends are immediately apparent. These quantities are much larger for the cations than for the neutrals, which is sensible in light of the ion-ion interaction in the case of the former. The replacement of the three H atoms on the tetrel atom by F causes a large enhancement as much as six-fold. This increase is especially large for the two heavier tetrel atoms Sn and Pb. Actually, it is the latter two tetrel atoms that consistently show the strongest binding with Sn having a slight edge.

Table 1. Binding energy (kcal/mol) between Cl$^-$ and ImTR$_3$ and ImHTR$_3^+$.

T	Neutral		Cation	
	ImTH$_3$	ImTF$_3$	ImHTH$_3^+$	ImHTF$_3^+$
Si	−1.79	−2.73	−5.74	−23.98
Ge	−1.87	−8.81	−5.26	−32.12
Sn	−3.57	−23.47	−10.14	−43.54
Pb	−2.78	−20.62	−7.87	−39.51

Representative geometries are depicted in Figure 1 for the Sn systems and show trends that parallel the energetics. The R(Sn·Cl) distances are shorter for the cationic receptors and are also shortened when the SnH$_3$ group is changed to SnF$_3$. Table 2 collects the R(T·Cl) distances for a full range of these complexes. As one would expect from the trends in the energetics, this distance is much shorter for the cations than for the neutral entities and the TF$_3$ systems hold the Cl in closer than does TH$_3$. The comparisons among the various tetrel atoms are more important. As the tetrel atom grows larger, one would expect a corresponding elongation of R(T·Cl). However, this trend would be opposed by the growing strength of the tetrel bond in the order Si < Ge << Pb < Sn so the pattern is not obvious to predict. Furthermore, there is little relation between tetrel atoms and R for the neutral ImTH$_3$ while R gets longer with heavier tetrel atom for TF$_3$. The conflict between the two trends is more complicated for the cations. The longest distance occurs for Ge for the TH$_3$ systems while there is a clearer trend of longer distances for larger T atoms for TF$_3$.

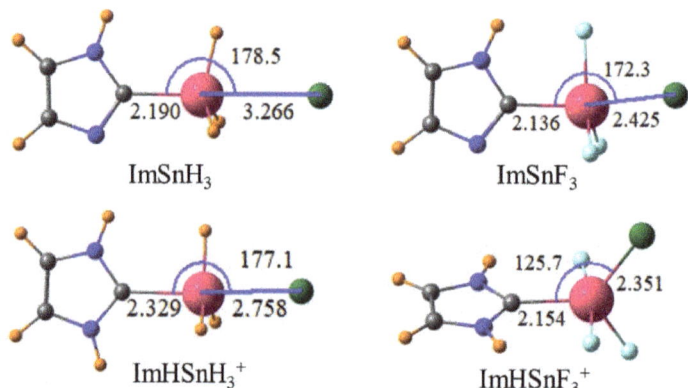

Figure 1. Geometries of indicated Lewis acids with Cl$^-$. Distances in Å, angles in degs.

Table 2. R(T·Cl) (Å) in optimized complexes.

	Neutral		Cation	
T	ImTH$_3$	ImTF$_3$	ImHTH$_3^+$	ImHTF$_3^+$
Si	3.441	2.293	2.557	2.112
Ge	3.491	2.310	3.032	2.178
Sn	3.266	2.425	2.758	2.351
Pb	3.435	2.467	2.988	2.419

There are several other interesting aspects of the geometry surrounding the T atom. Considering the SnH$_3$ systems on the left side of Figure 1, the θ(CSnH) angle in the neutral monomer is equal to 107.8°, which is nearly tetrahedral. It is reduced to 103.1° in the complex with Cl$^-$, i.e., the SnH$_3$ group flattens toward a trigonal bipyramid. This same group is already fairly flat in the cationic monomer with θ = 102.4°. However, upon complexation with Cl$^-$, the SnF$_3$ group undergoes a more radical change. Instead of adopting a position nearly opposite the C atom, the Cl moves well out of the imidazolium plane with θ(CSn·Cl) = 125.7°. Note that the R(Sn-Cl) distance of 2.351 Å is only 0.2 Å longer than the R(Sn-C) distance of 2.154 Å. The geometry around the Sn might fairly be described as a trigonal bipyramid with two apical F atoms and with Cl, C, and the third F occupying the three equatorial positions. This set of geometrical parameters and larger scale rearrangement for the ImHTF$_3^+$ complex is not limited to Sn but is characteristic of all four tetrel atoms.

More important than the binding energy itself is the free energy for the complexation reactions. ΔG contains not only zero point and thermal corrections but also entropic contributions. In part as a result of the transition from a pair of subunits to a single complex, the values of ΔG in Table 3 are less negative than ΔE in Table 1 and are even becoming positive in a number of instances. Some of the trends in ΔE survive the additional terms. For example, binding to the cationic receptors is stronger than to the neutrals and the replacement of TH$_3$ by TF$_3$ bolsters the strength of the interaction. The strongest binding occurs in all cases for Sn and Pb with a slight edge for the former. It might be noted that even within the confines of the strong dielectric environment of water, the binding of Cl$^-$ to the cationic ImHTF$_3^+$ species is a minimum of 15 kcal/mol and it rises to more than twice this amount for T=Sn and Pb. As mentioned above, the treatment of solvation here does not include specific interactions between the solvated system and discreet water molecules. The inclusion of this might have a bearing on these results.

Table 3. ΔG(298 K) (kcal/mol) for interactions between Cl$^-$ and ImTR$_3$ and ImHTR$_3^+$.

	Neutral		Cation	
	ImTH$_3$	ImTF$_3$	ImHTH$_3^+$	ImHTF$_3^+$
Si	+4.37	+5.42	+2.12	−14.95
Ge	+4.77	−1.23	+2.32	−23.91
Sn	+2.75	−13.31	−2.77	−36.53
Pb	+3.01	−12.44	−1.73	−32.25

3.2. More General Receptors

The success of cationic ImHSnF$_3^+$ as a receptor can be used as a starting point to explore modifications that might further enhance the binding. In the first place, one can imagine the ImHSnF$_3^+$ group being attached not to one of the imidazole C atoms but rather to N. The more electron-withdrawing power of the latter might strengthen the ability of the Sn atom to attract a nucleophile. Formation of this complex, pictured in Figure 2a, does enhance the binding by some 5 kcal/mol, which is indicated in Table 4. The geometry is basically unaltered by this change besides a small contraction of the R(Sn·Cl) distance. A second modification would be to add a third N atom to imidazolium to generate a triazolium species, which is shown in Figure 2b. It is this group which has served as the point of attachment for the anion-binding species in a number of experimental works [7,27–30,35,36,62,94–99]. Table 4 indicates that this change weakens the interaction by roughly 10%. On the other hand, switching the point of connection from C to N again raises ΔG to a point where it surpasses that of N-Im by a small amount with R(Sn·Cl) reduced to 2.341 Å, which is seen in Figure 2c.

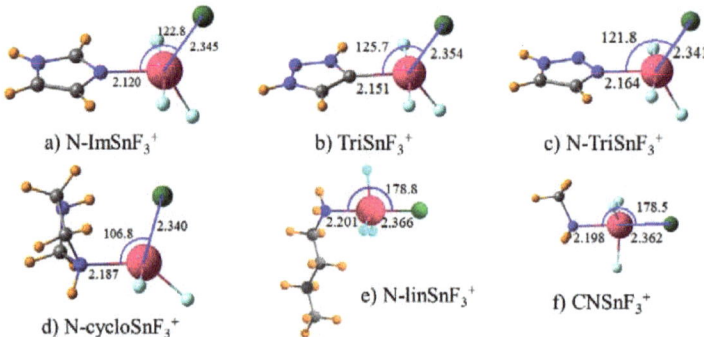

a) N-ImSnF$_3^+$ b) TriSnF$_3^+$ c) N-TriSnF$_3^+$

d) N-cycloSnF$_3^+$ e) N-linSnF$_3^+$ f) CNSnF$_3^+$

Figure 2. Geometries of indicated Lewis acids with Cl$^-$. Distances in Å, angles in degs.

Table 4. Energetics (kcal/mol) for interactions between Cl$^-$ and indicated receptors.

	ΔG	ΔE
Monocations		
ImSnF$_3^+$	−36.53	−43.54
N-ImSnF$_3^+$	−41.75	−49.99
TriSnF$_3^+$	−33.29	−41.30
N-TriSnF$_3^+$	−42.65	−51.67
N-cycloSnF$_3^+$	−42.39	−51.76
N-linSnF$_3^+$	−40.85	−49.53
CNSnF$_3^+$	−42.57	−50.76
φ-CNSnF$_3^+$	−38.03	−46.43
Dications		
φ-(CNSnF$_3^+$)$_2$	−46.00	−54.83
C$_5$ diamine	−53.63	−62.76
C$_4$ diamine	−62.21	−73.11
C$_3$ diamine	−60.42	−70.52
C$_2$ diamine	−63.30	−71.94
C$_1$ diamine	−53.33	−62.73
n MeNH$_2$SnF$_3^+$		
$n = 2$	−53.53	−75.13
$n = 3$	−66.38	−103.77

Since the receptors considered at this point all contain a heteroaromatic ring, the question arises as to the importance of this aromaticity to the binding. The five-membered imidazole ring was, therefore, fully saturated with H atoms, which leads to a nonaromatic ring by retaining the two N atoms. This loss of aromaticity does not reduce the chloride affinity. When attached to an N atom of the CH$_2$CH$_2$NHCH$_2$NH$^+$ ring (abbreviated as N-cyclo), the SnF$_3$ group in Figure 2d binds Cl$^-$ with approximately the same ΔG as the aromatic N-ImHSnF$_3^+$ counterpart in Figure 2a. Additionally, there is a slight enhancement in ΔE. This result begs the question as to whether the cyclic nature of the receptor is an important component at all or whether the second N atom is essential. The heterocycle of Figure 2d was, therefore, replaced with a simple amine CH$_3$(CH$_2$)$_3$NH$_2$ with the same number of five heavy atoms, which is represented in Figure 2e. This species, abbreviated as N-linHSnF$_3^+$, suffers only a very small loss of binding energy with ΔG still exceeding 40 kcal/mol. One may note a change in geometry around the Sn atom. In this simple amine, the Cl atom situates itself directly opposite the N atom, which leaves the three F atoms in equatorial sites. The next question relates to the length of the amine. If it is shortened from a *n*-butyl chain to a simple methyl group, how might that affect the binding. Such a shortening yields a small enhancement in the chloride affinity, which raises ΔG from 40.85 kcal/mol to 42.57 kcal/mol, which is indicated by the CNSnF$_3^+$ entry in Table 4 with the corresponding complex illustrated in Figure 2f.

There may be a particular advantage in the placement of the receptor on an aromatic system such as a phenyl ring, which is typical of those that have been considered experimentally and computationally in the past. In order to address this issue, the methylamine molecule was covalently attached to a phenyl group, which is illustrated in Figure 3a. Instead of augmenting the binding, this attachment had the opposite effect of reducing the binding energy by about 4 kcal/mol, or 10%, which was noted by the φ-CNSnF$_3^+$ entry in Table 4. It may be noted that this attachment to the phenyl ring induces a change in the geometry wherein the θ(NSn·Cl) angle decreases by 19° from 178.5° to 159.8° although the R(Sn·Cl) distance remains virtually unchanged. In addition of the attachment to a spacer such as a phenyl group, the commonly used receptors contain a pair of binding units in a bidentate arrangement. This sort of structure was mimicked by connecting two CH$_2$NH$_2$SnF$_3^+$ groups onto the same benzene ring. As illustrated in Figure 3b, the chloride ion occupies a near symmetric position, which is bound to both Sn atoms. Additionally, one F atom from each of the SnF$_3$ groups swings around so that they too are symmetrically disposed to the two Sn atoms. As indicated in Table 4, this bidentate receptor represents an enhancement in the binding with ΔG increasing from 38 kcal/mol to 46 kcal/mol. On the other hand, given the doubling of the positive charge on this receptor and the addition of a second tetrel bond, this 20% increase is a rather disappointing increment. Another reason for disappointment is that the bidentate geometry in Figure 3b has a more linear θ(NSn·F) angle of 173° compared to 160° in Figure 3a, which would ordinarily be more conducive to a strong noncovalent bond.

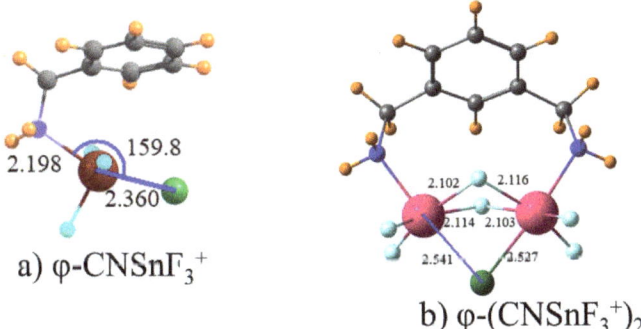

Figure 3. Geometries of indicated Lewis acids with Cl$^-$. Distances in Å, angles in degs.

Since aromaticity offers little advantage, there is little reason to connect the two SnF$_3^+$ groups through a phenyl ring spacer. Perhaps a bidentate arrangement in the same spirit could be offered by a simple set of methylene groups as an alkyl diamine. These systems were designed with varying numbers *n* of methylene groups in SnF$_3^+$NH$_2$(CH$_2$)$_n$NH$_2$SnF$_3^+$, permitted to react with Cl$^-$ and the resulting structures are depicted in Figure 4. All have the desired bipodal binding with the Cl nearly symmetrically disposed toward the two Sn atoms and with similar R(Sn·Cl) distances. The energetic data in Table 4 indicates that all of these receptors bind more strongly to Cl$^-$ than does the original receptor φ-(CNSnF$_3^+$)$_2$, i.e., the benzene connector offers no advantage. Of the various size diamine dications, C$_5$ and C$_1$ are the least favorable and C$_4$ and C$_2$ the most favorable.

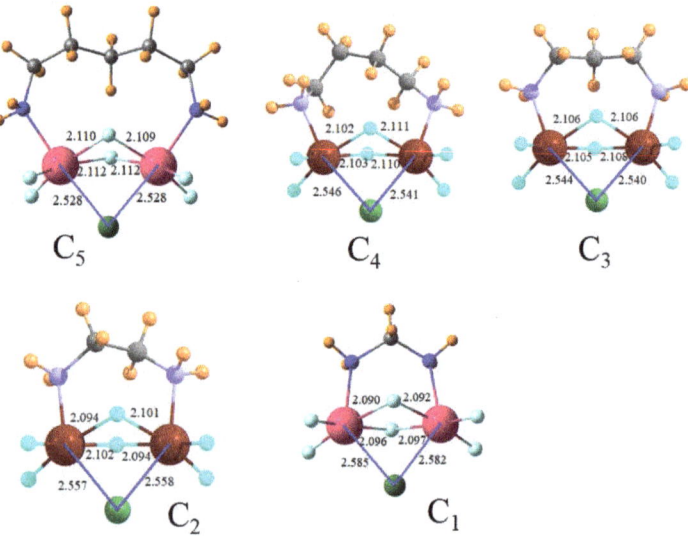

Figure 4. Geometries of cyclic F$_3$SnNH$_2$(CH$_2$)$_n$NH$_2$SnF$_3$ dications (C$_n$) with Cl$^-$. Distances in Å, angles in degrees.

The structure of each receptor in Figures 3b and 4, with its bidentate binding to the chloride, may impair the ability of each of the two tetrel bonds from achieving its full potential interaction energy. For example, the θ(NSn·Cl) bond would naturally incline toward 180° but this is not possible for a number of these complexes. In order to relieve this geometrical stress, the two cations within the single molecule were separated into a pair of mono-cations. In particular, the Cl$^-$ was allowed to bind to two individual MeNH$_2$SnF$_3^+$ ions and the resulting complex is pictured in Figure 5a. Despite the geometrical freedom, the two tetrel-bonding groups adopt a geometry very much like the single-molecule dications. Specifically, the two θ(NSn·Cl) angles in Figure 5a are 166° and 157°, which is somewhat deviant from linearity. These angles are not very different from those in the C3 and C4 diamines with angles of 157° and 160°, respectively. Additionally, perhaps more to the point, the freedom granted by the pair of mono-cations does not enhance the binding energy. Table 4 shows that ΔG is 53.5 kcal/mol, which is even lower than for most of the diamine dications (although ΔE does profit from a small enhancement). As a last point of interest in this regard, the addition of a third MeNH$_2$SnF$_3^+$ mono-cation increases the chloride binding but only by a small degree of 24%. This small increase may be due to steric crowding involving the third tetrel group. As evident in Figure 5b, the third R(Sn·Cl) distance is 3.745 Å, which is more than a full Å longer than the other two distances. The close proximity of the tetrel groups in Figure 5 was not imposed since optimizations were begun with these groups were nearly opposite one another.

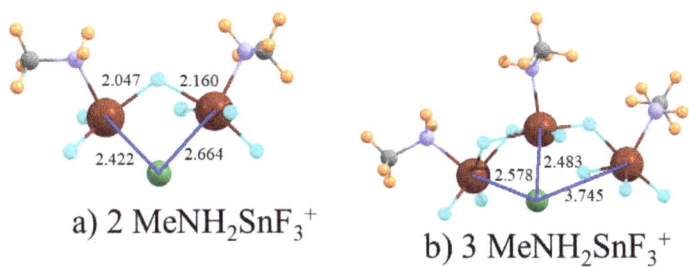

Figure 5. Geometries of Cl⁻ with (**a**) 2 and (**b**) 3 CH$_3$NH$_2$SnF$_3^+$ cations. Distances in Å.

3.3. Anions Other than Chloride

The forgoing analysis has been based on Cl⁻ as the universal anion. However, one of the important roles of a desirable anion receptor is its ability to distinguish among a sea of different anions. For this purpose, MeNH$_2$SnF$_3^+$ was chosen as the prototype monocationic receptor and the C$_2$ diamine SnF$_3^+$NH$_2$(CH$_2$)$_2$NH$_2$SnF$_3^+$ as dication. Both of these exhibit strong binding to the chloride. In addition to the four simple halides, other anions chosen for examination, due to their importance and prevalence, are OH⁻, NO$_3^-$, and HCO$_3^-$.

The binding energetics collected in Table 5 indicate that OH⁻ engages in the strongest interactions with either of the cationic receptors. In the case of the monocation, OH⁻ is followed by F⁻ and then by HCO$_3^-$. The latter two anions reverse places for the dication. There is little to distinguish NO$_3^-$ from the three larger halides whose binding follows the order of increasing size: Cl⁻ > Br⁻ > I⁻. As was observed in the earlier cases, ΔE is a bit more negative than ΔG.

Table 5. Energetics (kcal/mol) for interactions between anions and indicated mono and di-cationic receptors.

Anion	ΔG	ΔE
MeNH$_2$SnF$_3^+$		
F⁻	−64.54	−73.61
Cl⁻	−42.57	−50.76
Br⁻	−38.83	−46.64
I⁻	−36.46	−43.85
OH⁻	−78.62	−89.61
NO$_3^-$	−38.46	−52.81
HCO$_3^-$	−52.44	−67.65
C$_2$ Diamine Dication: SnF$_3^+$NH$_2$(CH$_2$)$_2$NH$_2$SnF$_3^+$		
F⁻	−87.88	−98.36
Cl⁻	−63.30	−71.94
Br⁻	−59.98	−68.20
I⁻	−58.04	−66.09
OH⁻	−112.98	−126.58
NO$_3^-$	−64.86	−79.14
HCO$_3^-$	−94.46	−109.52

The geometries of the various complexes with the halides are parallel to those for Cl⁻. The same may be said of the structures involving OH⁻, which can be seen in Figure 6. It is the O atoms of NO$_3^-$ and HCO$_3^-$ that directly interact with Sn and both are able to engage in bifurcated tetrel bonding with more than one O atom participating. Nonetheless, despite this possible advantage, it is the OH⁻ anion with its single O atom that is most strongly bound.

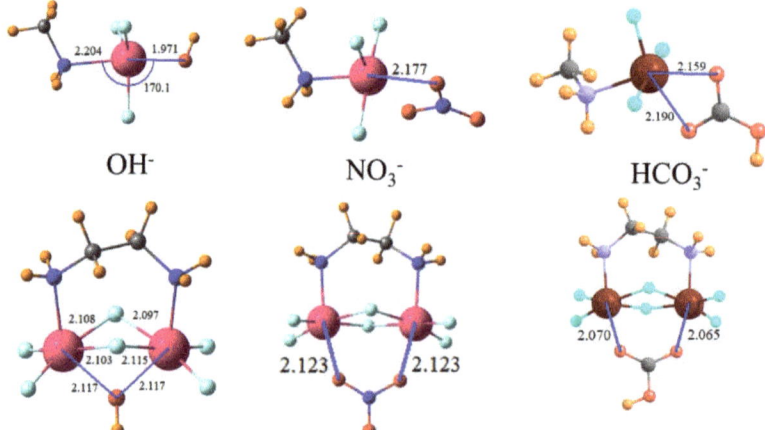

Figure 6. Geometries of indicated anion with $CH_3NH_2SnF_3^+$ cation in upper half and $F_3SnNH_2(CH_2)_2NH_2SnF_3^{+2}$ dication in lower half. Distances in Å.

3.4. Comparison of Receptors with Mobile Counterions

In order to extract any anion from solution, a receptor must compete with the anion's counterions. K^+ was chosen as a typical counter-ion, which might commonly occur. The concentration of the positive charge on a single atom might be anticipated to forge a very strong ion-ion interaction with each of the anions mentioned above. However, the binding energetics are comparatively quite small, which may be seen in Table 6. For example, Cl^- binds to K^+ with a ΔG of only -1.7 kcal/mol compared to the very much larger -42.6 kcal/mol for the tetrel-bonding $MeNH_2SnF_3^+$. Overall, the latter binds more strongly to the various anions than does K^+ by a factor between 10 and 40. The addition of a second K^+ can be used to compare with the dications. As seen in Table 6, this binding energy is no more than 10 kcal/mol, which compares with quantities between 60 kcal/mol and 113 kcal/mol for the dual tetrel bonded systems in Table 5.

Table 6. Energetics (kcal/mol) for interactions between anions and one or two K^+ cations.

	ΔG	ΔE
K^+		
F^-	−6.97	−11.73
Cl^-	−1.72	−5.95
Br^-	−0.90	−4.97
I^-	+0.07	−3.85
OH^-	−6.18	−12.61
NO_3^-	−1.75	−8.88
HCO_3^-	−3.85	−10.98
$2\,K^+$		
F^-	−10.03	−21.71
Cl^-	−1.04	−10.56
Br^-	−0.02	−8.73
I^-	+1.96	−6.60
OH^-	−10.07	−21.79
NO_3^-	−1.95	−15.92
HCO_3^-	−5.08	−18.82

These energy differences translate into a tremendous advantage for the tetrel-bonding species over the simple K^+ cations in the capture of these anions. If one expresses this advantage as the equilibrium constant $K = \exp(\delta\Delta G/RT)$ where $\delta\Delta G$ represents the difference in binding free energy between the former and the corresponding number of K^+ cations. The values obtained are listed in Table 7 at 25 °C. These advantages are very large from a minimum of 10^{27} all the way up to 10^{75}. Additionally, the dicationic receptors display a much larger advantage than the mono-cations.

Table 7. Preference of anions for tetrel-bonding species over one or two K^+ cations.

Anion	Monocation/K^+	Dication/$2K^+$
F^-	1.4×10^{42}	1.0×10^{57}
Cl^-	8.2×10^{29}	3.9×10^{45}
Br^-	6.0×10^{27}	8.2×10^{43}
I^-	5.7×10^{26}	8.7×10^{43}
OH^-	1.1×10^{53}	2.3×10^{75}
NO_3^-	7.7×10^{26}	1.2×10^{46}
HCO_3^-	3.8×10^{35}	2.9×10^{65}

3.5. Geometric Deformations of Monomers

It has been observed before [100,101] that substituents surrounding tetravalent tetrel atoms hinder the unimpeded approach of a nucleophile. If some of the substituents are bulky enough, they may prevent the formation of a tetrel bond entirely. However, even when smaller substituents are present, they must be pulled away to make room for the approaching nucleophile, which induces a certain amount of deformation energy into the Lewis acid molecule. This quantity has been shown to be as large as 20 kcal/mol and can be even larger [100] than the binding energy itself. This situation occurs in the tetrel-bonded complexes here as well. From the diagrams of the various complexes, one can see that the geometry changes around the tetrel atom are not a mere flattening out of the SnF_3 group to accommodate the chloride. It is true that the structure around the Sn atom adjusts from tetrahedral in the monomer to something more akin to a trigonal bipyramid within the complex. However, the apices of this bipyramid are not necessarily the C/N atom of the receptor and the Cl. In many of the optimized structures, these two atoms adopt equatorial positions along with one of the F atoms while the two remaining F atoms are positioned at the apices.

The deformation energies of the various cationic Lewis acids caused by their complexation with Cl^- are reported in the second column of Table 8 where it may be seen that there is a larger deformation energy for the first five mono-cations, in which all undergo the greater distortion required to rearrange so as to place F atoms at the apices. The deformation energies of the latter complexes all exceed 30 kcal/mol while the simpler rearrangements that retain the three F atoms in equatorial positions lie between 24 kcal/mol and 27 kcal/mol.

Rearrangements of the bipodal receptors are a bit simpler conceptually. The monomers contain a pair of Sn-F-Sn bridges not unlike the structures of the complexes pictured in the various figures. Therefore, the bulk of the rearrangement involves that necessary to make the two θ(N-Sn-Cl) angles as close to linearity as possible. In the ϕ-$(CNSnF_3^+)_2$ dication complex, for example, this angle differs from linearity by only 6°, which involves a deformation that requires 45 kcal/mol. The C_5 diamine achieves an 8° nonlinearity at a cost of only 38 kcal/mol, which suggests a bit more flexibility. The smaller diamines require a bit less deformation energy even if sacrificing greater nonlinearity: θ(N-Sn-Cl) = 20°, 23°, 34°, and 45°, respectively, for C_4, C_3, C_2, and C_1 diamines. Note that the binding energy in the first column of Table 8 does not suffer from this increasing nonlinearity.

Table 8. Binding, deformation, and interaction energy (kcal/mol) for interactions between Cl^- and indicated receptors.

Receptor	ΔE	E_{def}	E_{int} [a]
Monocations			
$ImSnF_3^+$	−43.54	32.82 [b]	−76.36
$N\text{-}ImSnF_3^+$	−49.99	43.40 [b]	−93.39
$TriSnF_3^+$	−41.30	34.33 [b]	−75.63
$N\text{-}TriSnF_3^+$	−51.67	30.36 [b]	−82.03
$N\text{-}cycloSnF_3^+$	−51.76	33.30 [b]	−85.06
$N\text{-}linSnF_3^+$	−49.53	24.18	−73.71
$CNSnF_3^+$	−50.76	24.52	−75.28
$\phi\text{-}CNSnF_3^+$	−46.43	27.15	−73.58
Dications			
$\phi\text{-}(CNSnF_3^+)_2$	−54.83	44.98	−99.81
C_5 diamine	−62.76	38.47	−101.23
C_4 diamine	−73.11	27.88	−100.99
C_3 diamine	−70.52	31.85	−102.37
C_2 diamine	−71.94	29.13	−101.07
C_1 diamine	−62.73	36.88	−99.61

[a] $E_{int} = \Delta E - E_{def}$; [b] two F atoms at apices of trigonal bipyramid in complex.

When the deformation energy is added to the total binding energy ΔE, the resulting E_{int} represents the interaction between Cl^- and the Lewis acid, once it has been deformed to the geometry, it adopts within the context of the full dimer. These quantities in the last column of Table 8 are quite large. They lie in the range of 73 kcal/mol to 93 kcal/mol for the mono cations especially large for $N\text{-}ImSnF_3^+$ wherein the SnF_3 group is attached to the N atom of imidazolium. E_{int} is even larger for the dications where it hovers consistently around 100 kcal/mol. Note that the interaction energy of Cl^- with a pre-deformed dicationic chelator, like the binding energy, remains quite a bit smaller than twice the analogous quantity for the monocations. These tetrel bond energies cannot be considered as simply additive.

4. Discussion

Of the various tetrel atoms tested, Sn forms the strongest interactions with a chloride anion, which is followed closely by Pb. The tetrel bond is strongly enhanced when the TH_3 group is perfluorinated to TF_3. The interaction is further strengthened if the molecule containing the tetrel atom is endowed with a full positive charge. With this information as a starting point, the imidazolium group to which the SnF_3 group is attached was varied in a methodical way to see if there were any ways to improve the binding. Binding is improved if this group is covalently attached to a Nitrogen atom of imidazolium rather than Carbon. On the other hand, replacement of imidazolium by triazolium had a slight weakening effect even though the tetrel bond is enhanced if the point of attachment is changed from C to N. The aromaticity of either of these two groups seems irrelevant since the replacement of imidazolium by its fully saturated five-membered ring analogue has no deleterious effect on the tetrel bond. Nor is the cyclic structure important, the binding is scarcely affected when a linear chain is used instead. The length of this chain on the N atom connection to SnF_3 is unimportant as well since its shortening from *n*-butyl to a simple methyl group produces only a small enhancement. There seems little advantage in placing this amine group on a phenyl connector since doing so weakens the tetrel bond by perhaps 10%.

A chelating arrangement whereby the Cl^- forms tetrel bonds to two SnF_3 groups simultaneously increases the total binding energy but by far less than a factor of two. For example, placing two $CH_3NH_2SnF_3^+$ groups on the phenyl ring produces only a magnification of the total binding energy by 21% when compared to that of a single such tetrel-bonding group. The size of this increase is not a result of geometric distortion since both θ(NSn·Cl) angles lie within 6° of linearity within this clathrate structure. Replacing the rigid phenyl ring by a more flexible $(CH_2)_n$ alkane chain improves the overall binding regardless of the length of this chain. The optimal length appears to be n = 2. Placing the two SnF_3 groups onto two separate molecules does not result in a stronger interaction, which suggests that steric constraints within the single dication molecule are not a detrimental factor. Just as adding a second group resulted in a magnification of only 1.2, a third such cation increases the binding free energy by the same factor. The modesty of the enhancement arising from a doubling of the positive charge on the receptor echoes recent [99,102] experimental findings.

It is worth reiterating that a very recent work [79] suggested that Cl^- is an excellent choice as the test anion since its calculated binding to a series of Lewis acids mimics the experimental trends arising from NMR measurements. While the binding of Cl^- is just a bit stronger than the larger halides as well as NO_3^-, HCO_3^- binds more strongly to the $MeNH_2SnF_3^+$ monocation. The smaller size of F^- with its concentrated negative charge leads to a larger binding free energy and OH^- even more so. The calculated trend of diminishing binding that accompanies the increasing size of the halide is consistent with experimental findings [99]. These same trends are in evidence when these anions engage in a bifurcated tetrel bond with a uni-molecular $SnF_3^+NH_2(CH_2)_2NH_2SnF_3^+$ dication even though the magnitudes are larger. These differences in binding energy can result in highly selective receptors. For example, the 24 kcal/mol difference in ΔG binding of F^- over Cl^- in Table 5 translates to a 10^{17} equilibrium preference of the former over the latter. Even smaller differences in ΔG reflect substantial selectivity. The 3.4 kcal/mol advantage of Cl^- over Br^- yields a 300-fold equilibrium ratio. However, the very strong binding of OH^- might preclude the use of these receptors in basic environments where hydroxide would likely displace other anions.

In order to preferentially bind with an anion in solution, a receptor must successfully compete with counterions. The tetrel-bonding receptors examined here are extremely effective in this regard. Their binding energies with the various monoanions are much greater than those of K^+ counterions despite the ability of the latter to move freely around each anion. The preference of any given anion for the monocationic tetrel-bonding receptors, over a K^+ counterion, expressed as an equilibrium ratio, varies between 10^{27}–10^{53}. This preference is even larger for the dications when compared to a pair of K^+ counterions, which rise up to as high as 10^{75}.

It will be observed that both Gibbs free energy (ΔG) and electronic energy (ΔE) has been provided for all of the complexation reactions here. The former corrects the latter for zero-point vibrational energies as well as entropic effects. The latter additions make ΔG less negative than ΔE, but the discrepancy is fairly uniform and is typically on the order of 7–10 kcal/mol, which is a bit larger for the dications. As an end result, both energetic quantities obey similar trends.

It should be stressed that the self-consistent reaction field approach used here to model immersion in a solvent represents only an approximation of the full solvation effects. This model treats the solvent as a dielectric continuum that reacts to, and stabilizes, the charge distribution within the solute in an iterative manner. In doing so, it essentially averages over the many configurations that the solvent molecules will adopt over the course of a measurement. However, specific interactions of any individual solvent molecule with the solute are not explicitly evaluated. For this reason, the calculated energetics in water should be treated as only approximations. Nonetheless, this procedure has the virtue of providing some measure of the relative stabilization caused by immersing the solute in the solvent milieu. The trends in the data that emerge are likely realistic and differences from one system to the next of more than a few kcal/mol can be treated as meaningful. For example, the very large equilibrium ratios in Table 7 between the preference of each anion for a tetrel-bonding receptor vs K^+ counterions are very unlikely to be reversed if other means of estimating solvation are employed.

Due to the high dielectric constant of water, solvation has quite a large impact on the binding energies. Taking the $ImGeTH_3$ complex with Cl^- as an example, the interaction energy in water of −1.9 kcal/mol rises to −14.6 kcal/mol in vacuo. The effect on the charged $ImGeTF_3^+$ receptor is even more extreme since ΔE grows from -32.1 kcal/mol to -144.4 kcal/mol. Very similar increases are observed in ΔG. One may also consider how solvation contributes to the huge advantage that the tetrel-bonding receptors enjoy over K^+ in the competition for an anion. Table 6 indicates a very weak interaction between K^+ and Cl^- in water with ΔG of only -1.72 kcal/mol, which is a major factor in the advantage of the tetrel-bonding receptor in the competition for this anion. The situation in the gas phase leads to much larger binding energies. Without the very substantial solvation energy of the cation, ΔG is greatly enlarged to -113.1 kcal/mol in vacuo. Taking the tetrel-bonding $MeNH_2SnF_3^+$ cation as a counterpoint, its binding energy with Cl^- of -42.57 kcal/mol in water increases to -181.1 kcal/mol in vacuum, which is an even larger increment. As a result, the 41 kcal/mol advantage that $MeNH_2SnF_3^+$ holds over K^+ in solution is increased to 68 kcal/mol without the moderating influence of water. Therefore, one may surmise that the stronger binding of tetrel-bonding species when compared to a small and compact counterion is intrinsic and is not an artifact of the solvation phenomena.

The reason for this reduced advantage in water derives from the solvation energies of the individual species. For exemplary purposes, one may consider the interactions of Cl^- with both $MeNH_2SnF_3^+$ and K^+. Considering first the monomers, the solvation energy of K^+ is larger by 9 kcal/mol than that of $MeNH_2SnF_3^+$ due to its smaller size and more compact charge. A similar advantage accrues to the $K^+ \cdot Cl^-$ ion pair vs. the larger tetrel-bonded complex where it increases by 24 kcal/mol. This greater stabilization advantage of the $K^+ \cdot Cl^-$ complex vs the separate ions increases its binding energy relative to the $MeNH_2SnF_3^+$ analogue. The net result is that the lesser binding energy of K^+ vs the tetrel bond in the gas phase is reduced by 15 kcal/mol in water.

It might finally be remarked that some of these interactions between the receptor and the anion are quite strong since they are in excess of 50 kcal/mol. When combined with the rather short $R(Sn \cdot X)$ distances, it would be legitimate to refer to many of these interactions as bordering on covalent with the Sn atom adopting a hyper-valent bonding character. The arrangement of the atoms around the Sn atom in Figure 2, for example, might best be described as pentavalent trigonal bipyramidal. An octahedral hexavalent environment, albeit a distorted one, could be invoked for a number of the bipodal receptors, which is shown in Figure 4.

In conclusion, tetrel bonding offers a highly attractive way of forming strong complexes with anions that can easily extract these anions from an aqueous environment containing counter-ions. The $-SnF_3$ group is particularly effective in this regard especially when the receptor contains a positive charge. A bipodal dicationic receptor has advantages over a mono-cation that can engage in only a single tetrel bond. It is hoped that the ideas presented here may guide researchers in the synthesis and testing of improved anion receptors.

Conflicts of Interest: The authors declare no conflict of interest.

References

1. Langton, M.J.; Serpell, C.J.; Beer, P.D. Anion Recognition in Water: Recent Advances from a Supramolecular and Macromolecular Perspective. *Angew. Chem. Int. Ed.* **2016**, *55*, 1974–1987. [CrossRef] [PubMed]
2. Pflugrath, J.W.; Quiocho, F.A. Sulphate sequestered in the sulphate-binding protein of Salmonella typhimurium is bound solely by hydrogen bonds. *Nature* **1985**, *314*, 257–260. [CrossRef] [PubMed]
3. Luecke, H.; Quiocho, F.A. High specificity of a phosphate transport protein determined by hydrogen bonds. *Nature* **1990**, *347*, 402–406. [CrossRef] [PubMed]
4. Koropatkin, N.M.; Pakrasi, H.B.; Smith, T.J. Atomic structure of a nitrate-binding protein crucial for photosynthetic productivity. *Proc. Natl. Acad. Sci. USA* **2006**, *103*, 9820–9825. [CrossRef] [PubMed]
5. Koropatkin, N.M.; Koppenaal, D.W.; Pakrasi, H.B.; Smith, T.J. The Structure of a Cyanobacterial Bicarbonate Transport Protein, CmpA. *J. Biol. Chem.* **2007**, *282*, 2606–2614. [CrossRef] [PubMed]
6. Gale, P.A. Preface. *Coord. Chem. Rev.* **2006**, *250*, 2917. [CrossRef]
7. Zurro, M.; Asmus, S.; Bamberger, J.; Beckendorf, S.; García Mancheño, O. Chiral Triazoles in Anion-Binding Catalysis: New Entry to Enantioselective Reissert-Type Reactions. *Chem. Eur. J.* **2016**, *22*, 3785–3793. [CrossRef] [PubMed]

8. Amendola, V.; Bergamaschi, G.; Boiocchi, M.; Legnani, L.; Presti, E.L.; Miljkovic, A.; Monzani, E.; Pancotti, F. Chloride-binding in organic-water mixtures: The powerful synergy of C-H donor groups within a bowl-shaped cavity. *Chem. Commun.* **2016**, *52*, 10910–10913. [CrossRef] [PubMed]
9. Toure, M.; Charles, L.; Chendo, C.; Viel, S.; Chuzel, O.; Parrain, J.-L. Straightforward and Controlled Shape Access to Efficient Macrocyclic Imidazolylboronium Anion Receptors. *Chem. Eur. J.* **2016**, *22*, 8937–8942. [CrossRef] [PubMed]
10. Cybulski, S.M.; Scheiner, S. Hydrogen bonding and proton transfers involving the carboxylate group. *J. Am. Chem. Soc.* **1989**, *111*, 23–31. [CrossRef]
11. Hillenbrand, E.A.; Scheiner, S. Effects of molecular charge and methyl substitution on proton transfers between oxygen atoms. *J. Am. Chem. Soc.* **1984**, *106*, 6266–6273. [CrossRef]
12. Cybulski, S.; Scheiner, S. Hydrogen bonding and proton transfers involving triply bonded atoms. Acetylene and hydrocyanic acid. *J. Am. Chem. Soc.* **1987**, *109*, 4199–4206. [CrossRef]
13. Steed, J.W. Anion-tuned supramolecular gels: A natural evolution from urea supramolecular chemistry. *Chem. Soc. Rev.* **2010**, *39*, 3686–3699. [CrossRef] [PubMed]
14. Li, A.-F.; Wang, J.-H.; Wang, F.; Jiang, Y.-B. Anion complexation and sensing using modified urea and thiourea-based receptors. *Chem. Soc. Rev.* **2010**, *39*, 3729–3745. [CrossRef] [PubMed]
15. Nehra, A.; Bandaru, S.; Yarramala, D.S.; Rao, C.P. Differential Recognition of Anions with Selectivity towards F^- by a Calix[6]arene–Thiourea Conjugate Investigated by Spectroscopy, Microscopy, and Computational Modeling by DFT. *Chem. Eur. J.* **2016**, *22*, 8903–8914. [CrossRef] [PubMed]
16. Kataev, E.A.; Müller, C.; Kolesnikov, G.V.; Khrustalev, V.N. Guanidinium-Based Artificial Receptors for Binding Orthophosphate in Aqueous Solution. *Eur. J. Org. Chem.* **2014**, *2014*, 2747–2753. [CrossRef]
17. Kuchelmeister, H.Y.; Schmuck, C. Nucleotide Recognition in Water by a Guanidinium-Based Artificial Tweezer Receptor. *Chem. Eur. J.* **2011**, *17*, 5311–5318. [CrossRef] [PubMed]
18. Riley, K.E.; Ford, C.L., Jr.; Demouchet, K. Comparison of hydrogen bonds, halogen bonds, CH··π interactions, and CX··π interactions using high-level ab initio methods. *Chem. Phys. Lett.* **2015**, *621*, 165–170. [CrossRef]
19. Riley, K.E.; Murray, J.S.; Fanfrlík, J.; Rezáč, J.; Solá, R.J.; Concha, M.C.; Ramos, F.M.; Politzer, P. Halogen bond tunability I: The effects of aromatic fluorine substitution on the strengths of halogen-bonding interactions involving chlorine, bromine, and iodine. *J. Mol. Model.* **2011**, *17*, 3309–3318. [CrossRef] [PubMed]
20. Alkorta, I.; Rozas, S.; Elguero, J. Charge-transfer complexes between dihalogen compounds and electron donors. *J. Phys. Chem. A* **1998**, *102*, 9278–9285. [CrossRef]
21. Cavallo, G.; Metrangolo, P.; Milani, R.; Pilati, T.; Priimagi, A.; Resnati, G.; Terraneo, G. The Halogen Bond. *Chem. Rev.* **2016**, *116*, 2478–2601. [CrossRef] [PubMed]
22. Alkorta, I.; Sanchez-Sanz, G.; Elguero, J.; Bene, J.E.D. FCl:PCX complexes: Old and new types of halogen bonds. *J. Phys. Chem. A* **2012**, *116*, 2300–2308. [CrossRef] [PubMed]
23. Metrangolo, P.; Neukirch, H.; Pilati, T.; Resnati, G. Halogen bonding based recognition processes: A world parallel to hydrogen bonding. *Acc. Chem. Res.* **2005**, *38*, 386–395. [CrossRef] [PubMed]
24. Politzer, P.; Murray, J.S. A unified view of halogen bonding, hydrogen bonding and other s-hole interactions. In *Noncovalent Forces*; Scheiner, S., Ed.; Springer: Dordrecht, The Netherlands, 2015; Volume 19, pp. 357–389.
25. Mullaney, B.R.; Thompson, A.L.; Beer, P.D. An All-Halogen Bonding Rotaxane for Selective Sensing of Halides in Aqueous Media. *Angew. Chem. Int. Ed.* **2014**, *53*, 11458–11462. [CrossRef] [PubMed]
26. Mele, A.; Metrangolo, P.; Neukirch, H.; Pilati, T.; Resnati, G. A Halogen-Bonding-Based Heteroditopic Receptor for Alkali Metal Halides. *J. Am. Chem. Soc.* **2005**, *127*, 14972–14973. [CrossRef] [PubMed]
27. Brown, A.; Beer, P.D. Halogen bonding anion recognition. *Chem. Commun.* **2016**, *52*, 8645–8658. [CrossRef] [PubMed]
28. Lim, J.Y.C.; Cunningham, M.J.; Davis, J.J.; Beer, P.D. Halogen bonding enhanced electrochemical halide anion sensing by redox-active ferrocene receptors. *Chem. Commun.* **2015**, *51*, 14640–14643. [CrossRef] [PubMed]
29. Caballero, A.; Swan, L.; Zapata, F.; Beer, P.D. Iodide-Induced Shuttling of a Halogen- and Hydrogen-Bonding Two-Station Rotaxane. *Angew. Chem. Int. Ed.* **2014**, *53*, 11854–11858. [CrossRef] [PubMed]
30. Tepper, R.; Schulze, B.; Jäger, M.; Friebe, C.; Scharf, D.H.; Görls, H.; Schubert, U.S. Anion Receptors Based on Halogen Bonding with Halo-1,2,3-triazoliums. *J. Org. Chem.* **2015**, *80*, 3139–3150. [CrossRef] [PubMed]
31. Wageling, N.B.; Neuhaus, G.F.; Rose, A.M.; Decato, D.A.; Berryman, O.B. Advantages of organic halogen bonding for halide recognition. *Supramol. Chem.* **2016**, *28*, 665–672. [CrossRef]

32. Molina, P.; Zapata, F.; Caballero, A. Anion Recognition Strategies Based on Combined Noncovalent Interactions. *Chem. Rev.* **2017**, *117*, 9907–9972. [CrossRef] [PubMed]
33. Serpell, C.J.; Kilah, N.L.; Costa, P.J.; Félix, V.; Beer, P.D. Halogen Bond Anion Templated Assembly of an Imidazolium Pseudorotaxane. *Angew. Chem. Int. Ed.* **2010**, *49*, 5322–5326. [CrossRef] [PubMed]
34. Caballero, A.; Zapata, F.; White, N.G.; Costa, P.J.; Félix, V.; Beer, P.D. A Halogen-Bonding Catenane for Anion Recognition and Sensing. *Angew. Chem. Int. Ed.* **2012**, *51*, 1876–1880. [CrossRef] [PubMed]
35. Gilday, L.C.; White, N.G.; Beer, P.D. Halogen- and hydrogen-bonding triazole-functionalised porphyrin-based receptors for anion recognition. *Dalton Trans.* **2013**, *42*, 15766–15773. [CrossRef] [PubMed]
36. Mercurio, J.M.; Knighton, R.C.; Cookson, J.; Beer, P.D. Halotriazolium Axle Functionalised [2]Rotaxanes for Anion Recognition: Investigating the Effects of Halogen-Bond Donor and Preorganisation. *Chem. Eur. J.* **2014**, *20*, 11740–11749. [CrossRef] [PubMed]
37. Chudzinski, M.G.; McClary, C.A.; Taylor, M.S. Anion receptors composed of hydrogen- and halogen-bond donor groups: Modulating selectivity with combinations of distinct noncovalent interactions. *J. Am. Chem. Soc.* **2011**, *133*, 10559–10567. [CrossRef] [PubMed]
38. Sarwar, M.G.; Dragisic, B.; Dimitrijevic, E.; Taylor, M.S. Halogen bonding between anions and iodoperfluoroorganics: Solution-phase thermodynamics and multidentate-receptor design. *Chem. Eur. J.* **2013**, *19*, 2050–2058. [CrossRef] [PubMed]
39. Chakraborty, S.; Dutta, R.; Ghosh, P. Halogen bonding assisted selective removal of bromide. *Chem. Commun.* **2015**, *51*, 14793–14796. [CrossRef] [PubMed]
40. Nepal, B.; Scheiner, S. Competitive Halide Binding by Halogen Versus Hydrogen Bonding: Bis-triazole Pyridinium. *Chem. Eur. J.* **2015**, *21*, 13330–13335. [CrossRef] [PubMed]
41. Nepal, B.; Scheiner, S. Substituent Effects on the Binding of Halides by Neutral and Dicationic Bis(triazolium) Receptors. *J. Phys. Chem. A* **2015**, *119*, 13064–13073. [CrossRef] [PubMed]
42. Nepal, B.; Scheiner, S. Building a Better Halide Receptor: Optimum Choice of Spacer, Binding Unit, and Halosubstitution. *ChemPhysChem* **2016**, *17*, 836–844. [CrossRef] [PubMed]
43. Scheiner, S. Highly Selective Halide Receptors Based on Chalcogen, Pnicogen, and Tetrel Bonds. *Chem. Eur. J.* **2016**, *22*, 18850–18858. [CrossRef] [PubMed]
44. Scheiner, S. Assembly of Effective Halide Receptors from Components. Comparing Hydrogen, Halogen, and Tetrel Bonds. *J. Phys. Chem. A* **2017**, *121*, 3606–3615. [CrossRef] [PubMed]
45. Scheiner, S. Halogen Bonds Formed between Substituted Imidazoliums and N Bases of Varying N-Hybridization. *Molecules* **2017**, *22*, 1634. [CrossRef] [PubMed]
46. Scheiner, S. Comparison of halide receptors based on H, halogen, chalcogen, pnicogen, and tetrel bonds. *Faraday Discuss.* **2017**, *203*, 213–226. [CrossRef] [PubMed]
47. Iwaoka, M.; Tomoda, S. Nature of the intramolecular Se···N nonbonded interaction of 2-selenobenzylamine derivatives. An experimental evaluation by ^{1}H, ^{77}Se, and ^{15}N-NMR spectroscopy. *J. Am. Chem. Soc.* **1996**, *118*, 8077–8084. [CrossRef]
48. Nagao, Y.; Hirata, T.; Goto, S.; Sano, S.; Kakehi, A.; Iizuka, K.; Shiro, M. Intramolecular nonbonded S···O interaction recognized in (acylimino)thiadiazoline derivatives as angiotensin II receptor antagonists and related compounds. *J. Am. Chem. Soc.* **1998**, *120*, 3104–3110. [CrossRef]
49. Sanz, P.; Mó, O.; Yáñez, M. Characterization of intramolecular hydrogen bonds and competitive chalcogen–chalcogen interactions on the basis of the topology of the charge density. *Phys. Chem. Chem. Phys.* **2003**, *5*, 2942–2947. [CrossRef]
50. Adhikari, U.; Scheiner, S. Sensitivity of pnicogen, chalcogen, halogen and H-bonds to angular distortions. *Chem. Phys. Lett.* **2012**, *532*, 31–35. [CrossRef]
51. Adhikari, U.; Scheiner, S. Effects of Charge and Substituent on the S···N Chalcogen Bond. *J. Phys. Chem. A* **2014**, *118*, 3183–3192. [CrossRef] [PubMed]
52. Nziko, V.D.P.N.; Scheiner, S. Chalcogen Bonding between Tetravalent SF$_4$ and Amines. *J. Phys. Chem. A* **2014**, *118*, 10849–10856. [CrossRef] [PubMed]
53. Fick, R.J.; Kroner, G.M.; Nepal, B.; Magnani, R.; Horowitz, S.; Houtz, R.L.; Scheiner, S.; Trievel, R.C. Sulfur–Oxygen Chalcogen Bonding Mediates AdoMet Recognition in the Lysine Methyltransferase SET7/9. *ACS Chem. Biol.* **2016**, *11*, 748–754. [CrossRef] [PubMed]
54. Esrafili, M.D.; Nurazar, R. Chalcogen bonds formed through π-holes: SO$_3$ complexes with nitrogen and phosphorus bases. *Mol. Phys.* **2016**, *114*, 276–282. [CrossRef]

55. Klinkhammer, K.W.; Pyykko, P. Ab initio interpretation of the closed-shell intermolecular E···E attraction in dipnicogen (H$_2$E-EH$_2$)$_2$ and (HE-EH)$_2$ hydride model dimers. *Inorg. Chem.* **1995**, *34*, 4134–4138. [CrossRef]
56. Moilanen, J.; Ganesamoorthy, C.; Balakrishna, M.S.; Tuononen, H.M. Weak interactions between trivalent pnictogen centers: Computational analysis of bonding in dimers X$_3$E···EX$_3$ (E = Pnictogen, X = Halogen). *Inorg. Chem.* **2009**, *48*, 6740–6747. [CrossRef] [PubMed]
57. Bene, J.E.D.; Alkorta, I.; Sanchez-Sanz, G.; Elguero, J. Structures, energies, bonding, and NMR properties of pnicogen complexes H$_2$XP:NXH$_2$ (X = H, CH$_3$, NH$_2$, OH, F, Cl). *J. Phys. Chem. A* **2011**, *115*, 13724–13731. [CrossRef] [PubMed]
58. Li, Q.-Z.; Li, R.; Liu, X.-F.; Li, W.-Z.; Cheng, J.-B. Concerted interaction between pnicogen and halogen bonds in XCl-FH$_2$P-NH$_3$ (X=F, OH, CN, NC, and FCC). *ChemPhysChem* **2012**, *13*, 1205–1212. [CrossRef] [PubMed]
59. Scheiner, S. The pnicogen bond: Its relation to hydrogen, halogen, and other noncovalent bonds. *Acc. Chem. Res.* **2013**, *46*, 280–288. [CrossRef] [PubMed]
60. Bauzá, A.; Mooibroek, T.J.; Frontera, A. σ-Hole Opposite to a Lone Pair: Unconventional Pnicogen Bonding Interactions between ZF$_3$ (Z=N, P, As, and Sb) Compounds and Several Donors. *ChemPhysChem* **2016**, *17*, 1608–1614. [CrossRef] [PubMed]
61. Benz, S.; Macchione, M.; Verolet, Q.; Mareda, J.; Sakai, N.; Matile, S. Anion Transport with Chalcogen Bonds. *J. Am. Chem. Soc.* **2016**, *138*, 9093–9096. [CrossRef] [PubMed]
62. Lim, J.Y.C.; Marques, I.; Thompson, A.L.; Christensen, K.E.; Félix, V.; Beer, P.D. Chalcogen Bonding Macrocycles and [2]Rotaxanes for Anion Recognition. *J. Am. Chem. Soc.* **2017**, *139*, 3122–3133. [CrossRef] [PubMed]
63. Garrett, G.E.; Carrera, E.I.; Seferos, D.S.; Taylor, M.S. Anion recognition by a bidentate chalcogen bond donor. *Chem. Commun.* **2016**, *52*, 9881–9884. [CrossRef] [PubMed]
64. Sánchez-Sanz, G.; Trujillo, C. Improvement of Anion Transport Systems by Modulation of Chalcogen Interactions: The influence of solvent. *J. Phys. Chem. A* **2018**, *122*, 1369–1377. [CrossRef] [PubMed]
65. Alkorta, I.; Rozas, I.; Elguero, J. Molecular Complexes between Silicon Derivatives and Electron-Rich Groups. *J. Phys. Chem. A* **2001**, *105*, 743–749. [CrossRef]
66. Bauzá, A.; Mooibroek, T.J.; Frontera, A. Tetrel-Bonding Interaction: Rediscovered Supramolecular Force? *Angew. Chem. Int. Ed.* **2013**, *52*, 12317–12321. [CrossRef] [PubMed]
67. Grabowski, S.J. Tetrel bond–σ-hole bond as a preliminary stage of the S$_N$2 reaction. *Phys. Chem. Chem. Phys.* **2014**, *16*, 1824–1834. [CrossRef] [PubMed]
68. Tang, Q.; Li, Q. Interplay between tetrel bonding and hydrogen bonding interactions in complexes involving F$_2$XO (X=C and Si) and HCN. *Comput. Theor. Chem.* **2014**, *1050*, 51–57. [CrossRef]
69. Azofra, L.M.; Scheiner, S. Tetrel, chalcogen, and CH··O hydrogen bonds in complexes pairing carbonyl-containing molecules with 1, 2, and 3 molecules of CO$_2$. *J. Chem. Phys.* **2015**, *142*, 034307. [CrossRef] [PubMed]
70. Scheiner, S. Comparison of CH···O, SH···O, Chalcogen, and Tetrel Bonds Formed by Neutral and Cationic Sulfur-Containing Compounds. *J. Phys. Chem. A* **2015**, *119*, 9189–9199. [CrossRef] [PubMed]
71. Del Bene, J.E.; Alkorta, I.; Elguero, J. Exploring the (H$_2$C=PH$_2$)$^+$:N-Base Potential Surfaces: Complexes Stabilized by Pnicogen, Hydrogen, and Tetrel Bonds. *J. Phys. Chem. A* **2015**, *119*, 11701–11710. [CrossRef] [PubMed]
72. Southern, S.A.; Bryce, D.L. NMR Investigations of Noncovalent Carbon Tetrel Bonds. Computational Assessment and Initial Experimental Observation. *J. Phys. Chem. A* **2015**, *119*, 11891–11899. [CrossRef] [PubMed]
73. Esrafili, M.D.; Vakili, M.; Javaheri, M.; Sobhi, H.R. Tuning of tetrel bonds interactions by substitution and cooperative effects in XH$_3$Si···NCH···HM (X = H, F, Cl, Br; M = Li, Na, BeH and MgH) complexes. *Mol. Phys.* **2016**, *114*, 1974–1982. [CrossRef]
74. Caballero, A.; White, N.G.; Beer, P.D. A bidentate halogen-bonding bromoimidazoliophane receptor for bromide ion recognition in aqueous media. *Angew. Chem. Int. Ed. Engl.* **2011**, *50*, 1845–1848. [CrossRef] [PubMed]
75. Walter, S.M.; Kniep, F.; Herdtweck, E.; Huber, S.M. Halogen-bond-induced activation of a carbon–heteroatom bond. *Angew. Chem. Int. Ed.* **2011**, *50*, 7187–7191. [CrossRef] [PubMed]

76. Walter, S.M.; Kniep, F.; Rout, L.; Schmidtchen, F.P.; Herdtweck, E.; Huber, S.M. Isothermal Calorimetric Titrations on Charge-Assisted Halogen Bonds: Role of Entropy, Counterions, Solvent, and Temperature. *J. Am. Chem. Soc.* **2012**, *134*, 8507–8512. [CrossRef] [PubMed]
77. Zapata, F.; Caballero, A.; White, N.G.; Claridge, T.D.W.; Costa, P.J.; Félix, V.; Beer, P.D. Fluorescent charge-assisted halogen-bonding macrocyclic halo-imidazolium receptors for anion recognition and sensing in aqueous media. *J. Am. Chem. Soc.* **2012**, *134*, 11533–11541. [CrossRef] [PubMed]
78. Sabater, P.; Zapata, F.; Caballero, A.; de la Visitación, N.; Alkorta, I.; Elguero, J.; Molina, P. Comparative Study of Charge-Assisted Hydrogen- and Halogen-Bonding Capabilities in Solution of Two-Armed Imidazolium Receptors toward Oxoanions. *J. Org. Chem.* **2016**, *81*, 7448–7458. [CrossRef] [PubMed]
79. Benz, S.; Poblador-Bahamonde, A.I.; Low-Ders, N.; Matile, S. Catalysis with Pnictogen, Chalcogen, and Halogen Bonds. *Angew. Chem. Int. Ed.* **2018**, *57*, 5408–5412. [CrossRef] [PubMed]
80. Frisch, M.J.; Trucks, G.W.; Schlegel, H.B.; Scuseria, G.E.; Robb, M.A.; Cheeseman, J.R.; Scalmani, G.; Barone, V.; Mennucci, B.; Petersson, G.A.; et al. *Gaussian 09, Revision B.01*; Gaussian, Inc.: Wallingford, CT, USA, 2009.
81. Zhao, Y.; Truhlar, D.G. The M06 suite of density functionals for main group thermochemistry, thermochemical kinetics, noncovalent interactions, excited states, and transition elements: Two new functionals and systematic testing of four M06-class functionals and 12 other functionals. *Theor. Chem. Acc.* **2008**, *120*, 215–241.
82. Feller, D. The role of databases in support of computational chemistry calculations. *J. Comput. Chem.* **1996**, *17*, 1571–1586. [CrossRef]
83. Schuchardt, K.L.; Didier, B.T.; Elsethagen, T.; Sun, L.; Gurumoorthi, V.; Chase, J.; Li, J.; Windus, T.L. Basis Set Exchange: A Community Database for Computational Sciences. *J. Chem. Inf. Model.* **2007**, *47*, 1045–1052. [CrossRef] [PubMed]
84. Boese, A.D. Density Functional Theory and Hydrogen Bonds: Are We There Yet? *ChemPhysChem* **2015**, *16*, 978–985. [CrossRef] [PubMed]
85. Forni, A.; Pieraccini, S.; Rendine, S.; Sironi, M. Halogen bonds with benzene: An assessment of DFT functionals. *J. Comput. Chem.* **2014**, *35*, 386–394. [CrossRef] [PubMed]
86. Bauzá, A.; Alkorta, I.; Frontera, A.; Elguero, J. On the Reliability of Pure and Hybrid DFT Methods for the Evaluation of Halogen, Chalcogen, and Pnicogen Bonds Involving Anionic and Neutral Electron Donors. *J. Chem. Theory Comput.* **2013**, *9*, 5201–5210. [CrossRef] [PubMed]
87. Mardirossian, N.; Head-Gordon, M. Characterizing and Understanding the Remarkably Slow Basis Set Convergence of Several Minnesota Density Functionals for Intermolecular Interaction Energies. *J. Chem. Theory Comput.* **2013**, *9*, 4453–4461. [CrossRef] [PubMed]
88. Elm, J.; Bildeb, M.; Mikkelsena, K.V. Assessment of binding energies of atmospherically relevant clusters. *Phys. Chem. Chem. Phys.* **2013**, *15*, 16442–16445. [CrossRef] [PubMed]
89. Rosokha, S.V.; Stern, C.L.; Ritzert, J.T. Experimental and computational probes of the nature of halogen bonding: Complexes of bromine-containing molecules with bromide anions. *Chem. Eur. J.* **2013**, *19*, 8774–8788. [CrossRef] [PubMed]
90. Liao, M.-S.; Lu, Y.; Scheiner, S. Performance assessment of density-functional methods for study of charge-transfer complexes. *J. Comput. Chem.* **2003**, *24*, 623–631. [CrossRef] [PubMed]
91. Boys, S.F.; Bernardi, F. The calculation of small molecular interactions by the differences of separate total energies. Some procedures with reduced errors. *Mol. Phys.* **1970**, *19*, 553–566. [CrossRef]
92. Latajka, Z.; Scheiner, S. Primary and secondary basis set superposition error at the SCF and MP2 levels: H_3N–Li^+ and H_2O–Li^+. *J. Chem. Phys.* **1987**, *87*, 1194–1204. [CrossRef]
93. Barone, V.; Cossi, M. Quantum calculation of molecular energies and energy gradients in solution by a conductor solvent model. *J. Phys. Chem. A* **1998**, *102*, 1995–2001. [CrossRef]
94. Jungbauer, S.H.; Huber, S.M. Cationic Multidentate Halogen-Bond Donors in Halide Abstraction Organocatalysis: Catalyst Optimization by Preorganization. *J. Am. Chem. Soc.* **2015**, *137*, 12110–12120. [CrossRef] [PubMed]
95. Mole, T.K.; Arter, W.E.; Marques, I.; Félix, V.; Beer, P.D. Neutral bimetallic rhenium(I)-containing halogen and hydrogen bonding acyclic receptors for anion recognition. *J. Organomet. Chem.* **2015**, *792*, 206–210. [CrossRef]
96. Barendt, T.A.; Docker, A.; Marques, I.; Félix, V.; Beer, P.D. Selective Nitrate Recognition by a Halogen-Bonding Four-Station [3]Rotaxane Molecular Shuttle. *Angew. Chem. Int. Ed.* **2016**, *55*, 11069–11076. [CrossRef] [PubMed]

97. Tepper, R.; Schulze, B.; Bellstedt, P.; Heidler, J.; Gorls, H.; Jager, M.; Schubert, U.S. Halogen-bond-based cooperative ion-pair recognition by a crown-ether-embedded 5-iodo-1,2,3-triazole. *Chem. Commun.* **2017**, *53*, 2260–2263. [CrossRef] [PubMed]
98. Borissov, A.; Lim, J.Y.C.; Brown, A.; Christensen, K.E.; Thompson, A.L.; Smith, M.D.; Beer, P.D. Neutral iodotriazole foldamers as tetradentate halogen bonding anion receptors. *Chem. Commun.* **2017**, *53*, 2483–2486. [CrossRef] [PubMed]
99. Dreger, A.; Engelage, E.; Mallick, B.; Beer, P.D.; Huber, S.M. The role of charge in 1,2,3-triazol(ium)-based halogen bonding activators. *Chem. Commun.* **2018**, *54*, 4013–4016. [CrossRef] [PubMed]
100. Scheiner, S. Steric Crowding in Tetrel Bonds. *J. Phys. Chem. A* **2018**, *122*, 2550–2562. [CrossRef] [PubMed]
101. Zierkiewicz, W.; Michalczyk, M.; Scheiner, S. Implications of monomer deformation for tetrel and pnicogen bonds. *Phys. Chem. Chem. Phys.* **2018**, *20*, 8832–8841. [CrossRef] [PubMed]
102. Heinen, F.; Engelage, E.; Dreger, A.; Weiss, R.; Huber, S.M. Iodine(III) Derivatives as Halogen Bonding Organocatalysts. *Angew. Chem. Int. Ed.* **2018**, *57*, 3830–3833. [CrossRef] [PubMed]

© 2018 by the author. Licensee MDPI, Basel, Switzerland. This article is an open access article distributed under the terms and conditions of the Creative Commons Attribution (CC BY) license (http://creativecommons.org/licenses/by/4.0/).

Article

An Ab Initio Investigation of the Geometries and Binding Strengths of Tetrel-, Pnictogen-, and Chalcogen-Bonded Complexes of CO_2, N_2O, and CS_2 with Simple Lewis Bases: Some Generalizations

Ibon Alkorta [1,*] and Anthony C. Legon [2,*]

1. Instituto de Química Médica (IQM-CSIC), Juan de la Cierva, 3, E-28006 Madrid, Spain
2. School of Chemistry, University of Bristol, Cantock's Close, Bristol BS8 1TS, UK
* Correspondence: ibon@iqm.csic.es (I.A.); a.c.legon@bristol.ac.uk (A.C.L.); Tel.: +44-117-331-7708 (A.C.L.)

Received: 20 August 2018; Accepted: 30 August 2018; Published: 4 September 2018

Abstract: Geometries, equilibrium dissociation energies (D_e), and intermolecular stretching, quadratic force constants (k_σ) are presented for the complexes $B \cdots CO_2$, $B \cdots N_2O$, and $B \cdots CS_2$, where B is one of the following Lewis bases: CO, HCCH, H_2S, HCN, H_2O, PH_3, and NH_3. The geometries and force constants were calculated at the CCSD(T)/aug-cc-pVTZ level of theory, while generation of D_e employed the CCSD(T)/CBS complete basis-set extrapolation. The non-covalent, intermolecular bond in the $B \cdots CO_2$ complexes involves the interaction of the electrophilic region around the C atom of CO_2 (as revealed by the molecular electrostatic surface potential (MESP) of CO_2) with non-bonding or π-bonding electron pairs of B. The conclusions for the $B \cdots N_2O$ series are similar, but with small geometrical distortions that can be rationalized in terms of secondary interactions. The $B \cdots CS_2$ series exhibits a different type of geometry that can be interpreted in terms of the interaction of the electrophilic region near one of the S atoms and centered on the C_∞ axis of CS_2 (as revealed by the MESP) with the n-pairs or π-pairs of B. The tetrel, pnictogen, and chalcogen bonds so established in $B \cdots CO_2$, $B \cdots N_2O$, and $B \cdots CS_2$, respectively, are rationalized in terms of some simple, electrostatically based rules previously enunciated for hydrogen- and halogen-bonded complexes, $B \cdots HX$ and $B \cdots XY$. It is also shown that the dissociation energy D_e is directly proportional to the force constant k_σ, with a constant of proportionality identical within experimental error to that found previously for many $B \cdots HX$ and $B \cdots XY$ complexes.

Keywords: intermolecular force constants; dissociation energies; CCSD(T)/aug-cc-pVTZ calculations; non-covalent bonds

1. Introduction

Investigation, both experimentally and theoretically, of non-covalent interactions among molecules is a topic of rapidly increasing interest. The hydrogen bond, known for almost a century, is of fundamental importance in chemistry and biology. The halogen bond is a weak interaction, in which interest within both disciplines grew rapidly in the last two decades. Modern definitions of the hydrogen bond [1] and the halogen bond [2], made under the auspices of the International Union of Pure and Applied Chemistry (IUPAC), arose naturally from the increased activity. Tetrel bonds, pnictogen bonds, and chalcogen bonds, close relatives of hydrogen and halogen bonds, were recognized as weak, non-covalent interactions in both the gas phase [3] and condensed phase [4] for several decades, but were named only in 2013 [5], 2011 [6], and 2009 [7], respectively. A task group set up by the IUPAC is currently working on the definitions of these three, newly named interactions (see: https://iupac.org/projects/project-details/?project_nr=2016-001-2-300).

It is now widely accepted [2,3,8] that each of these non-covalent bonds arises mainly from the interaction of an electrophilic region associated with an atom of the element E (where E is hydrogen, a halogen, or an element of group 14, 15, or 16) with the nucleophilic region (e.g., a non-bonding or π-bonding electron pair) in another molecule or the same molecule. Electrophilic and nucleophilic regions can be identified via the electrostatic potential near to the appropriate regions of the molecules [9]. A convenient modern and readily available way of identifying such regions is the molecular electrostatic surface potential (MESP), which is the potential energy of a non-perturbing, unit-positive point charge at the iso-surface on which the electron density is constant [10], and it is usually expressed as $0.00n$ e/bohr^3 ($n = 2$ here).

The closely related molecules CO_2, N_2O, and CS_2 form a series of interest in the context of non-covalent bonding. Each provides an electrophilic site by means of which either tetrel, pnictogen, or chalcogen bonds, respectively, could be formed. Both CO_2 and CS_2 are non-dipolar; thus, the molecular electric quadrupole moment is the first non-zero term in the expansion of the electric charge distribution; however, this moment is of opposite sign in the two molecules [11,12]. For CO_2, the sign corresponds to the partial charge description $^{\delta-}O = ^{2\delta+}C = O^{\delta-}$, while, for CS_2, the reverse arrangement $^{\delta+}S = ^{2\delta-}C = S^{\delta+}$ is implied. These charge distributions can be readily identified in the MESPs shown for each molecule (calculated at the 0.002 e/bohr^3 iso-surface) in Figure 1, which shows side-on and end-on views of the MESPs of CO_2, N_2O, and CS_2. Accordingly, we expect CO_2 to form tetrel bonds perpendicular to its C_∞ axis, via the electrophilic (blue) region at the C atom, with, e.g., the n-pair of a Lewis base. Conversely, CS_2 is likely to form chalcogen bonds via the electrophilic (blue) region that lies at each S atom and is centered on the C_∞ axis. Clearly, the charge distributions of CO_2 and N_2O, as represented by their MESPs in Figure 1, are very similar, as are the signs and magnitudes of their electric quadrupole moments [11,13]; however, N_2O also has a small electric dipole moment. Nitrous oxide is, therefore, expected to form a complex with a given Lewis base of similar geometry to that of its carbon dioxide counterpart, but with small distortions resulting from the lower symmetry and the non-zero electric dipole moment in the case of N_2O.

It this article, we present the geometries and interaction strengths of complexes of the type B$\cdots CO_2$, B$\cdots CS_2$, and B$\cdots N_2O$ for the series of Lewis bases, B = CO, HCCH, H_2S, HCN, H_2O, PH_3, and NH_3, as calculated ab initio at the CCSD(T)/aug-cc-pVTZ level of theory. The geometries so calculated can be compared with those established experimentally via gas-phase rotational or vibration–rotation spectra for some, but not all, of the complexes B$\cdots CO_2$ [14–21] and B$\cdots N_2O$ [21–29]; however, data for B$\cdots CS_2$ are sparse [30]. The interaction strength can be described in two possible ways. The first is the energy required for the reaction B$\cdots CO_2$ = B + CO_2, that is, the equilibrium dissociation energy D_e. The second is the intermolecular quadratic stretching force constant k_σ, which is proportional to the energy required for a unit infinitesimal displacement from equilibrium along the dissociation coordinate. It was shown elsewhere for hydrogen-bonded complexes B$\cdots$HX and halogen-bonded complexes B$\cdots$XY (X and Y are halogen atoms) that D_e is directly proportional to k_σ, with a constant of proportionality of $1.5(1) \times 10^{-3}$ m$^2\cdot$mol^{-1}, whether k_σ is obtained experimentally [31] from centrifugal distortion effects in the rotational spectra of the complexes or calculated ab initio [32].

Given the definitions of hydrogen and halogen bonds in terms of the interaction of nucleophilic regions of Lewis bases B with electrophilic regions near the atoms H of HX and X of XY, the aim of the work presented here is to examine by means of ab initio calculations (1) whether the complexes B$\cdots CO_2$, B$\cdots N_2O$, and B$\cdots CS_2$ involve tetrel, pnictogen, and chalcogen bonds, respectively, and (2) whether there is direct proportionality of D_e and k_σ for these complexes, and, if so, does the constant of proportionality found for hydrogen- and halogen-bonded complexes B$\cdots$HX and B$\cdots$XY also hold in these non-covalent bonds.

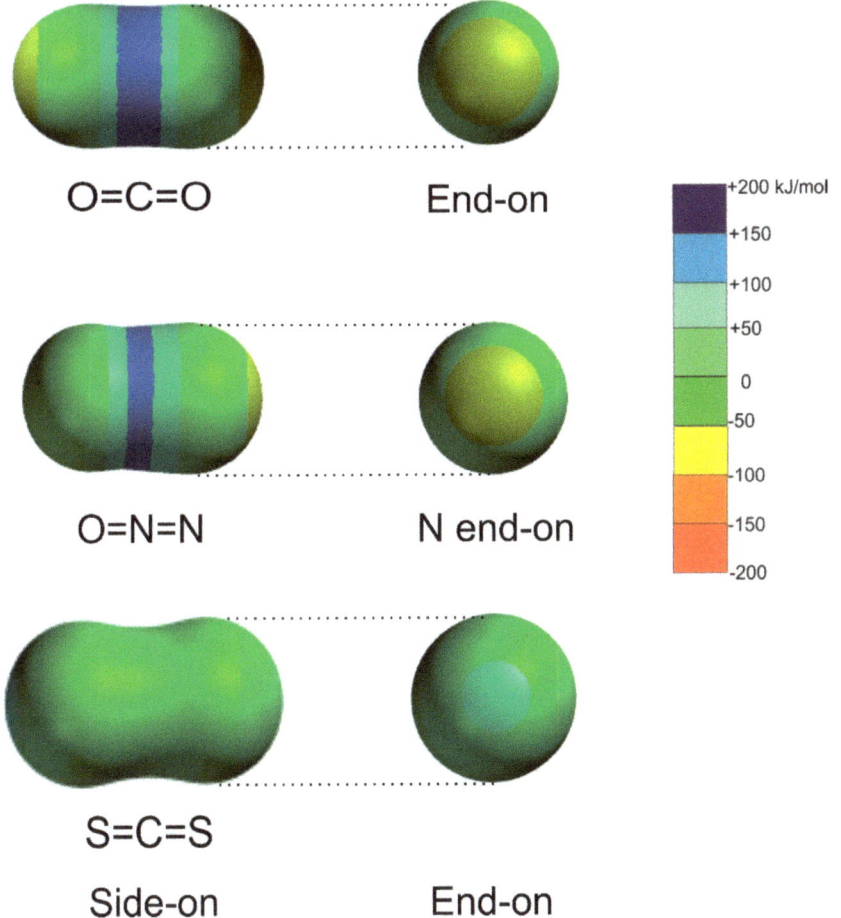

Figure 1. Molecular electrostatic surfaces potential (MESPs) for carbon dioxide, nitrous oxide, and carbon disulfide calculated for the 0.002 e/bohr^3 iso-surface at the MP2/6-311++G** level.

2. Theoretical Methods

We present here equilibrium geometries and values of D_e and k_σ (defined earlier) calculated ab initio for the members of three series of complexes, namely the series of B$\cdots$CO$_2$, B$\cdots$N$_2$O, and B$\cdots$CS$_2$, where B is one of the simple Lewis bases, CO, HCCH, H$_2$S, HCN, H$_2$O, PH$_3$, or NH$_3$. The geometry optimizations and the calculations of k_σ were conducted at the CCSD(T)/aug-cc-pVTZ level of theory [33,34]. To evaluate k_σ, the energy $E(r_e)$ at the equilibrium geometry was first obtained, and the energy $E(r)$ was then scanned for ± 20 pm about the appropriate equilibrium intermolecular distance r_e in increments $(r - r_e) = 5$ pm with optimization in all internal coordinates but r at each point. The curve of $E(r - r_e)$ as a function of $(r - r_e)$ was fitted to a third-order polynomial in $(r - r_e)$, and the second derivative was evaluated at $r = r_e$ to yield the quadratic force constant $k_\sigma = \left(\frac{\partial^2 E(r)}{\partial r^2}\right)_{r=r_e}$, which is the curvature at the minimum. All curves used in the evaluation of all k_σ presented here are available as supplementary information, as are the optimized geometries. Figure 2 shows a plot of $E(r - r_e)$ versus $(r - r_e)$ for the complex H$_3$N$\cdots$S=C=S, which is predicted by the ab initio calculations to possess C_{3v} symmetry at equilibrium, with the linear CS$_2$ molecule lying along the C_3 axis of NH_3, and therefore, with the inner S atom participating in a chalcogen bond to the n-electron pair of ammonia. Values of D_e with better accuracy were obtained using the method of extrapolation to a complete basis set [35] (CCSD(T)/CBS energy). For this purpose, the HF/aug-cc-pVnZ//CCSD(T)/aug-cc-pVTZ energies, with n = D, T, and Q, for the HF contribution and the CCSD(T)/aug-cc-pVn'Z//CCSD(T)/aug-cc-pVTZ, with n' = T and Q, for the correlation part were obtained for each system [36]. Finally, D_e was obtained as the difference of the CCSD(T)/CBS energy of the monomers and the complex. All the ab initio calculations were performed with the MOLPRO-2012 program [37]. The Z-matrices for optimized geometries are available as supplementary information. The molecular electrostatic surface potentials were generated using of the SPARTAN electronic structure package [38] at the MP2/6-311++G** level for CO_2, N_2O, CS_2, and PH_3.

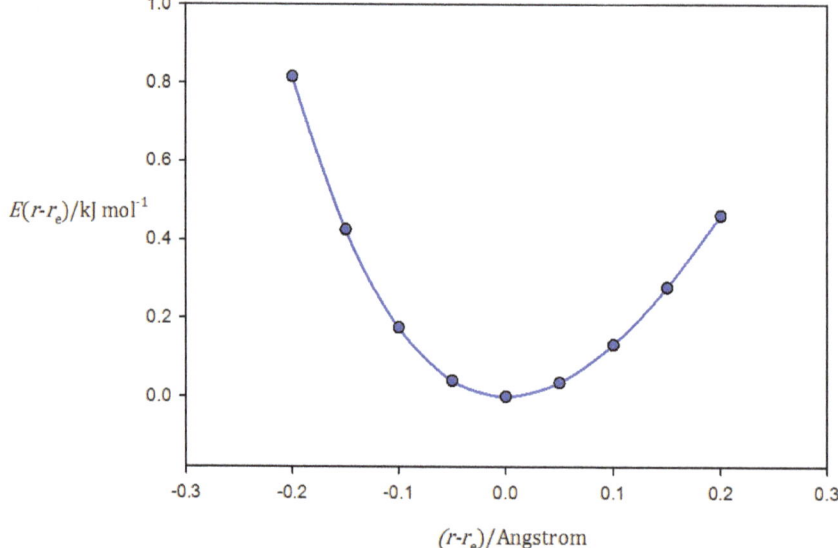

Figure 2. The variation in E ($r - r_e$) with $r - r_e$, used to calculate the intermolecular quadratic force k_σ (the curvature at the minimum) for $H_3N\cdots S=C=S$ at the CCSD(T)/aug-cc-pVTZ level of theory. The curve is a third-order polynomial fit to the calculated points (R^2 of fit = 0.9998). The polynomial was differentiated twice to obtain k_σ.

3. Results

3.1. Geometries of the $B\cdots CO_2$, $B\cdots N_2O$, and $B\cdots CS_2$ Complexes

Molecular diagrams showing the equilibrium geometry (drawn to scale) of each member of the $B\cdots CO_2$ series, where B = CO, HCCH, H_2S, HCN, H_2O, PH_3, and NH_3, are shown in Figure 3. The calculated (equilibrium) intermolecular distances are recorded in Table 1, together with their experimental counterparts (where the latter are available). The experimental distances were determined from microwave or high-resolution infrared spectroscopy conducted on supersonically expanded gas mixtures composed of the two component molecules diluted in an inert gas. The molecular shapes and intermolecular distances are, in each case, in reasonable agreement with those from experiment. It should be noted that the experimental distances are, in most cases, of the r_0 type, but are corrected for the contributions of the angular oscillations of the two components to the zero-point motion. There is no correction for the intermolecular radial contribution, however, and this normally leads to r_0 distances that are greater than the calculated equilibrium values. For the very floppy molecules considered here, the r_0 values are greater by the order of 0.05 to 0.1 Å.

Table 1. Calculated and observed intermolecular distances in B···CO_2 complexes.

Complex	Intermolecular Distance/Å		(Obs. − Calc.)/Å
	Calculated Ab Initio [a]	Observed	
OC···CO_2	$r(C···C) = 3.189$	3.277(1) [b]	0.088(1)
HCCH···CO_2	$r(\pi_{center}···C) = 3.201$	3.285(3) [c]	0.084(3)
HCN···CO_2 (T-shaped)	$r(N···C) = 2.962$	2.99(2) [d]	0.03(2)
CO_2···HCN (linear)	$r(O···H) = 2.236$	2.34 [e]	0.11
H_3N···CO_2	$r(N···C) = 2.922$	2.9875(2) [f]	0.066
H_2O···CO_2	$r(O···C) = 2.758$	2.836 [g]	0.078
H_2S···CO_2	$r(S···C) = 3.425$	3.449(1) [h]	0.024(1)
H_3P···CO_2	$r(P···C) = 3.528$	...	...

[a] See Figure 3 for the molecular diagrams (to scale) of the B···CO_2 complexes. [b] Reference [14]; [c] Reference [17]; [d] Reference [15,16]. [e] The distance reported here is the r_s value from Reference [39]; [f] Reference [20]; [g] Reference [19]; [h] Reference [18].

Table 2. Calculated and observed intermolecular distances in B···N_2O complexes.

Complex	Intermolecular Distance/Å		(Obs. − Calc.)/Å
	Calculated Ab Initio [a]	Observed	
OC···N_2O	$r(C···N_{center}) = 3.176$	3.36(1) [b]	0.18
HCCH···N_2O	$r(\pi_{center}···N_{center}) = 3.201$	3.296 [c]	0.095(1)
HCN···N_2O (T-shaped)	$r(C···N_{center}) = 3.002$	...	...
HCN···N_2O (parallel)	$r(C···N_{center}) = 3.271$	3.392 [d]	0.121
H_3N···N_2O	$r(N···N_{center}) = 3.021$	3.088 [e]	0.067
H_2O···N_2O	$r(O···N_{center}) = 2.855$	2.97(2) [f]	0.11(2)
H_2S···N_2O	$r(S···N_{center}) = 3.444$	...	...
H_3P···N_2O	$r(P···N_{center}) = 3.479$	...	...

[a] See later for the molecular diagrams (to scale) of the B···N_2O complexes. [b] r_s value estimated from data in Reference [24] is almost certainly an overestimate, as b_N is very small, and therefore, severely underestimated. [c] References [26,27]; [d] Reference [25]; [e] Reference [29]; [f] Reference [28].

It is clear from Figure 3 that the intermolecular bond is a tetrel bond in the sense that it involves the electrophilic region around C (the blue band that surrounds the C atom in the MESP of CO_2 shown in Figure 1) and either a non-bonding electron pair or a π-bonding electron pair as the nucleophilic site of the Lewis base B. In fact, the axis of the non-bonding electron pair coincides with the extension of the radius of the circle that defines the most electrophilic band around C in each of OC···CO_2, HCN···CO_2, H_3N···CO_2, and H_2S···CO_2, given that the n-pairs on S in H_2S lie at ~±90° to the plane of the H_2S nuclei, as established from earlier work on H_2S···HX and H_2S···XY (X and Y are halogen atoms) [9,40]. The fact that the ab-initio-derived configuration at O in H_2O···CO_2 is planar is not inconsistent with this conclusion. It was found for all H_2O···HX and H_2O···XY [9,40] investigated through rotational spectroscopy and/or ab initio calculations that, although the equilibrium configuration at O is non-planar, the barrier to planarity is low and lies below the zero-point energy level in most cases. The configuration is, therefore, rapidly inverting in the zero-point state and the molecule is effectively planar. For an interaction as weak as that in H_2O···CO_2, the barrier will probably be non-existent, as it is in H_2O···F_2 [41], for example. Some rules put forward originally for hydrogen-bonded complexes B···HX [9] and halogen-bonded complexes B···XY [40] can be easily modified to allow the geometries of the tetrel-bonded complexes shown in Figure 3 to be predicted. Thus, the modified rules become:

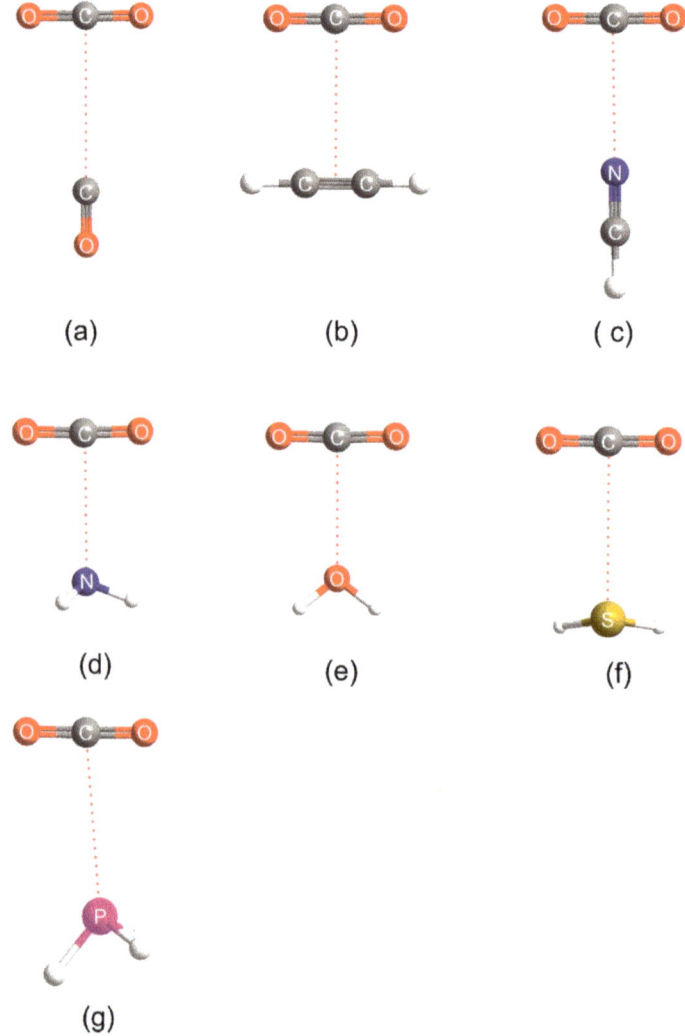

Figure 3. Molecular models drawn to scale of the geometries of B···CO_2 complexes calculated at the CCSD(T)/aug-cc-pVTZ level of theory, where B = CO, HCCH, HCN, NH_3, H_2O, H_2S, and PH_3 (**a**–**g**, respectively). Not shown is the linear, hydrogen-bonded isomer CO_2···HCN, which is 1.5 kJ·mol^{-1} higher in energy than the form in Ⓒ.

The equilibrium geometry of tetrel-bonded B···CO_2 complexes can be predicted by assuming that a radius of the most electrophilic ring around the C atom of CO_2 coincides with either (1) the axis of a non-bonding electron pair carried by B, or (2) the local symmetry axis of a π-bonding electron pair of B.

That is, in the original rules, "hydrogen-bonded complexes B···HX" is replaced by "tetrel-bonded complexes B···CO_2", and "the axis of HX" is replaced by "a radius of the most electrophilic ring around the C atom of CO_2".

The case of H_3P···CO_2 appears to be an exception to the rules, because the intermolecular bond does not lie exactly along the C_3 axis of phosphine. The reason for this becomes clear when the MESP of phosphine, shown in Figure 4, is examined. Approximately opposite the extension of each P–H bond is an electrophilic (blue) region which can interact with the nucleophilic (yellow-green) band around O of CO_2 (see Figure 1). This secondary interaction is, in fact, a pnictogen bond, and it is responsible for the distortion found in Figure 3g.

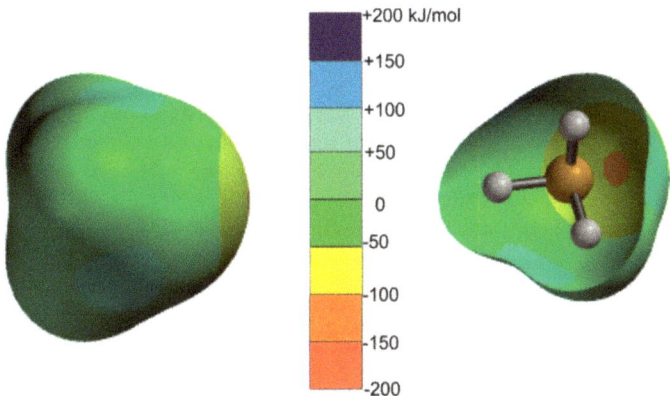

Figure 4. Molecular electrostatic surface potentials (MESPs) for phosphine calculated for the 0.002 e/bohr^3 iso-surface at the MP2/6-311++G** level. The surface in the right-hand diagram is cut away to reveal both the electrophilic (blue) regions near P on approximately the extension of the H–P bonds, and the nucleophilic (red dot) region on the C_3 axis.

The molecular geometries calculated ab initio for the corresponding B$\cdots$N$_2$O series are illustrated in Figure 5, and each has a similar, but not identical, shape to that of the corresponding member of the B$\cdots$CO$_2$ series, with the central N atom of N_2O acting as the primary electrophilic site. The lower symmetry of N_2O compared with that of CO_2 means, however, that the B$\cdots$N$_2$O complexes necessarily have lower symmetry and that secondary interactions become more important. The geometries shown in Figure 5 can be understood in terms of the rule set out in the preceding paragraph, that is, with the primary interaction involving the electrophilic (blue) band on the central N atom of N_2O with the n-pair or π-pair on the Lewis base B, but modified to allow a secondary interaction of the electrophilic region of B (i.e., C or H of HCN, H of HCCH, H of NH_3, H of H_2O, H of PH_3, or H of H_2S) with the nucleophilic region at O in N_2O (see Figure 1, end-on view). The conclusions for B$\cdots$CO$_2$ and B$\cdots$N$_2$O are, therefore, consistent with the previously noted similarity of the MESPs of CO_2 and N_2O displayed in Figure 1. The molecular shapes shown in Figure 5 correspond closely to those that are available experimentally (see Reference [3] for a convenient collection of experimentally determined shapes). The ab initio and experimental (where available) intermolecular distances for each B$\cdots$N$_2$O complex are included in Table 2.

Two geometries are given for HCN$\cdots$N$_2$O in Figure 5. Both correspond to minima in the energy, but are separated in energy by only 0.03 kJ·mol^{-1} at the CCSDT(T)/aug-cc-pVTZ level of theory and 0.45 kJ·mol^{-1} at the CCSD(T)/CBS level, with the parallel form (Figure 5c) lower in energy than the nearly perpendicular form (Figure 5b) in both cases. It is of interest to note that Miller and co-workers [25] found two isomers of this complex in their investigation of the high-resolution infrared spectrum of (N_2O, HCN) in a supersonically expanded gas mixture of the components diluted in helium. One was a parallel form (four such arrangements of N_2O and HCN were consistent with their observed rotational constants, including that found here by ab initio calculation), while the other was a hydrogen-bonded, linear isomer N=N=O$\cdots$HCN; however, these authors did not observe the T-shaped isomer shown in Figure 5b. Our calculations at the CCSD(T)/CBS level find the linear, hydrogen-bonded form N=N=O$\cdots$HCN to be higher in energy than the parallel isomer by 1.5 kJ·mol^{-1}. This observation suggests that, while the T-shaped isomer relaxes to the parallel form in the supersonic expansion, the higher-energy, hydrogen-bonded, linear isomer does not. Both linear, hydrogen-bonded [39,42] and T-shaped, tetrel-bonded [15,16] isomers of (CO$_2$, HCN) were observed experimentally. At the CCSD(T)/CBS level, O=C=O$\cdots$HCN is found to be 1.3 kJ·mol^{-1} higher in energy than the T-shaped isomer, in agreement with the experimental conclusions.

We emphasized in the introduction that the MESP of carbon disulfide is different from those of CO$_2$ and N$_2$O in that the most electrophilic (blue) site of CS$_2$ lies on the C$_\infty$ axis at the surface of each S atom (see Figure 1). As is clear from Figure 6, which displays the geometries of seven B$\cdots$CS$_2$ complexes calculated at the CCSD(T)/cc-aug-pVTZ level of theory, all complexes but H$_3$P$\cdots$CS$_2$ do indeed involve a chalcogen bond formed by the axial electrophilic region at one of the S atoms of CS$_2$ with an n- or π-electron pair of the Lewis base B. The calculated intermolecular distances are collected in Table 3. To the best of our knowledge, only H$_2$O$\cdots$CS$_2$ was investigated by means of its rotational spectrum [30]. The resulting value of r(O$\cdots$S) is included in Table 3. The angular geometries of the B$\cdots$CS$_2$ complexes displayed in Figure 6 can also be predicted by the rules set out elsewhere for hydrogen-bonded complexes B$\cdots$HX [9] or halogen-bonded complexes B$\cdots$XY [40], if they are modified by replacing, for example, "hydrogen-bonded complexes B$\cdots$HX" by "chalcogen-bonded complexes B$\cdots$CS$_2$" and the "HX axis" by "C$_\infty$ axis of CS$_2$" in the wording (see earlier). We note that there is a planar configuration at O found theoretically (see Figure 6) and experimentally [30] for H$_2$O$\cdots$CS$_2$, rather than the pyramidal configuration predicted by the rules. The explanation for this difference is identical to that given earlier for H$_2$O$\cdots$CO$_2$. On the other hand, the configuration at S in H$_2$S$\cdots$CS$_2$ is strongly pyramidal, with the intermolecular bond making an angle of approximately 90° with the plane of the H$_2$S nuclei, as found for almost all H$_2$S$\cdots$HX and H$_2$S$\cdots$XY complexes so far investigated [40]. However, there is a significant non-linearity of the S$\cdots$S=C nuclei. A possible reason for this non-linearity is that the intermolecular bond is very weak (D_e = 5.28 kJ·mol^{-1}, see Section 3.2) and the pair of equivalent electrophilic H atoms can undergo a secondary interaction with the weakly nucleophilic (yellow-green) region of CS$_2$ (see the MESP of CS$_2$ in Figure 1). The geometry of H$_3$P$\cdots$CS$_2$ involves a pnictogen bond and can be understood by reference to the MESP of phosphine in Figure 4. It seems that the primary interaction here involves one of the electrophilic (blue) regions near to P and approximately on the extension of each P–H bond (as seen in the cutaway version of the phosphine MESP in Figure 4) with the nucleophilic (yellow-green) region of CS$_2$. Evidently, this interaction is stronger than that of the terminal electrophilic (blue) region at S with the n-electron pair of phosphine (the red spot in the cutaway version of the MESP in Figure 4), leading to a primary P pnictogen bond.

Table 3. Calculated and observed intermolecular distances in B$\cdots$CS$_2$ complexes.

Complex	Intermolecular Distance/Å		(Obs. − Calc.)/Å
	Calculated Ab Initio [a]	Observed	
OC$\cdots$CS$_2$	r(C$\cdots$S) = 3.616	$\cdots$	$\cdots$
HCCH$\cdots$CS$_2$	r(π$_{center}$$\cdots$S) = 3.568	$\cdots$	$\cdots$
HCN$\cdots$CS$_2$	r(N$\cdots$S) = 3.285	$\cdots$	$\cdots$
H$_3$N$\cdots$CS$_2$	r(N$\cdots$S) = 3.304	$\cdots$	$\cdots$
H$_2$O$\cdots$CS$_2$	r(O$\cdots$S) = 3.132	3.197 [b]	0.065
H$_2$S$\cdots$CS$_2$	r(S$\cdots$S) = 3.773	$\cdots$	$\cdots$
H$_3$P$\cdots$CS$_2$	r(P$\cdots$S) = 3.798	$\cdots$	$\cdots$

[a] See Figure 6 for the molecular diagrams (to scale) of the B$\cdots$CS$_2$ complexes. [b] Reference [30].

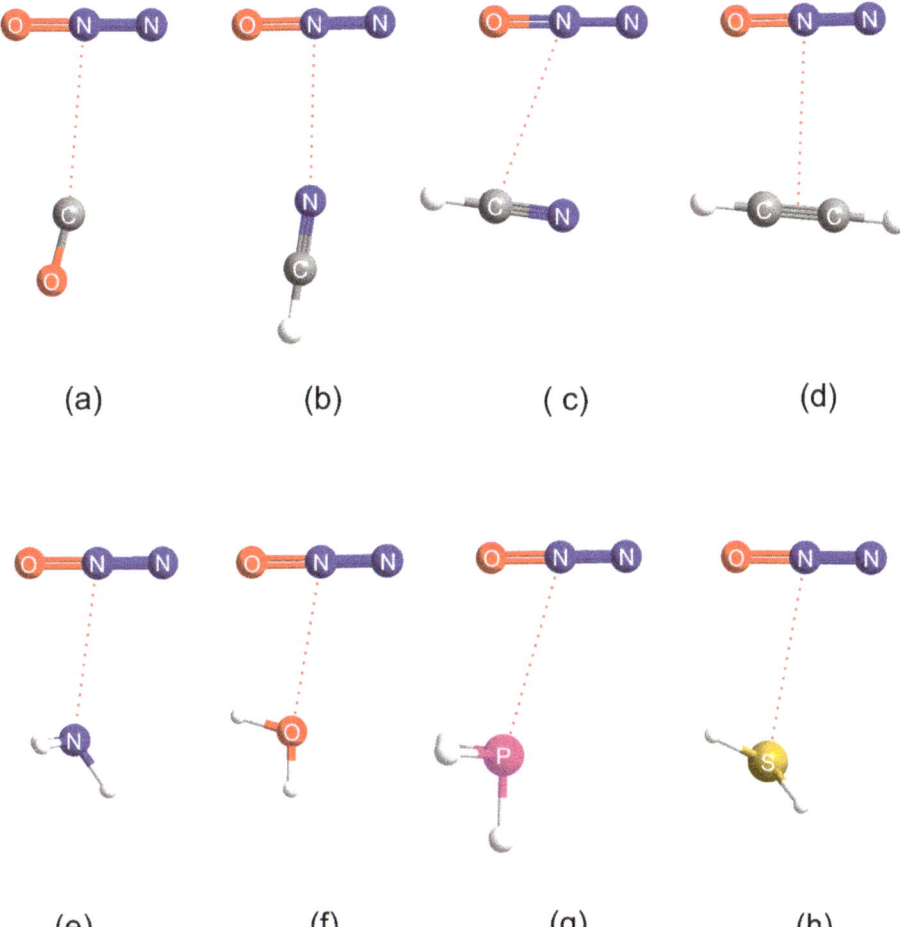

Figure 5. Molecular models drawn to scale of the geometries of B··· N$_2$O complexes calculated at the CCSD(T)/aug-cc-pVTZ level of theory, where B = CO, HCN, HCCH, NH$_3$, H$_2$O, PH$_3$, and H$_2$S (**a–h**, respectively; note that there are two models shown for HCN complexes). When B = HCN there are three low-energy conformers: the slipped parallel form at the global minimum, the T-shaped isomer higher in energy by only 0.03 kJ·mol^{-1}, and a linear, hydrogen-bonded conformer N$_2$O··· HCN (not shown) higher in energy by 1.3 kJ·mol^{-1} (see text for discussion).

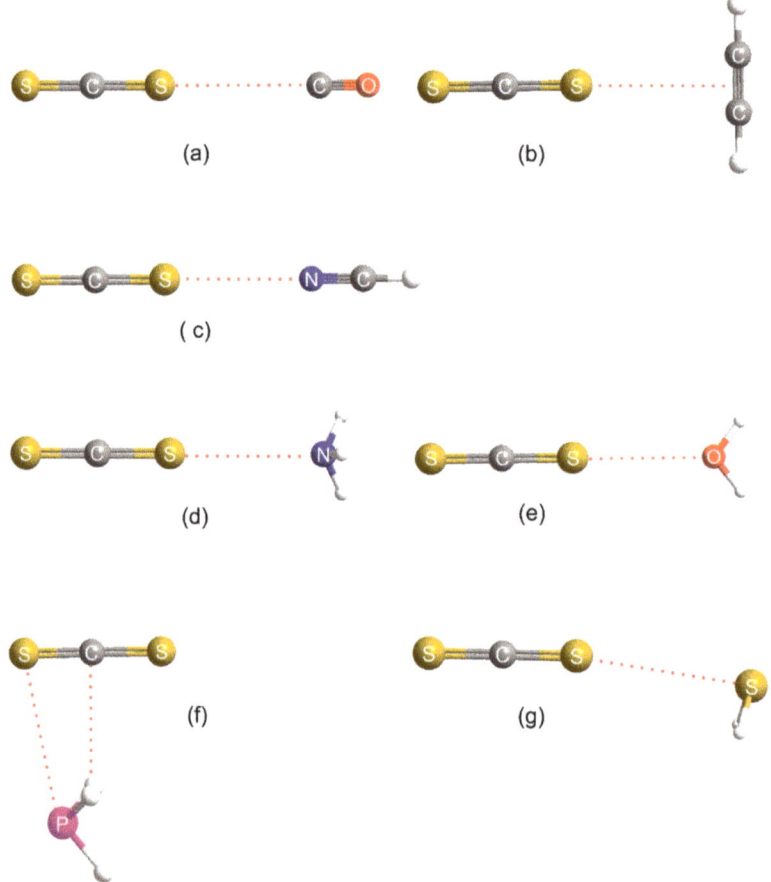

Figure 6. Molecular models drawn to scale of the geometries of B···CS$_2$ complexes calculated at the CCSD(T)/aug-cc-pVTZ level of theory, where B = CO, HCCH, HCN, NH$_3$, H$_2$O, PH$_3$ and H$_2$S (**a**–**g**, respectively).

3.2. The Relationship between D_e and k_σ in the B···CO$_2$, B···N$_2$O, and B···CS$_2$ Series

It was established [31] for a wide range of hydrogen-bonded complexes B···HX (X = F, Cl, Br, or I) and halogen-bonded complexes B···XY (X and Y are halogen atoms) that their dissociation energies D_e (as calculated ab initio at the CCSD(T)(F12c)/cc-pvdz-F12 level of theory) are directly proportional to their intermolecular stretching force constants k_σ (as determined experimentally from centrifugal distortion constants D_J or Δ_J obtained by measuring rotational spectra). The constant of proportionality was found to be 1.5(1) × 10^3 m^2·mol^{-1}. Later, it was shown for the B···HF, B···HCl, B···F$_2$, B···Cl$_2$, and B···ClF series, where B is a Lewis base, N$_2$, CO, HCCH, C$_2$H$_4$, HCN, H$_2$S, H$_2$O, PH$_3$, or NH$_3$, that the same constant of proportionality applies [32] when k_σ was calculated ab initio at the CCSD(T)/aug-cc-pVTZ level of theory and D_e was obtained via a CCSD(T)/CBS calculation, where CBS indicates a complete basis-set extrapolation using the aug-cc-pVnZ (n = T and Q) basis sets. The opportunity is taken here to investigate the corresponding relationship for the tetrel-bonded B···CO$_2$ complexes, the pnictogen-bonded B···N$_2$O complexes, and the chalcogen-bonded B···CS$_2$ complexes for the series of Lewis bases, B = CO, HCCH, H$_2$S, HCN, H$_2$O, PH$_3$, and NH$_3$, when both k_σ and D_e are calculated in the same way as described in Reference [32].

Values of D_e and k_σ so determined are recorded in Table 4, while Figure 7 shows a plot of D_e as the ordinate and k_σ as the abscissa for the B$\cdots$CO$_2$, B$\cdots$N$_2$O, and B$\cdots$CS$_2$ series investigated here, with color coding of the points as red, blue, and yellow, respectively. For consistency with HCN$\cdots$CO$_2$, of the isomers of HCN$\cdots$N$_2$O, only the data for the T-shaped form are included in Table 4 and Figure 7. The calculation of k_σ for the parallel isomer of N$_2$O$\cdots$HCCH was prevented by convergence problems, as well as for H$_2$S$\cdots$N$_2$O because, as the N$\cdots$S distance was varied, there was a switch to the hydrogen-bonded arrangement N$_2$O$\cdots$HSH. H$_3$P$\cdots$CS$_2$ was excluded because it does not involve a chalcogen bond, unlike the remaining B$\cdots$CS$_2$ complexes. The results of a linear regression fit of the points in Figure 7 are as follows: gradient = $1.44(20) \times 10^3$ m$^2 \cdot$mol^{-1} and intercept on the ordinate = $-0.32(124)$ kJ$\cdot$mol^{-1}. Thus, within the errors of the fit, D_e and k_σ are directly proportional, and the slope of the regression line agrees with those found previously for the B$\cdots$HF and B$\cdots$HCl series, and for the halogen-bonded series B$\cdots$F$_2$, B$\cdots$Cl$_2$, and B$\cdots$ClF [32] when calculations were conducted at identical levels of theory, namely $1.38(7) \times 10^3$ m$^2 \cdot$mol^{-1} and $1.49(5) \times 10^3$ m$^2 \cdot$mol^{-1}, respectively. Plots of D_e versus k_σ using D_e values calculated at the CCSD(T)(F12c)/cc-pVDZ-F12 level of theory and experimentally available k_σ [31], but with many more complexes in each of these two classes, gave almost identical slopes of $1.52(3) \times 10^3$ m$^2 \cdot$mol^{-1} and $1.47(3) \times 10^3$ m$^2 \cdot$mol^{-1}, respectively. Evidently, the same relationship between D_e and k_σ holds for hydrogen-bonded complexes B$\cdots$HX, halogen-bonded complexes B$\cdots$XY, the tetrel-bonded complexes B$\cdots$CO$_2$, the pnictogen-bonded complexes B$\cdots$N$_2$O, and the chalcogen-bonded complexes B$\cdots$CS$_2$. This fact is visually established by the plot of D_e versus k_σ shown in Figure 8. The figure includes all B$\cdots$HF, B$\cdots$HCl, B$\cdots$F$_2$, B$\cdots$Cl$_2$, and B$\cdots$ClF complexes reported in Reference [32] and all the B$\cdots$CO$_2$, B$\cdots$N$_2$O, and B$\cdots$CS$_2$ complexes included in Figure 7. Both sets of series were calculated in the same way, i.e., CCSD(T)/aug-cc-pVTZ for k_σ and CCSD(T)/CBS for D_e. The linear regression fit for all these data leads to $1.40(4) \times 10^3$ m$^2 \cdot$mol^{-1} for the slope and $-0.42(46)$ kJ$\cdot$mol^{-1} for the intercept.

Table 4. Intermolecular dissociation energies D_e and quadratic force constants k_σ for B$\cdots$CO$_2$, B$\cdots$N$_2$O, and B$\cdots$CS$_2$ complexes.

Lewis Base B	B$\cdots$CO$_2$		B$\cdots$N$_2$O		B$\cdots$CS$_2$	
	D_e/kJ$\cdot$mol^{-1}	k_σ/(N$\cdot$m^{-1})	D_e/kJ$\cdot$mol^{-1}	k_σ/(N$\cdot$m^{-1})	D_e/kJ$\cdot$mol^{-1}	k_σ/(N$\cdot$m^{-1})
OC	4.89	4.53	4.61	5.13	2.99	3.21
HCCH	8.81	7.08	8.14	...[a]	4.27	3.58
HCN	9.18	7.15	7.84	7.72	6.16	5.13
H$_2$O	13.77	10.09	12.47	7.82	8.01	4.15
H$_2$S	7.82	4.84	7.25	...[b]	5.28	3.46
H$_3$N	14.53	8.23	11.79	8.08	9.36	5.31
H$_3$P	6.26	4.85	5.92	4.85	6.75	...[c]

[a] Convergence problems when attempting to calculate the $E(r - r_e)$ versus $(r - r_e)$ curve to obtain k_σ. [b] When attempting to calculate k_σ, the geometry of the complex changes to the hydrogen-bonded isomer N$_2$O$\cdots$HSH as $(r - r_e)$ increases. [c] The main non-covalent interaction in this complex is between P of PH$_3$ and S of CS$_2$, and it is a pnictogen bond, not a chalcogen bond.

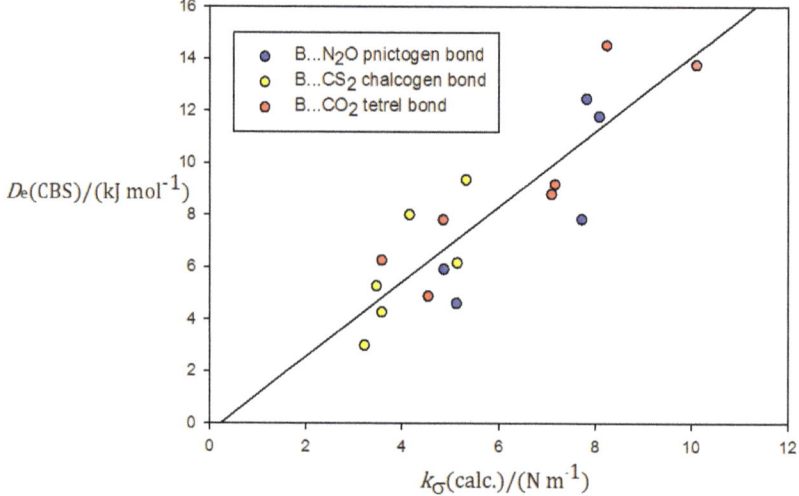

Figure 7. Plot of D_e calculated at the CCSD(T)/CBS level of theory (CBS indicates a complete basis-set extrapolation using the aug-cc-pVnZ (n = T and Q) basis sets) versus k_σ calculated at the CCSD(T)/aug-cc-pVTZ level for B$\cdots$CO$_2$, B$\cdots$N$_2$O, and B$\cdots$CS$_2$ complexes. See text for discussion.

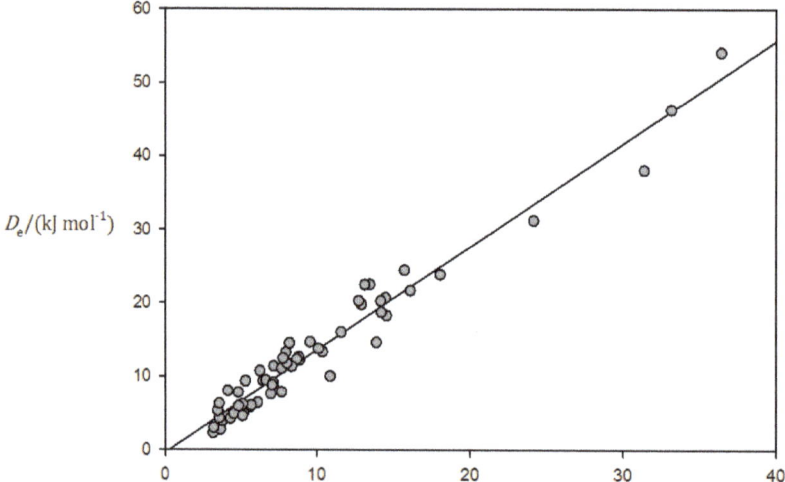

Figure 8. Plot of D_e calculated at the CCSD(T)/CBS level versus k_σ calculated at the CCSD(T)/aug-cc-pVTZ level for B$\cdots$CO$_2$, B$\cdots$N$_2$O, and B$\cdots$CS$_2$ complexes (this work; see also Figure 7), and B$\cdots$HF, B$\cdots$HCl, B$\cdots$F$_2$, B$\cdots$Cl$_2$, and B$\cdots$ClF complexes (see Reference [32] for the Lewis bases B involved and the values of D_e and k_σ for the B$\cdots$HX and B$\cdots$XY series).

4. Conclusions

The series of B$\cdots$CO$_2$, B$\cdots$N$_2$O, and B$\cdots$CS$_2$ complexes was investigated through ab initio calculations at the CCSD(T)/aug-pVTZ level of theory for the Lewis bases, B = CO, HCCH, H$_2$S, HCN, H$_2$O, PH$_3$, and NH$_3$. The atoms, except for some H, lie in a plane for all complexes. The intermolecular bonds in the B$\cdots$CO$_2$ complexes are formed by interaction of the electrophilic region around the C atom of CO$_2$ (see Figure 1) with n- or π-electron pairs (nucleophilic regions) carried by B and are, therefore, tetrel bonds. The geometry of each B···N$_2$O complex investigated (except perhaps for B = PH$_3$) is similar to that of the corresponding member of the B···CO$_2$ series. Thus, the primary non-covalent interaction involves the central N atom of N$_2$O with an n- or π-electron pair carried by B, but moderated by distortions that appear to arise from the secondary interaction of the electrophilic region of B (e.g., H atoms) with the O atom of N$_2$O. The B···CS$_2$ series is geometrically distinct from the other two in that (apart from B = PH$_3$) the primary non-covalent interaction is between the electrophilic region centered on the C$_\infty$ axis of CS$_2$ near to an S atom (see Figure 1) and an n- or π-electron pair of B, leading to a linear (or nearly linear in the case of B = H$_2$S) C=S···B system, and is, therefore, a chalcogen bond. These interpretations are electrostatic in origin and were applied previously to hydrogen bonds in B···HX complexes [9] and halogen bonds in B···XY complexes [40]. Consistent with the foregoing observations is the fact that the geometries of members of each of the three series, B···CO$_2$, B···N$_2$O, and B···CS$_2$, can be predicted by rules put forward some years ago for the same purpose for hydrogen-bonded complexes B···HX and halogen-bonded complexes B···XY. Moreover, this close relationship between hydrogen, halogen, tetrel, pnictogen, and chalcogen bonds is reflected in the recent generalized definition [43] proposed for non-covalent (E) bonds based on electrostatics, provided below.

An E bond occurs when there is evidence of a net attractive interaction between an electrophilic region associated with an E atom in a molecular entity and a nucleophilic region (e.g., an n-pair or π-pair of electrons) in another, or the same, molecular entity, where E is the general name for an element of Group 1, 11, 14, 15, 16, or 17 in the Periodic Table.

We note that some complexes investigated here can be described as of the σ-hole type, while others belong to the π-hole type.

Finally, we showed that the similarity between all of these types of non-covalent interaction extends to the direct proportionality of the dissociation energy D_e and the quadratic intermolecular stretching force constant k_σ, with a constant of proportionality $1.45(7) \times 10^3$ m^2·mol^{-1} describing all the series, B···HF, B···HCl, B···F$_2$, B···Cl$_2$, B···ClF, B···CO$_2$, B···N$_2$O, and B···CS$_2$, when the two measures of binding strength are calculated at the CCSD(T)/CBS and CCSD(T)/aug-cc-pVTZ levels of theory, respectively. As discussed in Reference [31], a Morse function is an example of a potential energy curve for which the dissociation energy and the force constant are directly proportional.

Supplementary Materials: The supplementary materials are available.

Author Contributions: I.A. and A.C.L. are both contributed to the design of experiments, formal analysis and the writing of draft.

Funding: This work was carried out with financial support from the Ministerio de Economía y Competitividad (Project No. CTQ2015-63997-C2-2-P) and the Comunidad Autónoma de Madrid (S2013/MIT2841, Fotocarbon).

Acknowledgments: A.C.L. thanks the School of Chemistry, University of Bristol for a Senior Research Fellowship.

Conflicts of Interest: The authors declare no conflict of interest.

References

1. Arunan, E.; Desiraju, G.R.; Klein, R.A.; Sadlej, J.; Scheiner, S.; Alkorta, I.; Clary, D.C.; Crabtree, R.H.; Dannenberg, J.J.; Hobza, P.; et al. Definition of the hydrogen bond (IUPAC Recommendations 2011). *Pure Appl. Chem.* **2011**, *83*, 1637–1641. [CrossRef]
2. Desiraju, G.R.; Ho, P.S.; Kloo, L.; Legon, A.C.; Marquardt, R.; Metrangolo, P.; Politzer, P.A.; Resnati, G.; Rissanen, K. Definition of the halogen bond (IUPAC Recommendations 2013). *Pure Appl. Chem.* **2013**, *85*, 1711–1713. [CrossRef]
3. Legon, A.C. Tetrel, pnictogen and chalcogen bonds identified in the gas phase before they had names: A systematic look at non-covalent interactions. *Phys. Chem. Chem. Phys.* **2017**, *19*, 14884–14896. [CrossRef] [PubMed]
4. Alcock, N.W. Secondary bonding to non-metallic elements. *Adv. Inorg. Chem. Radiochem.* **1972**, *15*, 1–58.

5. Bauzá, A.; Mooibroek, T.J.; Frontera, A. Tetrel-bonding interaction: Rediscovered supramolecular force? *Angew. Chem. Int. Ed.* **2013**, *52*, 12317–12321. [CrossRef] [PubMed]
6. Zahn, S.; Frank, R.; Hey-Hawkins, E.; Kirchner, B. Pnicogen bonds: A new molecular linker? *Chem. Eur. J.* **2011**, *17*, 6034–6038. [CrossRef] [PubMed]
7. Wang, W.; Ji, B.; Zhang, Y. Chalcogen bond: A sister noncovalent bond to halogen bond. *J. Phys. Chem. A* **2009**, *113*, 8132–8135. [CrossRef] [PubMed]
8. Cavallo, G.; Metrangolo, P.; Pilati, T.; Resnati, G.; Terraneo, G. Naming interactions from the electrophilic site. *Cryst. Growth Des.* **2014**, *14*, 2697–2702. [CrossRef]
9. Legon, A.C.; Millen, D.J. Angular geometries and other properties of hydrogen-bonded dimers: A simple electrostatic interpretation based on the success of the electron-pair model. *Chem. Soc. Rev.* **1987**, *16*, 467–498. [CrossRef]
10. Murray, J.S.; Lane, P.; Clark, T.; Politzer, P. Sigma-hole bonding: Molecules containing group VI atoms. *J. Mol. Model.* **2007**, *13*, 1033–1038. [CrossRef] [PubMed]
11. Graham, C.; Imrie, D.A.; Raab, R.E. Measurement of the electric quadrupole moments of CO_2, CO, N_2, Cl_2 and BF_3. *Mol. Phys.* **1998**, *93*, 49–56. [CrossRef]
12. Watson, J.N.; Craven, I.E.; Ritchie, G.L.D. Temperature dependence of electric field-gradient induced birefringence in carbon dioxide and carbon disulphide. *Chem. Phys. Lett.* **1997**, *274*, 1–6. [CrossRef]
13. Chetty, N.; Couling, V.W. Measurement of the electric quadrupole moment of N_2O. *J. Chem. Phys.* **2011**, *134*, 144307. [CrossRef] [PubMed]
14. Legon, A.C.; Suckley, A.P. Infrared diode-laser spectroscopy and Fourier-transform microwave spectroscopy of the (CO_2, CO) dimer in a pulsed jet. *J. Chem. Phys.* **1989**, *91*, 4440–4447. [CrossRef]
15. Leopold, K.R.; Fraser, G.T.; Klemperer, W. Rotational spectrum and structure of $HCN-CO_2$. *J. Chem. Phys.* **1984**, *80*, 1039–1046. [CrossRef]
16. Legon, A.C.; Suckley, A.P. Pulsed-jet, diode-laser IR spectroscopy of the $v = 1\rightarrow0$ transition in the CO_2 asymmetric stretching mode of (CO_2, HCN). *Chem. Phys. Lett.* **1989**, *157*, 5–10. [CrossRef]
17. Pritchard, D.G.; Nandi, R.N.; Muenter, J.S.; Howard, B.J. Vibration-rotation spectrum of the carbon dioxide-acetylene van der Waals complex in the 3μ region. *J. Chem. Phys.* **1988**, *89*, 1245–1250. [CrossRef]
18. Ricel, J.K.; Coudert, H.; Matsumura, K.; Suenram, R.D.; Stahl, W.; Pauley, D.J.; Kukolich, S.G. The rotational and tunnelling spectrum of the H_2S-CO_2 van der Waals complex. *J. Chem. Phys.* **1990**, *92*, 6408–6419. [CrossRef]
19. Peterson, K.I.; Klemperer, W. Structure and internal rotation of H_2O-CO_2, $HDO-CO_2$, and D_2O-CO_2 van der Waals complexes. *J. Chem. Phys.* **1984**, *80*, 2439–2445. [CrossRef]
20. Fraser, G.T.; Leopold, K.R.; Klemperer, W. The rotational spectrum, internal rotation, and structure of NH_3-CO_2. *J. Chem. Phys.* **1984**, *81*, 2577–2584. [CrossRef]
21. Fraser, G.T.; Nelson, D.D.; Charo, A.; Klemperer, W. Microwave and infrared characterization of several weakly bound NH_3 complexes. *J. Chem. Phys.* **1985**, *82*, 2535–2546. [CrossRef]
22. Qian, H.-B.; Howard, B.J. High Resolution infrared spectroscopy and structure of $CO-N_2O$. *J. Mol. Spectrosc.* **1997**, *184*, 156–161. [CrossRef]
23. Xu, Y.; McKellar, A.R.W. The C–O Stretching band of the $CO-N_2O$ van der waals complex. *J. Mol. Spectrosc.* **1996**, *180*, 164–169. [CrossRef]
24. Nagari, M.S.; Xu, Y.; Jäger, W. Rotational spectroscopic investigation of the weak interaction between CO and N_2O. *J. Mol. Spectrosc.* **1999**, *197*, 244–253. [CrossRef] [PubMed]
25. Dayton, D.C.; Pedersen, L.G.; Miller, R.E. Structural determinations for two isomeric forms of N_2O-HCN. *J. Phys. Chem.* **1991**, *96*, 1087–1095. [CrossRef]
26. Hu, T.A.; Ling, H.S.; Muenter, J.S. Vibration-rotation spectrum of the acetylene-nitrous oxide van der Waals complex in the 3 micron region. *J. Chem. Phys.* **1991**, *95*, 1537–1542. [CrossRef]
27. Peebles, R.A.; Peebles, S.A.; Kuczkowski, R.L. Isotopic studies, structure and modeling of the nitrous oxide-acetylene complex. *J. Phys. Chem. A* **1999**, *103*, 10813–10818. [CrossRef]
28. Zolandz, D.; Yaron, D.; Peterson, K.I.; Klemperer, W. Water in weak interactions: The structure of the water–nitrous oxide complex. *J. Chem. Phys.* **1992**, *97*, 2861–2868. [CrossRef]
29. Fraser, G.T.; Nelson, D.D.; Gerfen, G.J.; Klemperer, W. The rotational spectrum, barrier to internal rotation, and structure of NH_3-N_2O. *J. Chem. Phys.* **1985**, *83*, 5442–5449. [CrossRef]

30. Ogata, T.; Lovas, F.J. Microwave fourier transform spectrum of the water-carbon disulfide complex. *J. Mol. Spectrosc.* **1993**, *162*, 505–512. [CrossRef]
31. Legon, A.C. A reduced radial potential energy function for the halogen bond and the hydrogen bond in complexes B···XY and B···HX, where X and Y are halogen atoms. *Phys. Chem. Chem. Phys.* **2014**, *16*, 12415–12421. [CrossRef] [PubMed]
32. Alkorta, I.; Legon, A.C. Strengths of non-covalent interactions in hydrogen-bonded complexes B···HX and halogen-bonded complexes B···XY (X, Y = F, Cl): An ab initio investigation. *New J. Chem.* **2018**, *42*, 10548–10554. [CrossRef]
33. Purvis, G.D., III; Bartlett, R.J. A full coupled-cluster singles and doubles model—The inclusion of disconnected triples. *J. Chem. Phys.* **1982**, *76*, 1910–1918. [CrossRef]
34. Dunning, T.H., Jr. Gaussian basis sets for use in correlated molecular calculations. I. The atoms boron through neon and hydrogen. *J. Chem. Phys.* **1989**, *90*, 1007–1023. [CrossRef]
35. Feller, D. The use of systematic sequences of wave functions for estimating the complete basis set, full configuration interaction limit in water. *J. Chem. Phys.* **1993**, *98*, 7059–7071. [CrossRef]
36. Halkier, A.; Helgaker, T.; Jorgensen, P.; Klopper, W.; Olsen, J. Basis-set convergence of the energy in molecular Hartree–Fock calculations. *Chem. Phys. Lett.* **1999**, *302*, 437–446. [CrossRef]
37. Werner, H.-J.; Knowles, P.J.; Knizia, G.; Manby, F.R.; Schütz, M.; Celani, P.; Korona, T.; Lindh, R.; Mitrushenkov, A.; Rauhut, G.; et al. *MOLPRO*, version 2012.1. Available online: http://www.molpro.net (accessed on 3 September 2018).
38. Deppmeier, B.J.; Driessen, A.J.; Hehre, T.S.; Hehre, W.J.; Johnson, J.A.; Klunzinger, P.E.; Leonard, J.M.; Pham, I.N.; Pietro, W.J.; Yu, J.; et al. *SPARTAN'14 Mechanics Program, Release 1.1.8*; Wavefunction Inc.; SPARTAN Inc.: Irvine, CA, USA, 2014.
39. Klots, T.D.; Ruoff, R.S.; Gutowsky, H.S. Rotational spectrum and structure of the linear CO_2–HCN dimer: Dependence of isomer formation on carrier gas. *J. Chem. Phys.* **1989**, *90*, 4216–4221. [CrossRef]
40. Legon, A.C. Pre-reactive complexes of dihalogens XY with Lewis bases B in the gas phase: A systematic case for the 'halogen' analogue B···XY of the hydrogen bond B···HX. *Angew. Chem. Int. Ed. Engl.* **1999**, *38*, 2686–2714. [CrossRef]
41. Cooke, S.A.; Cotti, G.; Evans, C.M.; Holloway, J.H.; Kisiel, Z.; Legon, A.C.; Thumwood, J.M.A. Pre-reactive complexes in mixtures of water vapour with halogens: Characterisation of H_2O···ClF and H_2O···F_2 by a combination of rotational spectroscopy and ab initio calculations. *Chem. Eur. J.* **2001**, *7*, 2295–2305. [CrossRef]
42. Dayton, D.C.; Pedersen, L.G.; Miller, R.E. Infrared spectroscopy and ab initio-theory of the structural isomers of CO_2–HCN. *J. Chem. Phys.* **1990**, *93*, 4560–4570. [CrossRef]
43. Legon, A.C.; Walker, N.R. What's in a name? 'Coinage-metal' non-covalent bonds and their definition. *Phys. Chem. Chem. Phys.* **2018**, *20*, 19332–19338. [CrossRef] [PubMed]

Sample Availability: No samples are available from the authors.

© 2018 by the authors. Licensee MDPI, Basel, Switzerland. This article is an open access article distributed under the terms and conditions of the Creative Commons Attribution (CC BY) license (http://creativecommons.org/licenses/by/4.0/).

Article

Comparative Strengths of Tetrel, Pnicogen, Chalcogen, and Halogen Bonds and Contributing Factors

Wenbo Dong [1], Qingzhong Li [1,*] and Steve Scheiner [2,*]

1 The Laboratory of Theoretical and Computational Chemistry, School of Chemistry and Chemical Engineering, Yantai University, Yantai 264005, China; dongwenbo1994@163.com
2 Department of Chemistry and Biochemistry, Utah State University, Logan, UT 84322-0300, USA
* Correspondence: lqz@ytu.edu.cn (Q.L.); steve.scheiner@usu.edu (S.S.);
Tel.: +86-535-690-2063 (Q.L.); +1-435-797-7419 (S.S.)

Received: 29 June 2018; Accepted: 9 July 2018; Published: 10 July 2018

Abstract: Ab initio calculations are employed to assess the relative strengths of various noncovalent bonds. Tetrel, pnicogen, chalcogen, and halogen atoms are represented by third-row atoms Ge, As, Se, and Br, respectively. Each atom was placed in a series of molecular bonding situations, beginning with all H atoms, then progressing to methyl substitutions, and F substituents placed in various locations around the central atom. Each Lewis acid was allowed to engage in a complex with NH_3 as a common nucleophile, and the strength and other aspects of the dimer were assessed. In the context of fully hydrogenated acids, the strengths of the various bonds varied in the pattern of chalcogen > halogen > pnicogen ≈ tetrel. Methyl substitution weakened all bonds, but not in a uniform manner, resulting in a greatly weakened halogen bond. Fluorosubstitution strengthened the interactions, increasing its effect as the number of F atoms rises. The effect was strongest when the F atom lay directly opposite the base, resulting in a halogen > chalcogen > pnicogen > tetrel order of bond strength. Replacing third-row atoms by their second-row counterparts weakened the bonds, but not uniformly. Tetrel bonds were weakest for the fully hydrogenated acids and surpassed pnicogen bonds when F had been added to the acid.

Keywords: halogen bond; chalcogen bond; pnicogen bond; tetrel bond

1. Introduction

A revolution of sorts, albeit a gradual one, occurred in the field of noncovalent interactions as it became progressively more apparent that the venerable H-bond was not completely unique. That is, the bridging proton in H-bonds could be replaced by a variety of other atoms, with little if any loss in noncovalent bond energy. The first class of atoms that fit this criterion was the halogens [1–10]. The ability of these very electronegative atoms to replace an H atom was deemed counterintuitive at first, as the polarity of the R-H bond, placing a partial positive charge on the proton, was considered a prime ingredient of the classic H-bond. This issue was resolved when it was found that the charge distribution around the halogen atom in an analogous R-X (X = halogen) bond is highly asymmetric. While there is an equatorial band of negative electrostatic potential surrounding the X atom in a R-X bond, there is also a positive polar region situated directly opposite the R atom. This positive area, frequently referred to as a σ-hole, can attract an approaching nucleophile in precisely the same way the partially positive charge surrounding the proton of a H-bond can [11,12]. Of course, the interaction, whether an H or halogen bond, is not entirely electrostatic as it contains other elements, such as charge transfer and dispersion, but this charge distribution voided the argument that a halogen atom must necessarily repel an incoming nucleophile.

As further work proceeded, it soon became apparent that this same phenomenon can be extended to more than just the halogen family of elements. Chalcogen atoms of the S/Se group could engage [13–21] in very similar bonding to a nucleophile. As these atoms are commonly involved in covalent bonding with two substituents, one would expect a pair of such σ-holes, one opposite each of these two substituents. Indeed, chalcogen bonding reflected this pattern, as there were two such sites that could engage in these interactions. Along similar lines of thinking, there ought to be three σ-holes surrounding a trivalent pnicogen atom, each of which is in principle capable of forming a so-called pnicogen bond, which has in fact been observed [22–27]. Tetrel atoms of the Si/Ge family, too, can participate [28–36] in analogously named bonds, and their most common tetravalent covalent bonding situation provides four separate sites for potential tetrel bonds.

These various noncovalent bond analogues of the H-bond share a number of characteristics [37–54]. For example, regardless of the particular family of atoms, as one moves down a column of the periodic table, the atom becomes progressively less electronegative and more polarizable. Both of these factors tend to amplify the σ-hole, and reinforce charge transfer from the nucleophile, and to thus strengthen the corresponding noncovalent bond. These same factors can also be intensified if electron-withdrawing substituents are added to the atom in question. Scores of previous results have accordingly demonstrated the ability of such substituents to strengthen the noncovalent bond to the approaching nucleophile. This effect is most pronounced when the substituent lies directly opposite the nucleophile, where it can better accommodate additional charge accumulation from the electron donor. Due to a strong electrostatic component in all of these bonds, they are dramatically strengthened if the Lewis acid is positively charged, with a like reinforcement upon interaction with an anion.

One issue that has borne only very moderate study is the comparison of these different noncovalent bonds with an eye toward those factors that differentiate one from another. While a substantial amount of work has compared each separate sort of bond with the prototype H-bond [55–70], much less is known about the strength of one with respect to another. On the face of it, one might think that the least electronegative atom is prone to form the most intense σ-holes, and that is in fact what is seen within any given family of atoms, e.g., the halogens or the chalcogens. This premise would lead to the general conclusion that the noncovalent bonding strength ought to increase in the order halogen < chalcogen < pnicogen < tetrel. However, a scan of the literature would suggest this is not the case. As another issue, in order to approach the pertinent atom, the nucleophile must avoid steric and repulsive electrostatic interactions with any of the substituents or lone electron pairs surrounding this central atom. As one proceeds along the halogen-to-tetrel series, the number of lone pairs diminishes while there is an increase in the number of substituents. This trend will also factor into the relative strengths of the different sorts of bonds.

At this juncture, then, there is a need for a thorough and fair comparison of the different sorts of noncovalent bonds. Can one make the general claim that one type will be stronger than another, given like covalent bonding situations? If this is found to be the case, then can one explain this trend on simple chemical grounds? In order to establish any such pattern, it is essential that the calculations be performed at a level that can be considered fully reliable, particularly if the differences in binding strength are small. The present work attempts to answer these questions using high-level ab initio calculations.

2. Systems and Theoretical Methods

In order to establish a valid and consistent baseline, focus was first placed on the third row of the periodic table where Br was compared with Se, As, and Ge, as representative of halogen, chalcogen, pnicogen, and tetrel atoms, respectively. Each was placed in a variety of molecular environments, starting with all H substituents as a base point. In the next phase, one H atom was replaced by a methyl group, as a small representative alkyl chain to which many of these atoms would normally be attached. As electron-withdrawing groups are known to amplify each of these sorts of noncovalent bonds, F atoms were added in various modes. This atom could be added in a position directly opposite the nucleophile where it was thought to have the largest amplifying effect. As an alternative, to take advantage of its electron-withdrawing capability, but without its distortion of the σ* antibonding orbital that acts as a sink to electron density transferred from the nucleophile, the F atom could be placed in a peripheral position, viz. bonded to the atom in question, but not directly opposite the nucleophile. In order to assure that the results were not limited to only third-row atoms, additional calculations were carried out with their second-row counterparts. NH_3 was taken as the universal nucleophilic partner, as its small size and presence of a single lone pair avoided the complications that might result from interactions other than the ones of interest.

The complexes and monomers were fully optimized using second-order Møller–Plesset perturbation theory (MP2) with the aug-cc-pVTZ basis set [71,72]. Harmonic vibrational frequencies were then computed at the same level in order to verify that the structures obtained correspond to minima with no imaginary frequencies and to obtain vibrational frequencies. Optimization and frequency calculations were carried out using the Gaussian 09 program [73]. Optimized coordinates of monomers and complexes are supplied in the Supplementary Materials section.

Interaction energy (E_{int}) and binding energy (E_b) were used as a measure of the strength of the interactions; they were calculated as the difference between the energy of the complex relative to the monomers in the complex geometry, and the optimized monomers, respectively. E_{int} was also computed with the CCSD(T) (Coupled Cluster with Single and Double Excitations, with Iterative Triples) method for the optimized MP2 structure. Both terms were corrected for basis set superposition error (BSSE) using the counterpoise procedure [74,75] outlined by Boys and Bernardi.

Molecular electrostatic potentials (MEPs) were computed on the 0.001 au electron density contour at the MP2/aug-cc-pVTZ level with the Wave Function Analysis-Surface Analysis-Suite (WFA-SAS) program [76]. The value of the MEP maximum of the σ-hole of the Lewis acid monomer facing the base was evaluated. The Natural Bond Orbital (NBO) method [77] was utilized to extract atomic charges and intermolecular orbital interactions between occupied and empty orbitals using the NBO-3.1 program, included within the Gaussian-09 program.

The bonding characteristics were analyzed by means of Atoms-in-Molecules (AIM) theory [78]. The relevant bond critical point (BCP) and topological parameters including electron density, Laplacian, and total electron energy density were obtained using the AIM2000 program [79]. To help understand the origin of the binding within each complex, the interaction energy was decomposed into five physically meaningful terms: electrostatic (E^{ele}), exchange (E^{ex}), repulsion (E^{rep}), polarization (E^{pol}), and dispersion (E^{disp}). This decomposition was performed using the localized molecular orbital-energy decomposition analysis (LMO-EDA) method [80] at the MP2/aug-cc-pVTZ level via the GAMESS program [81].

3. Results

3.1. Energies and Geometries

The diagrams in Figure 1 correspond to the unsubstituted Lewis acids where all atoms bonded to the central A atom are H. As displayed in the topmost section of Table 1, the MP2 interaction energies varied between −6.8 and −9.1 kJ/mol, with the tetrel and chalcogen bonds being the weakest and strongest, respectively. Raising the level of calculation up to CCSD(T) had a small effect, changing these quantities by 0.2–0.4 kJ/mol. This higher-level treatment of correlation weakened all the bonds, with the exception of the tetrel bond, making it slightly stronger than the pnicogen bond. The MP2 binding energies in the third column of Table 1 were only slightly less negative than E_{int}, a result of small deformation energies of the two monomers upon forming the complex. The intermolecular R(A⋯N) distances conformed roughly with the energy trends, although the halogen bond was the shortest, despite its being weaker than the chalcogen bond. This may have been partly due to the smaller covalent radius of Br. In terms of mutual orientations, both the halogen and tetrel bonds were fully linear, while the chalcogen and pnicogen bonds deviated by nearly 20° from linearity.

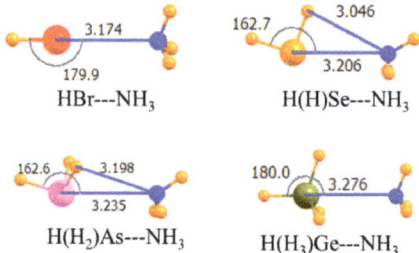

Figure 1. Optimized geometries of complexes of non-fluorinated Lewis acids with NH$_3$. Intermolecular distances are in Å, angles in deg.

Table 1. Interaction energy (E$_{int}$), binding energy (E$_b$), intermolecular distance (R, Å), and angle θ(R-A···N)(deg) where R represents the atom directly opposite NH$_3$. Energies are in kJ/mol.

Lewis Acid	E$_{int,MP2}$	E$_{int,CCSD(T)}$	E$_{b,MP2}$	R	θ(R-A···N)
		H-H$_n$A			
HBr	−7.79	−7.57	−7.77	3.174	179.9
H(H)Se	−9.08	−8.68	−8.98	3.206	162.7
H(H$_2$)As	−7.10	−6.81	−7.02	3.235	162.6
H(H$_3$)Ge	−6.83	−7.07	−6.42	3.276	180.0
		Me-H$_n$A			
MeBr	−5.01	−4.57	−4.99	3.233	160.4
Me(H)Se	−7.90	−7.50	−7.82	3.234	166.6
Me(H$_2$)As	−6.27	−6.09	−6.19	3.268	168.5
Me(H$_3$)Ge	−5.29	−5.71	−4.90	3.329	179.7
		H-F$_n$A			
H(F)Se	−14.43	−14.08	−13.89	3.023	161.6
H(F$_2$)As	−18.35	−18.36	−17.11	2.882	155.0
H(F$_3$)Ge	−120.73	−122.35	−35.04	2.101	180.0
		Me-F$_n$A			
Me(F)Se	−11.86	−11.66	−11.46	3.114	162.7
Me(F$_2$)As	−14.44	−14.73	−13.64	3.021	155.3
Me(F$_3$)Ge	−111.65	−111.27	−25.77	2.115	180.0
		F-H$_n$A			
FBr	−67.87	−59.18	−61.71	2.293	180.0
F(H)Se	−49.25	−44.35	−45.97	2.423	169.6
F(H$_2$)As	−34.57	−32.30	−32.76	2.597	165.3
F(H$_3$)Ge	−30.93	−30.59	−25.28	2.640	179.9

Replacing the H atom that lay opposite the NH$_3$ with a methyl group led to the geometries depicted in Figure 2. The only substantive reorientation induced by this methyl substitution involved the Br halogen wherein the NH$_3$ came off of the C-Br axis by 20°, and its C$_3$ axis turned away from the Br. This methylation weakened all of the bonds, but by varying amounts. This weakening was an expected consequence of the electron-releasing properties of this alkyl group. Using the CCSD(T) data as a reference, the pnicogen bond was weakened by only 0.7 kJ/mol, while the largest decrease of 3.0 kJ/mol occurred for the halogen bond, consistent with its nonlinearity. The chalcogen bond remained the strongest of the four, and it was the halogen bond that was weakest for these methyl-substituted Lewis acids. This same pattern was carried over into the binding energies, which included geometric deformations of the monomers. Commensurate with the weakening of all bonds, the intermolecular distances were all a bit longer for Me-H$_n$A. Note also that this substitution induced a 20° nonlinearity into the halogen bond, while slightly enhancing the linearity of the chalcogen and pnicogen bonds.

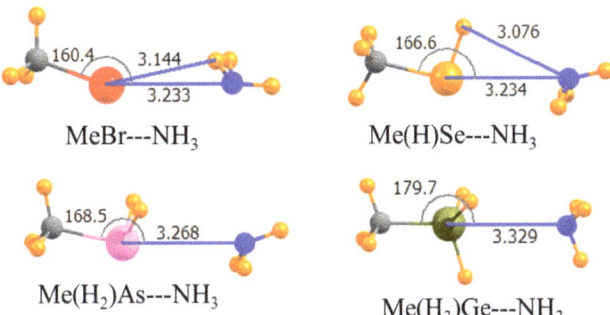

Figure 2. Optimized geometries of complexes of methyl Lewis acids with NH$_3$. Intermolecular distances are in Å, angles in deg.

The introduction of electron-withdrawing F atoms onto any given A atom is known to enhance its Lewis acidity. The next section of data in Table 1 relates to replacing all H atoms but the one that lay directly opposite the NH$_3$ by F. The molecular structures in Figure 3 indicated very nonlinear chalcogen and pnicogen bonds, with θ(HA···N) equal to 162° and 155°, respectively. Despite this nonlinearity, this substitution substantially increased the strength of both these bonds. The interaction energy of the former rose by 6.6 kJ/mol, and the latter by 12.3 kJ/mol. The larger increment in the case of the pnicogen bond was likely due to the introduction of two F atoms rather than the single F atom for the chalcogen bond. Upon adding a third F atom, there was a dramatic change. The very large increase in the case of the tetrel bond, more than 100 kJ/mol, resulted from its transition into what might be better termed a covalent bond, or at least partially covalent. Note that the R(Ge···N) distance was only 2.1 Å, a contraction of more than a full Å in comparison to the previous two cases. This very close encounter cannot be established without substantial monomer deformation. That is, the three F atoms must have peeled back away from the approaching N as the HGeF$_3$ molecule lost its initial pseudo-tetrahedral shape in forming the trigonal bipyramid that encompassed the fifth NH$_3$ ligand. This distortion cost some 85 kJ/mol. The final binding energy E_b of the tetrel bond was 35 kJ/mol, more than double that of the pnicogen bond, which is itself stronger than the chalcogen bond. As in the earlier case, switching out the H atom on the Lewis acid opposite the NH$_3$ with a methyl group, as in Figure 4, weakened all of the interactions. This decrement in the binding energy was only 2–3 kJ/mol for the chalcogen and pnicogen bonds, but amounted to 9 kJ/mol for the tetrel analogue.

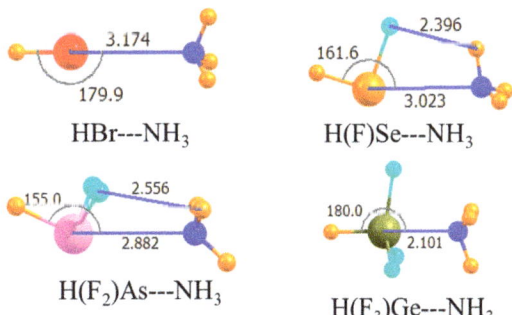

Figure 3. Optimized geometries of complexes involving a partially fluorinated Lewis acid with H opposite N. Intermolecular distances are in Å, angles in deg.

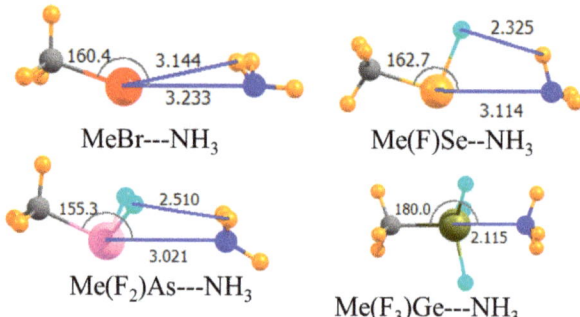

Figure 4. Optimized geometries of complexes involving a fluorinated Lewis acid with CH₃ opposite N. Intermolecular distances are in Å, angles in deg.

The placement of an electron-withdrawing F atom directly opposite the Lewis base is known to have an even stronger effect than when it is peripherally located. The molecular structures in Figure 5 show again that the halogen and tetrel bonds remained linear, reflecting the symmetry of the Lewis acids, while the less symmetrical chalcogen and pnicogen bonds were substantially distorted from linearity, with the NH₃ moving up and closer to the H atom(s) of the acid. The bottom section of Table 1 shows the very large binding magnifications quite clearly. Note first that when the F atom was so positioned on the Ge atom, Ge did not engage in a covalent bond with the NH₃. Indeed, the tetrel bond was the weakest of the array, which followed the pattern halogen > chalcogen > pnicogen > tetrel. This order was a very clear one, with fairly large differences between one bond type and another. Moreover, this pattern was valid not only for E_b, but for the interaction energy as well. One also saw a clear correlation in that stronger bonds were connected by the shortest intermolecular separations. It might be noted parenthetically that with the F atom opposite the NH₃, Ge did not engage in a covalent bond with the base, with R(Ge···N) = 2.64 Å.

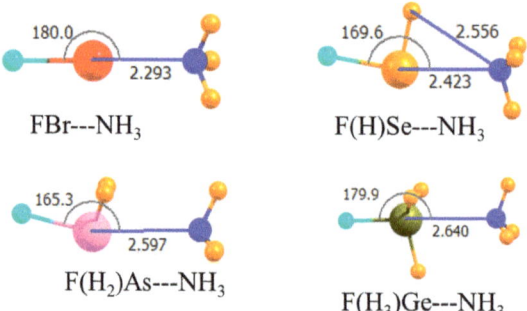

Figure 5. Optimized geometries of complexes involving a partially mono-fluorinated Lewis acid with one F atom opposite N. Intermolecular distances are in Å, angles in deg.

3.2. Analysis of the Wave Functions

There are a number of factors that contribute to the strength of noncovalent bonds of this type. As a component of the electrostatic attraction, one typically observes the presence of a so-called σ-hole on the A atom of the Lewis acid. This hole occurs directly opposite one of the covalent bonds in which the A atom is engaged, and attracts the Lewis base. The intensity of this hole is commonly measured by the value of the molecular electrostatic potential (MEP) on a particular isodensity surface, usually taken arbitrarily as ρ = 0.001 au. This maximum, labeled $V_{s,max}$, is reported in the second column of Table 2 for each of the Lewis acid monomers. There are certain points of similarity between these MEP maxima and the interaction energies in Table 1. Taking the F-H$_n$A acids as an example, the values of $V_{s,max}$ followed the same halogen > chalcogen > pnicogen > tetrel order as does E_{int}, and both exhibited the same (opposite) pattern of tetrel > pnicogen > chalcogen for the H-F$_n$A acids. However, there are inconsistencies as well. For example, the replacement of the H atom of HBr by a methyl group reduced the interaction energy while intensifying its σ-hole. Even though H(H$_3$)Ge formed the weakest of this class of bonds, it also presented the largest $V_{s,max}$.

Table 2. MEP maximum ($V_{s,max}$) of the σ-hole of the acid monomer facing the base, total charge transfer (CT) from the Lewis base to the acid, NBO values of E(2) from the N lone pair of NH$_3$ to σ*(A-R) antibonding orbitals where A is the central atom of the Lewis acid and R is (a) the atom directly opposite N and (b) the peripheral atom(s). Also shown is change in A-R bond length (Δr) and vibrational frequency (Δv) of the A-R stretch, where R lies opposite N.

Lewis Acid	$V_{s,max}$ au	CT me	E(2) [a] kcal/mol	E(2) [b] kcal/mol	Δr(A-R) Å	Δv(A-R) cm^{-1}
H-H$_n$A						
HBr	0.027	2	11.37	-	0.004	−26.8
H(H)Se	0.030	4	8.99	0.54	0.003	−107.6
H(H$_2$)As	0.028	7	10.70	2.51	0.006	−5.4
H(H$_3$)Ge	0.032	9	12.37	6.65	0.008	+1.2
Me-H$_n$A						
MeBr	0.030	1	3.26	-	0.001	−1.3
Me(H)Se	0.018	2	5.10	0.25	0.001	−0.7
Me(H$_2$)As	0.019	4	7.77	1.67	0.003	−2.2
Me(H$_3$)Ge	0.024	8	11.87	5.02	0.007	−10.1
H-F$_n$A						
H(F)Se	0.036	8	14.00	4.26	0.007	−38.1
H(F$_2$)As	0.046	27	15.34	14.46	0.009	−44.8
H(F$_3$)Ge	0.069	175	86.19	416.95	0.017	−2.2
Me-F$_n$A						
Me(F)Se	0.022	5	8.95	2.63	0.005	−6.7
Me(F$_2$)As	0.033	7	8.03	7.65	0.009	−35.7
Me(F$_3$)Ge	0.055	172	80.47	417.54	0.021	−101.6
F-H$_n$A						
FBr	0.093	143	255.40	-	0.070	−114.9
F(H)Se·	0.089	90	152.82	11.03	0.049	−107.6
F(H$_2$)As	0.079	53	82.81	14.96	0.033	−60.2
F(H$_3$)Ge	0.077	46	59.11	39.12	0.027	−63.3

[a] N$_{lp}$→σ*(A-R), R directly opposite N; [b] N$_{lp}$→σ*(A-R), R peripheral atom, sum of all such transfer energies.

In addition to Coulombic attraction between the two monomers, the noncovalent bond depends on a certain amount of intermolecular charge transfer. One way to measure this quantity is as a summation of NBO atomic charges on the atoms of each subunit. This total charge transfer is displayed in the third column of Table 2. Like $V_{s,max}$, CT also correlated generally with the interaction energies. The replacement of one H atom by CH$_3$ depressed both quantities, and both were substantially elevated by F-substitution. CT correctly predicted the energetic ordering of the noncovalent bond strengths of the F-H$_n$A acids. However, like the MEP maxima, CT failed to correlate with the interaction energies of the non-fluorinated species.

Charge transfer can be understood not only as that between the two molecules as a whole, but also between individual molecular orbitals. The bulk of the charge originates in the N lone pair of NH$_3$ that is pointing toward the acid. Its principal sink is the σ* antibonding orbital of the A-R$_a$ bond wherein R$_a$ lies opposite the N atom. The energetic consequence of this particular charge transfer was calculated by the NBO procedure, and is reported as $E(2)^a$ in Table 2. A second portion of the charge originating in the N lone pair made its way into the other σ*(A-R_b) antibonding orbitals, where R_b refers to the peripheral substituents on A, those not opposite the N atom. The cumulative sum of these transfers is tabulated as $E(2)^b$ in the fifth column of Table 2. Like the full CT, these individual components only partially reflected the energetics. Methyl substitution correctly reduced E(2), while fluorination led to a marked enhancement. $E(2)^a$ followed the same trend, as does the energetics for the F-H_nA series. However, E(2) was largest for H(H_3)Ge and smallest for H(H)Se, opposite to the trend in the interaction energies. Clearly, then, while consideration of the MEP and aspects of charge transfer bear some relation to the energetics, neither could be treated as fully predictive.

On the other hand, this charge transfer into the σ*(A-R_a) antibonding orbital afforded a reasonable indicator of the properties of this bond. The accumulation of additional charge in this antibonding orbital caused the A-R_a bond to weaken, and hence to stretch. This elongation is displayed in the penultimate column of Table 2 and appeared to rise and fall in parallel with $E(2)^a$. For example, the longest stretch for most of the acids in the top sections of Table 2 occurred for Ge, as does the largest value of $E(2)^a$. The exception to this rule corresponded to those acids in the bottom section of Table 2 where a F atom sat directly opposite the NH_3 base. In these cases, $E(2)^a$ followed the decreasing trend Br > Se > As > Ge, as did Δr. Indeed, there was a tight correlation between these two quantities. The correlation coefficient for a linear relationship between Δr and $E(2)^a$ was 0.974, which improved to 0.984 upon eliminating the two Ge systems that engage in a covalent bond.

The last column of Table 2 displays the stretching frequency of this same covalent bond, which shifted to the red in most instances, consistent with its elongation. However, correlations with other parameters were much weaker. For example, even though the charge transfer was quite modest in the complex between H(H)Se and NH_3, and the binding energy was rather small, the pertinent Se-H bond shifted by a full 108 cm^{-1} to the red, the largest shift of any of these complexes. The A-F red shifts in the F(H_2)As and F(H_3)Ge complexes were the reverse of the energetic quantities of these two dimers. One must recall, however, that unlike a particular bond length, the normal vibrational modes did not isolate any one particular bond. Instead they coupled together a number of different bonds, some stretched while others contracted, and included some degree of bending as well. Given this complicated character, it was not surprising to see poor correlation between frequency shifts and other parameters.

The properties of the AIM bond critical point generally offer an accurate barometer of the strength of a noncovalent bond. The three most widely used such quantities are collected in Table 3, and they bore some similarities with energetics. Methyl substitution reduced these values, while they increase upon fluorosubstitution. With respect to the F-H_nA acids, both ρ and $\nabla^2\rho$ correctly reproduced the halogen > chalcogen > pnicogen > tetrel energetic trend. However, AIM did not accurately reflect some of the other trends. Taking the unsubstituted H-H_nA acids as a case in point, neither ρ nor H displayed much differentiation from one type of bond to the next. The values of $\nabla^2\rho$ were somewhat different from one another, but seemed to exaggerate the strength of the halogen bond.

It is worthwhile to inquire as to how these four sorts of noncovalent bonds compare in terms of their basic contributing factors. Decomposition of the total interaction energy into electrostatic (E^{ele}), exchange (E^{ex}), repulsion (E^{rep}), polarization (E^{pol}), and dispersion (E^{disp}) provides a fingerprint of sorts for each interaction. These components are listed in Table 4 and show first for the nonfluorinated acids, that the electrostatic attraction was roughly twice that of dispersion, but there was little to distinguish one sort of bond from another in these two elements. On the other hand, exchange was roughly three times larger for the pnicogen and tetrel bonds, as compared to halogen and chalcogen. It was this outsized exchange energy which appeared to be a hallmark of the latter two types of bonds, regardless of substitution. Polarization energy represented the smallest component. With the exception of dispersion, which undergoes a small uptick, all of the other components were lowered upon methyl substitution. Even larger increments accompanied the replacement of one or more H atoms by F. This fluorosubstitution raised the polarization energy to the point where it exceeded dispersion, and could become competitive with the exchange energy for the halogen and chalcogen bonds.

Table 3. Electron density (ρ), Laplacian ($\nabla^2\rho$), and energy density (H) at the intermolecular BCP in the complexes (all are in au).

Lewis Acid	ρ	$\nabla^2\rho$	H
H-H$_n$A			
HBr	0.010	0.038	0.002
H(H)Se	0.009	0.034	0.001
H(H$_2$)As	0.009	0.029	0.001
H(H$_3$)Ge	0.008	0.025	0.001
Me-H$_n$A			
MeBr	0.008	0.034	0.002
Me(H)Se	0.008	0.032	0.002
Me(H$_2$)As	0.008	0.027	0.001
Me(H$_3$)Ge	0.007	0.023	0.001
H-F$_n$A			
H(F)Se	0.014	0.046	0.001
H(F$_2$)As	0.018	0.046	0.000
H(F$_3$)Ge	0.077	0.216	−0.030
Me-F$_n$A			
Me(F)Se	0.011	0.039	0.002
Me(F$_2$)As	0.013	0.037	0.001
Me(F$_3$)Ge	0.075	0.209	−0.028
F-H$_n$A			
FBr	0.061	0.132	−0.015
F(H)Se	0.044	0.105	−0.007
F(H$_2$)As	0.029	0.074	−0.002
F(H$_3$)Ge	0.023	0.072	0.000

Table 4. Electrostatic (E^{ele}), exchange (E^{ex}), repulsion (E^{rep}), polarization (E^{pol}), dispersion (E^{disp}), and interaction energies (E_{int}); all are in kJ/mol.

Lewis Acid	E^{ele}	E^{ex}	E^{rep}	E^{pol}	E^{disp}	E_{int}
H-H$_n$A						
HBr	−16.30	−10.29	49.53	−4.81	−7.27	−7.90
H(H)Se	−16.85	−10.29	48.61	−4.14	−7.69	−9.11
H(H$_2$)As	−16.18	−32.02	52.88	−4.43	−7.40	−7.15
H(H$_3$)Ge	−16.80	−32.65	53.63	−4.64	−6.40	−6.86
Me-H$_n$A						
MeBr	−8.15	−7.53	35.57	−2.34	−8.78	−5.02
Me(H)Se	−12.79	−9.04	42.39	−3.18	−8.74	−7.86
Me(H$_2$)As	−13.00	−28.72	47.23	−3.55	−8.23	−6.31
Me(H$_3$)Ge	−13.29	−29.59	48.03	−4.14	−6.40	−5.35
H-F$_n$A						
H(F)Se	−34.19	−19.00	93.92	−9.91	−10.66	−14.50
H(F$_2$)As	−55.05	−83.77	149.44	−17.56	−11.58	−18.52
H(F$_3$)Ge	−363.74	−371.98	782.50	−165.65	−2.17	−121.05
Me-F$_n$A						
Me(F)Se	−25.67	−15.70	76.37	−7.44	−10.87	−11.95
Me(F$_2$)As	−36.91	−59.31	103.54	−10.41	−11.50	−14.59
Me(F$_3$)Ge	−352.88	−369.72	773.13	−158.21	−4.31	−111.94
F-H$_n$A						
FBr	−197.84	−97.95	554.14	−120.05	−27.67	−68.09
F(H)Se	−146.89	−75.42	407.59	−74.03	−23.12	−49.49
F(H$_2$)As	−98.61	−147.64	269.90	−40.67	−17.85	−34.86
F(H$_3$)Ge	−89.20	−125.82	228.60	−31.27	−13.42	−31.10

3.3. 2nd Row Atoms

It would be injudicious to base all conclusions concerning the comparisons between the various sorts of noncovalent bonds upon atoms in a single row (the third) of the periodic table. Thus, similar calculations were performed for the analogous atoms of the preceding row. The energetic and geometrical data for the complexes of these Lewis acids with NH_3 are presented in Table 5, which may be directly compared with those in Table 1. It would be expected that the smaller size of the second-row atoms, coupled with their greater electronegativity, ought to have weakened their complexes with NH_3. This trend was indeed observed, but with some exceptions. For example, the halogen bond of HCl was weaker than that of HBr, but the tetrel bond of $H(H_3)Si$ was stronger than its third row congener; this same trend was noted after methyl substitution as well. Another issue arises with S. This atom was electronegative enough that it would not engage in a chalcogen bond with NH_3. Instead, the H_2S molecule rotated around to form a SH$\cdots$N H-bond, as does MeSH.

Table 5. Interaction energy (E_{int}), binding energy (E_b), intermolecular distance (R, Å), and angle θ(R-A$\cdots$N)(deg) where R represents the atom directly opposite N. Energies are in kJ/mol.

Lewis Acid	$E_{int,MP2}$	$E_{int,CCSD(T)}$	$E_{b,MP2}$	R	θ(R-A$\cdots$N)
H-H_nA					
HCl	−3.77	−3.73	−3.77	3.254	156.5
H(H)S [a]	—	—	—	—	—
H(H_2)P	−7.01	−6.47	−6.94	3.281	166.3
H(H_3)Si	−8.75	−8.87	−8.05	3.187	180.0
Me-H_nA					
MeCl	−4.01	−3.81	−3.98	3.409	145.2
Me(H)S [a]	—	—	—	—	—
Me(H_2)P	−6.57	−6.42	−6.50	3.294	171.2
Me(H_3)Si	−6.38	−6.74	−5.77	3.257	180.0
H-F_nA					
H(F)S	−11.12	−9.11	−8.82	3.176	164.0
H(F_2)P	−11.41	−11.79	−10.93	3.051	159.0
H(F_3)Si	−106.23	−108.75	−17.19	2.099	180.0
Me-F_nA					
Me(F)S	−7.09	−7.29	−6.92	3.315	164.2
Me(F_2)P	−8.25	−8.84	−7.95	3.247	157.0
Me(F_3)Si	−12.06	−13.51	−7.57	3.086	179.8
F-H_nA					
FCl	−53.34	−43.90	−45.78	2.231	180.0
F(H)S	−37.36	−33.14	−34.98	2.435	171.0
F(H_2)P	−28.34	−26.42	−27.12	2.604	167.7
F(H_3)Si	−35.26	−34.76	−25.01	2.489	180.0

[a] Does not form a chalcogen bond.

With respect to fluorinated species, the switch from third to second-row atoms produced the expected weakening of the interaction. But another distinction between second and third-row atoms was noted for the fluorinated Me(F_3)Ge and Me(F_3)Si acids. Whereas the former pulls in the NH_3 to form a short covalent Ge-N bond, the same was not true for its Si analogue wherein R(Si$\cdots$N) remained longer than 3 Å. The trend in binding energies of the F-H_nA acids remained halogen > chalcogen > pnicogen > tetrel, as it was for the third-row atoms. However, an exception occured in the consideration of the interaction energy. The very large (10 kJ/mol) geometrical distortion energy in FH_3Si was sufficient to enlarge its interaction energy to exceed that of the pnicogen bond involving FH_2P.

While the smaller size of the second-row atoms would tend toward shorter intermolecular separations, the weakening of the interactions should have acted to push the two molecules further apart. The values of R in Table 5 are thus not entirely different from those for the third-row atoms in Table 1. Angular aspects were also generally similar with a few exceptions. The halogen bond of HCl was distorted from linearity by some 23°, and MeCl was also less linear than its MeBr analogue.

4. Discussion

In the fully hydrogenated series of Lewis acids, the strengths of the various bonds varied in the pattern chalcogen > halogen > pnicogen > tetrel, although the last two were reversed if the level of calculation was raised from MP2 to CCSD(T). Note that this pattern did not obey the simple order of electronegativity of the central atom, which would have placed the tetrel bond as strongest. Replacement of one H atom by a methyl group (opposite the base) weakened all bonds, but had the strongest effect on the halogen bond, which becomes the weakest of the four. Fluorosubstitution very substantially strengthened the four sorts of bonds as the F atom pulled electron density away from the central atom. Leaving one H atom on the acid to lie opposite the base, and replacing all others by F strengthened the interactions, and this effect rose with the number of these peripheral F atoms. Indeed, the three F atoms of the HF_3Ge acid enhanced the interaction to the point where the interaction energy exceeded 120 kcal/mol as the very short tetrel bond acquired a covalent character. The very close approach forced the acid to deform to accommodate the base, but even so, the binding energy of the bonds followed the tetrel > pnicogen > chalcogen pattern, opposite to that observed prior to fluorosubstitution. Replacement of the sole H atom by a methyl group again weakened the interactions but leaves the ordering intact. A much more profound strengthening effect occurred if it was the H directly opposite the base that was replaced by F. These interaction energies magnified the unsubstituted interaction energies by a factor between 5 and 9, leading to quantities approaching 70 kcal/mol. It was the halogen bond that was enhanced the most and the tetrel bond the least, so that the order of these monofluorinated bonds washalogen > chalcogen > pnicogen > tetrel. This pattern applied to both interaction and binding energies alike.

The exchange attraction energy was much larger for pnicogen and tetrel bonds than for halogen and chalcogen bonds, making it the largest contributor to the former bond types. With these two exceptions, electrostatics provided the major contribution, surpassing both polarization and dispersion energies by a wide margin. Electrostatic components did not differentiate between the four types of bonds unless there was fluorosubstitution of the Lewis acid. When the F atom was situated directly opposite the base, there was a clear decreasing trend of E^{ele}: halogen > chalcogen > pnicogen > tetrel. It was also in these same configurations where polarization energy made a major contribution, but was otherwise generally smaller than the dispersion energy. The value of the molecular electrostatic potential at the site of the sigma hole conformed fairly well to the full electrostatic potential. For example, methyl substitution diminished both E^{ele} and $V_{s,max}$, and both quantities grew upon fluorosubstitution. The intensity of the σ-hole was also a good indicator of E^{ele} in the four sorts of bonds. On the other hand, neither $V_{s,max}$, nor the more complete E^{ele}, accurately reproduced trends in the full interaction energy. Charge transfer, either measured as the total from one molecule to the other, or specifically from the N lone pair MO of the base to a σ* antibonding orbital of the acid, also provided some guidance in terms of the full interaction energy. For example, both measures correctly predicted the order of bonding for the fluorinated acids. But there are certain inconsistencies as well in that charge transfer was largest for the tetrel bond, but it was the chalcogen bond that was strongest, when considering unsubstituted systems. Similar limitations applied to the density at the AIM bond critical point, which also showed little distinction between the four sorts of bonds in the absence of F substituents.

There was a general pattern where replacing third-row central atoms with their second-row analogues weakened the various bonds, but this trend was not fully consistent. Taking the non-substituted acids as a case in point, replacement of Br by Cl did substantially weaken the halogen bond. However, there was little effect on the As to P substitution of the pnicogen bond, and the tetrel bond was strengthened when Ge was replaced by Si. There was another issue that must be considered as well. While Se was not electronegative enough to engage in a SeH···N H-bond with NH_3, the more electronegative S chalcogen atom would form such an H-bond. In fact, the SH···N H-bond was strong enough that it eliminated the S···N chalcogen bond as a secondary minimum on the potential energy surface. The weakening effects of a smaller central atom were more consistently noted, however, when F atoms were added to the Lewis acid. In the case of peripheral substitutions, there was little distinction between chalcogen and pnicogen bonds for the second-row atoms. Whereas the halogen bond remained strongest when F was placed opposite the base, and chalcogen second, there was a sort of reversal in the other two bonds. While the binding energy of the pnicogen bond exceeded that of the tetrel bond, these two reversed when considering only the interaction energy between pre-deformed monomers. This latter issue arose due to the particularly large deformation energy of the FH_3Si molecule.

There are a number of works in the recent literature that have some bearing on the comparisons of these various sorts of noncovalent bonds. In a general sense, halogen bonding was found [82] preferable to pnicogen bonding when combined with an amine base, but the order reversed for an aromatic π-system. On the other hand, pnicogen bonding is more stable than halogen bonding in dimers formed by HArF and XH_2P [83]. In broadening the conversation to include chalcogen bonds, Shukla and Chopra [84] investigated how the substituents on the PH_2R and SeHR molecules determine the structure of the dimer, and thereby the presence of either a pnicogen or chalcogen bond, given the Se···P contact in both. Li and coworkers [85] compared chalcogen with halogen bonds for the fluorosubstituted $F_2C=Se$ and found the latter to be stronger. However, the comparison was clouded by the different geometries adopted by the two sorts of dimers, one employing a σ-hole and the other a π-hole above the Se atom.

The Esrafili group has produced some relevant work as well. One study [86] compared chalcogen with pnicogen bonds, both of which could occur, depending upon the molecular orientations, in $RHS:PH_2R$ dimers. The authors found only small differences, with interaction energies generally in the range between 8 and 18 kJ/mol. Halogen bonds were compared with their pnicogen counterparts [87] in the context of hypervalent ZOF_2X molecules wherein NH_3 could interact with either the pnicogen (Z) or halogen (X) atom. A shift from pnicogen to halogen bond preference was observed, consistent with the fact that the halogen bond strengthens as X becomes larger, and the opposite occurs for the X-P···N pnicogen bond. Similar trends were seen in the comparison of halogen with chalcogen bonds for the related YO_2X_2 acids [88].

Jiao et al. [89] found halogen bonds considerably stronger than pnicogen bonds, but only in a specialized set of dimers, namely dihalogens combined with phosphine derivatives PH_2R···BrX. Grabowski and Sokalski [36] considered NH_3 as the Lewis base, along with C_2H_2 and Cl^-. In the case of the former, combined with acids wherein a single F substituent was disposed opposite the base, the order noted for third-row atoms was halogen > chalcogen > pnicogen > tetrel, the same order as obtained here, as are the numerical values. This order persisted for second-row atoms. Shifting gears toward an anionic electron donor, Matile at al [90] have very recently calculated binding energies of various highly fluorinated Lewis acids to Cl^-, and found pnicogen bonding stronger than chalcogen bonding, and that fourth row atoms engaged in stronger bonds than their third-row analogues. In all cases, it was a C atom (of a phenyl ring) that is situated opposite the base, rather than F.

There have also been a number of works from this laboratory that relate to the issue of these comparisons. An early set of calculations [91] suggested a pnicogen > halogen > chalcogen bond strength order, but this work was limited to unsubstituted hydrides that were only weakly bound. HSX molecules, capable of both halogen and chalcogen bonds, yielded mixed results depending upon the nature of the X atom [92]. Whether considering a pnicogen, chalcogen, or halogen bond, all display similar sensitivity to stretching [60,93], but greater sensitivity to angular deformation [94], than do H-bonds. However, there is little to distinguish one from another, and all three are subject to similar substituent effects [95], which follow the general pattern $CH_3 < NH_2 < CF_3 < OH < Cl < NO_2 < F$.

Addition of positive charge on the electron acceptor strengthens all of these interactions, but has more of an effect upon a S···O chalcogen than a C···O tetrel bond, leading to a preference for the former [64]. Within the specialized context of bipodal receptors that engage in a pair of simultaneous noncovalent bonds with a halide [96–98], tetrel bonding has a clear edge over halogen, chalcogen, and pnicogen bonds. In another specialized context of hypervalency [99], pnicogen bonds show a clear edge over both chalcogen and halogen bonds.

5. Conclusions

In the absence of any replacements of H atoms, the chalcogen bond is the strongest followed in order by halogen, pnicogen, and tetrel. Methyl substitution on the Lewis acid weakens all bonds, particularly the halogen bond, which is the weakest of the four in this bonding environment. All bonds are strengthened by fluorosubstitution (peripheral to, rather than opposite the base), which leads to the bonding order: tetrel > pnicogen > chalcogen. The most dramatic bond enhancement arises from replacement of the atom opposite the base by F, and yet a different order of halogen > chalcogen > pnicogen > tetrel. If the third-row Lewis acid atoms are replaced by their second-row analogues, there is a general weakening of the noncovalent bonds, but this change is not consistent from one sort of bond to the next.

Supplementary Materials: The following are available online. Optimized coordinates of complexes and monomers.

Author Contributions: S.S. conceived of the idea for this project and wrote a first draft of the manuscript; W.D. carried out the calculations and compiled the data; Q.L. supervised the calculations and helped with a final draft.

Funding: This research was funded by the National Natural Science Foundation of China grant number 21573188.

Acknowledgments: APC was sponsored by MDPI.

Conflicts of Interest: The author declares no conflict of interest.

References

1. Lommerse, J.P.M.; Stone, A.J.; Taylor, R.; Allen, F.H. The nature and geometry of intermolecular interactions between halogens and oxygen or nitrogen. *J. Am. Chem. Soc.* **1996**, *118*, 3108–3116. [CrossRef]
2. Alkorta, I.; Rozas, S.; Elguero, J. Charge-transfer complexes between dihalogen compounds and electron donors. *J. Phys. Chem. A* **1998**, *102*, 9278–9285. [CrossRef]
3. Farina, A.; Meille, S.V.; Messina, M.T.; Metrangolo, P.; Resnati, G.; Vecchio, G. Resolution of racemic 1,2-dibromohexafluoropropane through halogen-bonded supramolecular helices. *Angew. Chem. Int. Ed. Engl.* **1999**, *38*, 2433–2436. [CrossRef]
4. Wash, P.L.; Ma, S.; Obst, U.; Rebek, J. Nitrogen-halogen intermolecular forces in solution. *J. Am. Chem. Soc.* **1999**, *121*, 7973–7974. [CrossRef]
5. Legon, A.C. Prereactive complexes of dihalogens XY with Lewis bases B in the gas phase: A systematic case for the halogen analogue B···XY of the hydrogen bond B···HX. *Angew. Chem. Int. Ed. Engl.* **1999**, *38*, 2686–2714. [CrossRef]
6. Caronna, T.; Liantonio, R.; Logothetis, T.A.; Metrangolo, P.; Pilati, T.; Resnati, G. Halogen bonding and π···π stacking control reactivity in the solid state. *J. Am. Chem. Soc.* **2004**, *126*, 4500–4501. [CrossRef] [PubMed]
7. Auffinger, P.; Hays, F.A.; Westhof, E.; Ho, P.S. Halogen bonds in biological molecules. *Proc. Natl. Acad. Sci. USA* **2004**, *101*, 16789–16794. [CrossRef] [PubMed]
8. Glaser, R.; Chen, N.; Wu, H.; Knotts, N.; Kaupp, M. ^{13}C NMR study of halogen bonding of haloarenes: Measurements of solvent effects and theoretical analysis. *J. Am. Chem. Soc.* **2004**, *126*, 4412–4419. [CrossRef] [PubMed]
9. Grabowski, S.J.; Bilewicz, E. Cooperativity halogen bonding effect—Ab initio calculations on H_2CO···$(ClF)_n$ complexes. *Chem. Phys. Lett.* **2006**, *427*, 51–55. [CrossRef]
10. Riley, K.E.; Merz, K.M. Insights into the strength and origin of halogen bonding: The halobenzene-formaldehyde dimer. *J. Phys. Chem. A* **2007**, *111*, 1688–1694. [CrossRef] [PubMed]
11. Politzer, P.; Lane, P.; Concha, M.C.; Ma, Y.; Murray, J.S. An overview of halogen bonding. *J. Mol. Model.* **2007**, *13*, 305–311. [CrossRef] [PubMed]

12. Clark, T.; Hennemann, M.; Murray, J.S.; Politzer, P. Halogen bonding: The σ-hole. *J. Mol. Model.* **2007**, *13*, 291–296. [CrossRef] [PubMed]
13. Rosenfield, R.E.; Parthasarathy, R.; Dunitz, J.D. Directional preferences of nonbonded atomic contacts with divalent sulfur. 1. Electrophiles and nucleophiles. *J. Am. Chem. Soc.* **1977**, *99*, 4860–4862. [CrossRef]
14. Row, T.N.G.; Parthasarathy, R. Directional preferences of nonbonded atomic contacts with divalent sulfur in terms of its orbital orientations. 2. Sulfur···sulfur interactions and nonspherical shape of sulfur in crystals. *J. Am. Chem. Soc.* **1981**, *103*, 477–479. [CrossRef]
15. Desiraju, G.R.; Nalini, V. Database analysis of crystal-structure-determininginteractions involving sulphur: Implications for the design of organic metals. *J. Mater. Chem.* **1991**, *1*, 201–203. [CrossRef]
16. Burling, F.T.; Goldstein, B.M. Computational studies of nonbonded sulfur-oxygen and selenium-oxygen interactions in the thiazole and selenazole nucleosides. *J. Am. Chem. Soc.* **1992**, *114*, 2313–2320. [CrossRef]
17. Iwaoka, M.; Tomoda, S. Nature of the intramolecular Se···N nonbonded interaction of 2-selenobenzylamine derivatives. An experimental evaluation by ^{1}H, ^{77}Se, and ^{15}N NMR spectroscopy. *J. Am. Chem. Soc.* **1996**, *118*, 8077–8084. [CrossRef]
18. Werz, D.B.; Gleiter, R.; Rominger, F. Nanotube formation favored by chalcogen-chalcogen interactions. *J. Am. Chem. Soc.* **2002**, *124*, 10638–10639. [CrossRef] [PubMed]
19. Sanz, P.; Mó, O.; Yáñez, M. Characterization of intramolecular hydrogen bonds and competitive chalcogen–chalcogen interactions on the basis of the topology of the charge density. *Phys. Chem. Chem. Phys.* **2003**, *5*, 2942–2947. [CrossRef]
20. Bleiholder, C.; Werz, D.B.; Koppel, H.; Gleiter, R. Theoretical investigations on chalcogen-chalcogen interactions: What makes these nonbonded interactions bonding? *J. Am. Chem. Soc.* **2006**, *128*, 2666–2674. [CrossRef] [PubMed]
21. Nziko, V.d.P.N.; Scheiner, S. Chalcogen bonding between tetravalent SF$_4$ and amines. *J. Phys. Chem. A* **2014**, *118*, 10849–10856. [CrossRef] [PubMed]
22. Klinkhammer, K.W.; Pyykko, P. Ab initio interpretation of the closed-shell intermolecular E···E attraction in dipnicogen (H$_2$E-EH$_2$)$_2$ and (HE-EH)$_2$ hydride model dimers. *Inorg. Chem.* **1995**, *34*, 4134–4138. [CrossRef]
23. Deiters, J.A.; Holmes, R.R. Ab initio treatment of a phosphorus coordinate, trigonal bipyramidal to pentafluoride-pyridine reaction square pyramidal to octahedral. *Phosphorus Sulfur Sillicon Relat. Elem.* **1997**, *123*, 329–340. [CrossRef]
24. Scheiner, S. The pnicogen bond: Its relation to hydrogen, halogen, and other noncovalent bonds. *Acc. Chem. Res.* **2013**, *46*, 280–288. [CrossRef] [PubMed]
25. Hill, N.J.; Levason, W.; Reid, G. Arsenic (III) halide complexes with phosphine and arsine co-ligands: Synthesis, spectroscopic and structural properties. *J. Chem. Soc. Dalton Trans.* **2002**, 1188–1192. [CrossRef]
26. Scheiner, S. Effects of multiple substitution upon the P···N noncovalent interaction. *Chem. Phys.* **2011**, *387*, 79–84. [CrossRef]
27. Scheiner, S. Can two trivalent N atoms engage in a direct N···N noncovalent interaction? *Chem. Phys. Lett.* **2011**, *514*, 32–35. [CrossRef]
28. Rossi, A.R.; Jasinski, J.M. Theoretical studies of neutral silane-ammonia adducts. *Chem. Phys. Lett.* **1990**, *169*, 399–404. [CrossRef]
29. Ruoff, R.S.; Emilsson, T.; Jaman, A.I.; Germann, T.C.; Gutowsky, H.S. Rotational spectra, dipole moment, and structure of the SiF$_4$–NH$_3$dimer. *J. Chem. Phys.* **1992**, *96*, 3441–3446. [CrossRef]
30. Schoeller, W.W.; Rozhenko, A. Pentacoordination at fluoro-substituted silanes by weak Lewis donor addition. *Eur. J. Inorg. Chem* **2000**, *2000*, 375–381. [CrossRef]
31. Politzer, P.; Murray, J.S.; Lane, P.; Concha, M.C. Electrostatically driven complexes of SiF$_4$ with amines. *Int. J. Quantum Chem.* **2009**, *109*, 3773–3780. [CrossRef]
32. Bauzá, A.; Mooibroek, T.J.; Frontera, A. Tetrel bonding interactions. *Chem. Rec.* **2016**, *16*, 473–487. [CrossRef] [PubMed]
33. Marín-Luna, M.; Alkorta, I.; Elguero, J. A theoretical study of the H$_n$F$_{4-n}$Si:N-base (n = 1–4) tetrel-bonded complexes. *Theor. Chem. Acc.* **2017**, *136*, 41. [CrossRef]
34. Geboes, Y.; De Proft, F.; Herrebout, W.A. Effect of fluorination on the competition of halogen bonding and hydrogen bonding: Complexes of fluoroiodomethane with dimethyl ether and trimethylamine. *J. Phys. Chem. A* **2017**, *121*, 4180–4188. [CrossRef] [PubMed]

35. An, X.; Yang, X.; Xiao, B.; Cheng, J.; Li, Q. Comparison of hydrogen and halogen bonds between dimethyl sulfoxide and hypohalous acid: Competition and Cooperativity. *Mol. Phys.* **2017**, *115*, 1614–1623. [CrossRef]
36. Grabowski, S.J.; Sokalski, W.A. Are various σ-hole bonds steered by the same mechanisms? *ChemPhysChem* **2017**, *18*, 1569–1577. [CrossRef] [PubMed]
37. Karpfen, A. Theoretical characterization of the trends in halogen bonding. In *Halogen Bonding. Fundamentals and Applications*; Metrangolo, P., Resnati, G., Eds.; Springer: Berlin, Germany, 2008; pp. 1–15.
38. Moilanen, J.; Ganesamoorthy, C.; Balakrishna, M.S.; Tuononen, H.M. Weak interactions between trivalent pnictogen centers: Computational analysis of bonding in dimers $X_3E\cdots EX_3$ (E = Pnictogen, X = Halogen). *Inorg. Chem.* **2009**, *48*, 6740–6747. [CrossRef] [PubMed]
39. Legon, A.C. The halogen bond: An interim perspective. *Phys. Chem. Chem. Phys.* **2010**, *12*, 7736–7747. [CrossRef] [PubMed]
40. Zahn, S.; Frank, R.; Hey-Hawkins, E.; Kirchner, B. Pnicogen bonds: A new molecular linker? *Chem. Eur. J.* **2011**, *17*, 6034–6038. [CrossRef] [PubMed]
41. Adhikari, U.; Scheiner, S. Comparison of $P\cdots D$ (D = P, N) with other noncovalent bonds in molecular aggregates. *J. Chem. Phys.* **2011**, *135*, 184306. [CrossRef] [PubMed]
42. Sánchez-Sanz, G.; Trujillo, C.; Alkorta, I.; Elguero, J. Intermolecular weak interactions in HTeXH dimers (X = O, S, Se, Te): Hydrogen bonds, chalcogen–chalcogen contacts and chiral discrimination. *ChemPhysChem* **2012**, *13*, 496–503. [CrossRef] [PubMed]
43. Adhikari, U.; Scheiner, S. Effects of carbon chain substituent on the $P\cdots N$ noncovalent bond. *Chem. Phys. Lett.* **2012**, *536*, 30–33. [CrossRef]
44. Sánchez-Sanz, G.; Trujillo, C.; Solimannejad, M.; Alkorta, I.; Elguero, J. Orthogonal interactions between nitryl derivatives and electron donors: Pnicogen bonds. *Phys. Chem. Chem. Phys.* **2013**, *15*, 14310–14318. [CrossRef] [PubMed]
45. Riley, K.E.; Hobza, P. The relative roles of electrostatics and dispersion in the stabilization of halogen bonds. *Phys. Chem. Chem. Phys.* **2013**, *15*, 17742–17751. [CrossRef] [PubMed]
46. Sedlak, R.; Stasyuk, O.A.; Fonseca Guerra, C.; Řezáč, J.; Růžička, A.; Hobza, P. New insight into the nature of bonding in the dimers of Lappert's stannylene and its Ge analogs: A quantum mechanical study. *J. Chem. Theory Comput.* **2016**, *12*, 1696–1704. [CrossRef] [PubMed]
47. Bauzá, A.; Mooibroek, T.J.; Frontera, A. σ-Hole opposite to a lone pair: Unconventional pnicogen bonding interactions between ZF_3 (Z = N, P, As, and Sb) compounds and several donors. *ChemPhysChem* **2016**, *17*, 1608–1614. [CrossRef] [PubMed]
48. Shukla, R.; Chopra, D. Understanding the effect of substitution on the formation of $S\cdots F$ chalcogen bond. *J. Chem. Sci.* **2016**, *128*, 1589–1596. [CrossRef]
49. Esrafili, M.D.; Kiani, H.; Mohammadian-Sabet, F. Tuning of carbon bonds by substituent effects: An ab initio study. *Mol. Phys.* **2016**, *114*, 3658–3668. [CrossRef]
50. Legon, A.C. Tetrel, pnictogen and chalcogen bonds identified in the gas phase before they had names: Asystematic look at non-covalent interactions. *Phys. Chem. Chem. Phys.* **2017**, *19*, 14884–14896. [CrossRef] [PubMed]
51. Robertson, C.C.; Wright, J.S.; Carrington, E.J.; Perutz, R.N.; Hunter, C.A.; Brammer, L. Hydrogen bonding vs halogen bonding: The solvent decides. *Chem. Sci.* **2017**, *8*, 5392–5398. [CrossRef] [PubMed]
52. Liu, M.-X.; Li, Q.-Z.; Scheiner, S. Comparison of tetrel bonds in neutraland protonated complexes of pyridineTF$_3$ and furanTF$_3$ (T = C, Si, and Ge) with NH_3. *Phys. Chem. Chem. Phys.* **2017**, *19*, 5550–5559. [CrossRef] [PubMed]
53. Grabowski, S.J. Hydrogen bonds, and σ-hole and π-hole bonds—Mechanisms protecting doublet and octet electron structures. *Phys. Chem. Chem. Phys.* **2017**, *19*, 29742–29759. [CrossRef] [PubMed]
54. Scheiner, S. Detailed comparison of the pnicogen bond with chalcogen, halogen and hydrogen bonds. *Int. J. Quantum Chem.* **2013**, *113*, 1609–1620. [CrossRef]
55. Aakeröy, C.B.; Fasulo, M.; Schultheiss, N.; Desper, J.; Moore, C. Structural competition between hydrogen bonds and halogen bonds. *J. Am. Chem. Soc.* **2007**, *129*, 13772–13773. [CrossRef] [PubMed]
56. Solimannejad, M.; Malekani, M.; Alkorta, I. Cooperativity between the hydrogen bonding and halogen bonding in $F_3CX\cdots NCH(CNH)\cdots NCH(CNH)$ complexes (X = Cl, Br). *Mol. Phys.* **2011**, *109*, 1641–1648. [CrossRef]

57. Del Bene, J.E.; Alkorta, I.; Sanchez-Sanz, G.; Elguero, J. ^{31}P–^{31}P Spin–spin coupling constants for pnicogen homodimers. *Chem. Phys. Lett.* **2011**, *512*, 184–187. [CrossRef]
58. Grabowski, S.J. QTAIM characteristics of halogen bond and related interactions. *J. Phys. Chem. A* **2012**, *116*, 1838–1845. [CrossRef] [PubMed]
59. Grabowski, S.J. Hydrogen and halogen bonds are ruled by the same mechanisms. *Phys. Chem. Chem. Phys.* **2013**, *15*, 7249–7259. [CrossRef] [PubMed]
60. Scheiner, S. Sensitivity of noncovalent bonds to intermolecular separation: Hydrogen, halogen, chalcogen, and pnicogen bonds. *CrystEngComm* **2013**, *15*, 3119–3124. [CrossRef]
61. Riley, K.E.; Rezác, J.; Hobza, P. Competition between halogen, dihalogen and hydrogen bonds in bromo- and iodomethanol dimers. *J. Mol. Model.* **2013**, *19*, 2879–2883. [CrossRef] [PubMed]
62. Tang, Q.; Li, Q. Interplay between tetrel bonding and hydrogen bonding interactions in complexes involving F_2XO (X = C and Si) and HCN. *Comput. Theor. Chem.* **2014**, *1050*, 51–57. [CrossRef]
63. McDowell, S.A.C.; Holder, Z.L. Computational study of non-covalent interactions in oxirane···XF complexes (X = H, F, Cl, Br, Li) and their F-/Li-substituted analogues. *Mol. Phys.* **2015**, *113*, 3757–3766. [CrossRef]
64. Scheiner, S. Comparison of CH···O, SH···O, chalcogen, and tetrel bonds formed by neutral and cationic sulfur-containing compounds. *J. Phys. Chem. A* **2015**, *119*, 9189–9199. [CrossRef] [PubMed]
65. Alkorta, I.; Del Bene, J.; Elguero, J. $H_2XP:OH_2$ Complexes: Hydrogen vs. pnicogen bonds. *Crystals* **2016**, *6*, 19. [CrossRef]
66. Fanfrlík, J.; Holub, J.; Růžičková, Z.; Řezáč, J.; Lane, P.D.; Wann, D.A.; Hnyk, D.; Růžička, A.; Hobza, P. Competition between halogen, hydrogen and dihydrogen bonding in brominated carboranes. *ChemPhysChem* **2016**, *17*, 3373–3376. [CrossRef] [PubMed]
67. Azofra, L.M.; Scheiner, S. Tetrel, chalcogen, and CH···O hydrogen bonds in complexes pairing carbonyl-containing molecules with 1, 2, and 3 molecules of CO_2. *J. Chem. Phys.* **2015**, *142*, 034307. [CrossRef] [PubMed]
68. Domagała, M.; Lutyńska, A.; Palusiak, M. Halogen bond versus hydrogen bond: The many-body interactions approach. *Int. J. Quantum Chem.* **2017**, *117*, e25348. [CrossRef]
69. Sánchez-Sanz, G.; Alkorta, I.; Elguero, J. Theoretical study of intramolecular interactions in peri-substituted naphthalenes: Chalcogen and hydrogen bonds. *Molecules* **2017**, *22*, 227. [CrossRef] [PubMed]
70. Zhao, C.; Lu, Y.; Zhu, Z.; Liu, H. Theoretical exploration of halogen bonding interactions in the complexes of novel nitroxide radical probes and comparison with hydrogen bonds. *J. Phys. Chem. A* **2018**, *122*, 5058–5068. [CrossRef] [PubMed]
71. Dunning, T.H.J. Gaussian basis sets for use in correlated molecular calculations. I. The atoms boron through neon and hydrogen. *J. Chem. Phys.* **1989**, *90*, 1007–1023. [CrossRef]
72. Woon, D.E.; Dunning, T.H., Jr. Gaussian basis sets for use in correlated molecular calculations. V. Core-valence basis sets for boron through neon. *J. Chem. Phys.* **1995**, *103*, 4572–4585. [CrossRef]
73. Frisch, M.J.; Trucks, G.W.; Schlegel, H.B.; Scuseria, G.E.; Robb, M.A.; Cheeseman, J.R.; Scalmani, G.; Barone, V.; Mennucci, B.; Petersson, G.A.; et al. *Gaussian 09*; Revision B.01; Gaussian Inc.: Wallingford, CT, USA, 2009.
74. Boys, S.F.; Bernardi, F. The calculation of small molecular interactions by the differences of separate total energies. Some procedures with reduced errors. *Mol. Phys.* **1970**, *19*, 553–566. [CrossRef]
75. Latajka, Z.; Scheiner, S. Primary and secondary basis set superposition error at the SCF and MP2 levels: $H_3N–Li^+$ and $H_2O–Li^+$. *J. Chem. Phys.* **1987**, *87*, 1194–1204. [CrossRef]
76. Bulat, F.A.; Toro-Labbé, A.; Brinck, T.; Murray, J.S.; Politzer, P. Quantitative analysis of molecular surfaces: Areas, volumes, electrostatic potentials and average local ionization energies. *J. Mol. Model.* **2010**, *16*, 1679–1691. [CrossRef] [PubMed]
77. Reed, A.E.; Curtiss, L.A.; Weinhold, F. Intermolecular interactions from a natural bond orbital, donor-acceptor viewpoint. *Chem. Rev.* **1988**, *88*, 899–926. [CrossRef]
78. Bader, R.F.W. *Atoms in Molecules, A Quantum Theory*; Clarendon Press: Oxford, UK, 1990; p. 438.
79. Bader, R.F.W. *AIM2000*; 2.0; McMaster University: Hamilton, ON, Canada, 2000.
80. Su, P.; Li, H. Energy decomposition analysis of covalent bonds and intermolecular interactions. *J. Chem. Phys.* **2009**, *131*, 014102. [CrossRef] [PubMed]
81. Schmidt, M.W.; Baldridge, K.K.; Boatz, J.A.; Elbert, S.T.; Gordon, M.S.; Jensen, J.H.; Koseki, S.; Matsunaga, N.; Nguyen, K.A.; Su, S.; et al. General atomic and molecular electronic structure system. *J. Comput. Chem.* **1993**, *14*, 1347–1363. [CrossRef]

82. Bauzá, A.; Quiñonero, D.; Deyà, P.M.; Frontera, A. Halogen bonding versus chalcogen and pnicogen bonding: A combined Cambridge structural database and theoretical study. *CrystEngComm* **2013**, *15*, 3137–3144. [CrossRef]
83. Liu, X.; Cheng, J.; Li, Q.; Li, W. Competition of hydrogen, halogen, and pnicogen bonds in the complexes of HArF with XH_2P (X = F, Cl, and Br). *Spectrochim. Acta A Mol. Biomol. Spectrosc.* **2013**, *101*, 172–177. [CrossRef] [PubMed]
84. Shukla, R.; Chopra, D. "Pnicogen bonds" or "chalcogen bonds": Exploiting the effect of substitution on the formation of P···Se noncovalent bonds. *Phys. Chem. Chem. Phys.* **2016**, *18*, 13820–13829. [CrossRef] [PubMed]
85. Guo, X.; An, X.; Li, Q. Se···N Chalcogen bond and Se···X halogen bond involving $F_2C=Se$: Influence of hybridization, substitution, and cooperativity. *J. Phys. Chem. A* **2015**, *119*, 3518–3527. [CrossRef] [PubMed]
86. Esrafili, M.D.; Akhgarpour, H. An ab initio study on competition between pnicogen and chalcogen bond interactions in binary $XHS:PH_2X$ complexes (X = F, Cl, CCH, COH, CH_3, OH, OCH_3 and NH_2). *Mol. Phys.* **2016**, *114*, 1847–1855. [CrossRef]
87. Esrafili, M.D.; Mohammadirad, N. Characterization of σ-hole interactions in 1:1 and 1:2 complexes of YOF_2X (X = F, Cl, Br, I; Y = P, As) with ammonia: Competition between halogen and pnicogen bonds. *Struct. Chem.* **2016**, *27*, 939–946. [CrossRef]
88. Esrafili, M.D.; Mohammadian-Sabet, F. σ-Hole bond tunability in $YO_2X_2:NH_3$ and $YO_2X_2:H_2O$ complexes (X = F, Cl, Br; Y = S, Se): Trends and theoretical aspects. *Struct. Chem.* **2016**, *27*, 617–625. [CrossRef]
89. Jiao, Y.; Liu, Y.; Zhao, W.; Wang, Z.; Ding, X.; Liu, H.; Lu, T. Theoretical study on the interactions of halogen-bonds and pnicogen-bonds in phosphine derivatives with Br_2, BrCl, and BrF. *Int. J. Quantum Chem.* **2017**, *117*, e25443. [CrossRef]
90. Benz, S.; Poblador-Bahamonde, A.I.; Low-Ders, N.; Matile, S. Catalysis with pnictogen, chalcogen, and halogen bonds. *Angew. Chem. Int. Ed.* **2018**, *57*, 5408–5412. [CrossRef] [PubMed]
91. Scheiner, S. On the properties of X···N noncovalent interactions for first-, second- and third-row X atoms. *J. Chem. Phys.* **2011**, *134*, 164313. [CrossRef] [PubMed]
92. Adhikari, U.; Scheiner, S. The S···N noncovalent interaction: Comparison with hydrogen and halogen bonds. *Chem. Phys. Lett.* **2011**, *514*, 36–39. [CrossRef]
93. Nepal, B.; Scheiner, S. Long-range behavior of noncovalent bonds. Neutral and charged H-bonds, pnicogen, chalcogen, and halogen bonds. *Chem. Phys.* **2015**, *456*, 34–40. [CrossRef]
94. Adhikari, U.; Scheiner, S. Sensitivity of pnicogen, chalcogen, halogen and H-bonds to angular distortions. *Chem. Phys. Lett.* **2012**, *532*, 31–35. [CrossRef]
95. Adhikari, U.; Scheiner, S. Substituent effects on Cl···N, S···N, and P···N noncovalent bonds. *J. Phys. Chem. A* **2012**, *116*, 3487–3497. [CrossRef] [PubMed]
96. Scheiner, S. Assembly of effective halide receptors from components. Comparing hydrogen, halogen, and tetrel bonds. *J. Phys. Chem. A* **2017**, *121*, 3606–3615. [CrossRef] [PubMed]
97. Scheiner, S. Highly selective halide receptors based on chalcogen, pnicogen, and tetrel bonds. *Chem. Eur. J.* **2016**, *22*, 18850–18858. [CrossRef] [PubMed]
98. Scheiner, S. Comparison of halide receptors based on H, halogen, chalcogen, pnicogen, and tetrel bonds. *Faraday Discuss. Chem. Soc.* **2017**, *203*, 213–226. [CrossRef] [PubMed]
99. Scheiner, S.; Lu, J. Halogen, chalcogen, and pnicogen bonding involving hypervalent atoms. *Chem. Eur. J.* **2018**, *24*, 8167–8177. [CrossRef] [PubMed]

Sample Availability: Sample Availability: Not available.

© 2018 by the authors. Licensee MDPI, Basel, Switzerland. This article is an open access article distributed under the terms and conditions of the Creative Commons Attribution (CC BY) license (http://creativecommons.org/licenses/by/4.0/).

Article

Complexes of CO_2 with the Azoles: Tetrel Bonds, Hydrogen Bonds and Other Secondary Interactions

Janet E. Del Bene [1,*], José Elguero [2] and Ibon Alkorta [2,*]

1 Department of Chemistry, Youngstown State University, Youngstown, OH 44555, USA
2 Instituto de Química Médica (IQM-CSIC), Juan de la Cierva, 3, E-28006 Madrid, Spain; iqmbe17@iqm.csic.es
* Correspondence: jedelbene@ysu.edu (J.E.D.B.); ibon@iqm.csic.es (I.A.);
  Tel.: +1-330-609-5593 (J.E.D.B.); +34-91-562-29-00 (I.A.)

Academic Editor: Steve Scheiner
Received: 31 March 2018; Accepted: 11 April 2018; Published: 14 April 2018

Abstract: Ab initio MP2/aug'-cc-pVTZ calculations have been performed to investigate the complexes of CO_2 with the azoles pyrrole, pyrazole, imidazole, 1,2,3- and 1,2,4-triazole, tetrazole and pentazole. Three types of complexes have been found on the CO_2:azole potential surfaces. These include ten complexes stabilized by tetrel bonds that have the azole molecule in the symmetry plane of the complex; seven tetrel-bonded complexes in which the CO_2 molecule is perpendicular to the symmetry plane; and four hydrogen-bonded complexes. Eight of the planar complexes are stabilized by $Nx \cdots C$ tetrel bonds and by a secondary interaction involving an adjacent Ny-H bond and an O atom of CO_2. The seven perpendicular CO_2:azole complexes form between CO_2 and two adjacent N atoms of the ring, both of which are electron-pair donors. In three of the four hydrogen-bonded complexes, the proton-donor Nz-H bond of the ring is bonded to two C-H bonds, thereby precluding the planar and perpendicular complexes. The fourth hydrogen-bonded complex forms with the strongest acid pentazole. Binding energies, charge-transfer energies and changes in CO_2 stretching and bending frequencies upon complex formation provide consistent descriptions of these complexes. Coupling constants across tetrel bonds are negligibly small, but $^{2h}J(Ny\text{-}C)$ across $Nz\text{-}H \cdots C$ hydrogen bonds are larger and increase as the number of N atoms in the ring increases.

Keywords: tetrel bond; hydrogen bond; carbon dioxide; azoles; IR spectra; spin-spin coupling constants

1. Introduction

Carbonic anhydrases belong to a family of enzymes that catalyze the reversible reaction that converts carbon dioxide and water to bicarbonate ion and a proton [1,2]. The proposed mechanism of action involves hydrogen-bond formation between the threonine N-H (Thr199) and a CO_2 oxygen atom [3–11]. It is also known than azoles are good inhibitors of carbonic anhydrase [1,5,8,12–14]. Azoles are five-membered heteroaromatic compounds that contain in their rings from one nitrogen atom (pyrrole) to five nitrogen atoms (pentazole) [15]. The X-ray crystal structure of 1H-1,2,4-triazole bound through N2 to the N-H of Thr199 has been reported [16]. The interaction of CO_2 with azoles is also important in carbon dioxide capture by metal-organic frameworks [17–20], zeolitic tetrazolate frameworks [21] and microporous organic polymers [22,23].

It is of interest to explore interactions that are important in biochemical reactions, such as those between azines and azoles with CO_2. In a previous paper, we presented the structures, binding energies and other properties of CO_2: azine complexes stabilized by $N \cdots C$ tetrel bonds [24]. The azoles have pyridine-like N atoms in positions 2, 3, 4 and 5 that may act as electron-pair donors as do the azines. However, the azoles also have pyrrole-like N atoms in positions 1, 2 and 4, as illustrated in Scheme 1. These N-H bonds may act as proton donors for hydrogen-bond formation.

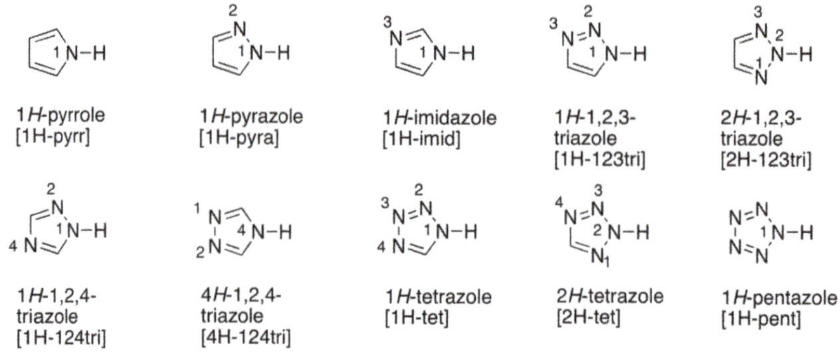

Scheme 1. The ten azoles and their codes.

A series of papers have reported experimental data on complexes relevant to those investigated in the present study. The structures of these complexes are illustrated in Scheme 2. The X-ray crystal structure of the complex between 3-amino-1H-1,2,4-triazole and CO_2 (**1**) shows a tetrel bond between the N of the amino group and the carbon of CO_2 (CSD refcodes: WALBOC and YUZCED) [25,26]. A Zn-1,2,4-triazole derivative forms a tetrel-bonded complex with CO_2 in the solid state using the free N2 atom of the triazole (**2**) (CSD refcodes: PEGBUA, PEGCAH, and PEGCEL) [27]. The hydrogen-bonded complex between a 2H-tetrazole and CO_2 has been proposed to explain the behavior of a microporous organic polymer (CSD refcode: TZPIM) [22]. Finally, complex **3** is stabilized by an N-C covalent bond (CSD refcode: EPIVOQ) [28]. A related transition structure has been proposed that is stabilized by an N···C tetrel bond.

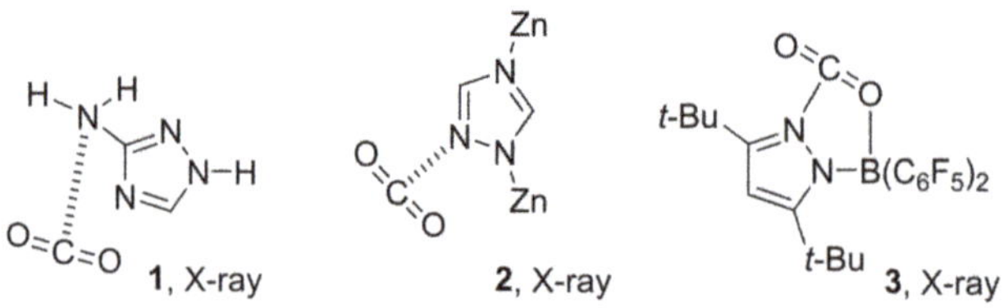

Scheme 2. Relevant experimentally-determined complexes.

There are also three papers in the literature that report theoretical studies of complexes related to those investigated in this paper. Vogiatzis, Mavrandonakis, Klopper and Froudakis have reported MP2/aug-cc-pVTZ structures of complexes of CO_2 with 1H-imid-23 and 2H-tet-34, with the CO_2 molecule lying in the plane of the azole [29]. These authors noted the presence of stabilizing interactions involving an Hα atom of the azole and CO_2. Prakash, Mathivon, Benoit, Chambaud and Hochlaf studied 1H-imid-23 and 1H-imid with CO_2 and also investigated π stacking of imidazole [30]. Hernández-Marín and Lemus-Santana studied the complexes between CO_2 and imidazole, 2-methylimidazole, benzimidazole and pyrazine using DFT methods [31]. Finally, Vidal-Vidal, Nieto Faza and Silva López extended studies of π-complexes to those with 1H-pyrrole, 1H-pyrazole, 1H-1,2,3-triazole, 2H-1,2,3-triazole, 4H-1,2,4-triazole, 2H-tetrazole and 1H-pentazole [32].

As a continuation of our work on intermolecular interactions, we present in this paper the results of a systematic study of complexes of the ten azoles with CO_2. These include two types of tetrel-bonded complexes and a set of hydrogen-bonded complexes. We report the structures of these complexes, their binding energies and charge-transfer energies, selected IR stretching and bending frequencies and changes in these frequencies upon complexation, as well as spin-spin coupling constants across tetrel and hydrogen bonds. It is the purpose of this paper to present and discuss these results.

2. Methods

The structures of the isolated CO_2 monomer, the azoles pyrrole, pyrazole, imidazole, 1,2,3- and 1,2,4-triazole, tetrazole and pentazole, and the complexes CO_2:azole were optimized at second-order Møller–Plesset perturbation theory (MP2) [33–36] with the aug'-cc-pVTZ basis set [37]. This basis set was derived from the Dunning aug-cc-pVTZ basis set [38,39] by removing diffuse functions from H atoms. Frequencies were computed to establish that these optimized structures correspond to equilibrium structures on their potential surfaces and to examine the changes in selected vibrational frequencies upon complex formation. Optimization and frequency calculations were performed using the Gaussian 09 program [40]. The binding energies ($-\Delta E$) of the complexes were computed as the negative of the reaction energy for the formation of the complex from CO_2 and the corresponding azole.

The electron density properties at bond critical points (BCPs) of complexes have been analyzed using the atoms in molecules (AIM) methodology [41–44] employing the AIMAll [45] program. The topological analysis of the electron density produces the molecular graph of each complex that identifies the location of electron density features of interest, including the electron density (ϱ) maxima associated with the various nuclei and saddle points that which correspond to BCPs. The zero gradient line that connects a BCP with two nuclei is the bond path. The natural bond orbital (NBO) method [46] has been employed to obtain the stabilizing charge-transfer interactions in complexes using the NBO-6 program [47]. Since MP2 orbitals are nonexistent, charge-transfer interactions have been computed using the B3LYP functional with the aug'-cc-pVTZ basis set at the MP2/aug'-cc-pVTZ complex geometries. This allows for the inclusion of at least some electron correlation effects.

Equation of motion coupled-cluster singles and doubles (EOM-CCSD) spin-spin coupling constants were evaluated in the CI (configuration interaction)-like approximation [48,49] with all electrons correlated. For these calculations, the Ahlrichs [50] qzp basis set was placed on ^{13}C, ^{15}N and ^{17}O, and the qz2p basis set on the ^{1}H atom bonded to N. The Dunning cc-pVDZ basis was used for the remaining H atoms. Total coupling constants were evaluated as the sum of the paramagnetic spin orbit (PSO), diamagnetic spin orbit (DSO), Fermi contact (FC) and spin dipole (SD) terms. Coupling constant calculations were performed using ACES II [51] on the HPC cluster Oakley at the Ohio Supercomputer Center.

3. Results and Discussion

3.1. Overview of the CO_2:Azole Complexes

Table 1 contains the names of the complexes, their binding energies and symmetries. The azoles are listed in Table 1 according to increasing number of nitrogen atoms. For each azole, the complexes are listed in order of decreasing binding energy. Three types of complexes have been found on the CO_2:azole surfaces, namely tetrel-bonded complexes in which the CO_2 molecule lies in the symmetry plane of the complex; tetrel-bonded complexes in which the CO_2 molecule is perpendicular to the symmetry plane; and hydrogen-bonded complexes. Planar tetrel-bonded complexes are identified as zH-azole-xy, where z refers to the location of the Nz-H covalent bond in the ring, azole identifies the particular azole molecule and xy indicates the N atom that forms the tetrel bond and an adjacent N-H or C-H that may interact with CO_2. The 2H-123tri-12 complex illustrated in Figure 1 is representative of planar tetrel-bonded complexes. For complexes in which the CO_2 molecule is perpendicular to the symmetry plane, the designation is similar, with xy referring to adjacent N atoms that have lone pairs of electrons and p indicating a perpendicular complex. 4H-124tri-12p in which N1 and N2 donate lone pairs for tetrel-bond formation is typical of these complexes and is also illustrated in Figure 1. Complexes in the third set are not tetrel-bonded, but hydrogen bonded, and are identified as zH-azole. The N1-H···O hydrogen-bonded complex 1H-imid is illustrated in Figure 1.

Table 1. Names of CO_2:azole complexes, their binding energies ($-\Delta E$, kJ·mol^{-1}) and symmetries.

Azole	Complex	$-\Delta E$	Sym
pyrrole	1H-pyrr	10.1	C_{2v}
pyrazole	1H-pyra-12	22.7	C_s
imidazole	1H-imid-23	19.9	C_s
	1H-imid	11.2	C_s
triazoles	1H-123tri-12	21.8	C_s
	2H-123tri-12	20.3	C_s
	1H-123tri-23p	15.8	C_s
	1H-124tri-12	21.0	C_s
	4H-124tri-12p	18.1	C_{2v}
	1H-124tri-45	17.9	C_s
	4H-124tri	12.9	C_{2v}
tetrazoles	1H-tet-12	20.3	C_s
	2H-tet-23	19.8	C_s
	2H-tet-12	18.9	C_s
	1H-tet-34p	15.1	C_s
	2H-tet-34p	13.7	C_s
	1H-tet-23p	13.3	C_s
pentazole	1H-pent-12	19.0	C_s
	1H-pent	17.6	C_{2v}
	1H-pent-34p	12.2	C_{2v}
	1H-pent-23p	11.2	C_s

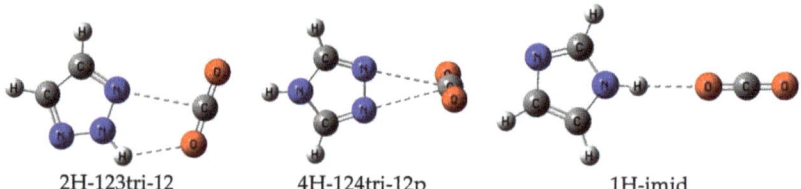

2H-123tri-12   4H-124tri-12p   1H-imid

Figure 1. Representative tetrel- and hydrogen-bonded CO_2:azole complexes.

3.2. Planar Complexes Stabilized by Tetrel Bonds

The structures, total energies and molecular graphs of planar complexes stabilized by tetrel bonds are reported in Table S1 of the Supporting Information. Table 2 reports their binding energies, charge-transfer energies and Nx-C, Ny-O′ and O′-H, or Cy-O′ and O′-H distances, with O′ the adjacent atom of CO_2. The binding energies of these complexes vary by less than 5 kJ·mol^{-1}, from 17.9 kJ·mol^{-1} for 1H-124tri-45 to 22.7 kJ·mol^{-1} for 1H-pyra-12. The Nx-C distances range from 2.781 Å in 1H-imid-23 to 3.027 Å in 1H-pent-12. However, the binding energies do not correlate with the Nx-C distances, as can be seen from the scattergram of Figure 2. To gain insight into Figure 2, it is advantageous to subdivide the planar tetrel-bonded complexes into three groups: those in which Cy-H is adjacent to Nx; those in which Ny-H is adjacent to Nx and have an Nx-C distance that is shorter than the Ny-O′ distance; and those that also have Ny-H adjacent Nx, but have an Ny-O′ distance that is shorter than the Nx-C distance.

Table 2. Binding energies (−ΔE) and charge-transfer energies (kJ·mol^{-1}), Nx-C, Ny-O', Cy-O' and H-O' distances (R, Å) for planar CO_2:azole complexes stabilized by Nx⋯C tetrel bonds.

Azole	Complex	−ΔE	R(Nx-C)	R(Ny-O'); R(Cy-O')[a]	R(NyH-O'); R(CyH-O')[a,b]	Primary CT[c]	Secondary CT[d]
pyrazole	1H-pyra-12	22.7	N2: 2.801	N1: 2.939	2.285	10.7	2.0
imidazole	1H-imid-23	19.9	N3: 2.781	C2: 3.171	2.732	13.4	1.0[d]
triazoles	1H-123tri-12	21.8	N2: 2.852	N1: 2.918	2.250	7.6	6.1
	2H-123tri-12	20.3	N1: 2.859	N2: 2.936	2.298	7.5	2.1
	1H-124tri-12	21.0	N2: 2.859	N1: 2.933	2.275	7.4	2.3
	1H-124tri-45	17.9	N4: 2.832	C5: 3.156	2.707	10.2	1.1[e]
tetrazoles	1H-tet-12	20.3	N2: 2.933	N1: 2.094	2.222	5.0	4.5
	2H-tet-23	19.8	N3: 2.917	N2: 2.904	2.252	5.3	6.3
	2H-tet-12	18.9	N1: 2.933	N2: 2.917	2.264	5.3	2.6
pentazole	1H-pent-12	19.0	N2: 3.027	N1: 2.878	2.197	3.1	4.7

[a] Ny-H or Cy-H is adjacent to Nx. [b] H-O' distances involving Cy-H are given in italics. [c] Nx$_{lp}$→σ*C-O. [d] O'$_{lp}$→σ*Ny-H. [e] The charge-transfer is O'$_{lp}$→σ*Cy-N1.

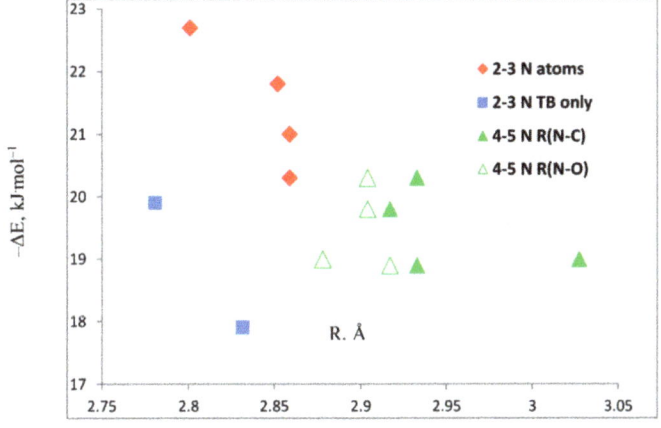

Figure 2. Binding energies versus distance for the planar tetrel-bonded CO_2:azole complexes. The solid symbols refer to the Nx-C distance; open symbols refer to the Ny-O' distance in complexes with tetrazole and pentazole.

The complexes 1H-imid-23 and 1H-124tri-45 are the two complexes that have a Cy-H bond adjacent to Nx. From Table 2 and Figure 2, it can be seen that although 1H-imid-23 has the shortest Nx-C distance, its binding energy is less than the binding energies of the planar tetrel-bonded complexes with pyrazole and the triazoles that have an Ny-H bond adjacent Nx. 1H-124tri-45 has the smallest binding energy among the planar tetrel-bonded complexes, even though its Nx-C distance is shorter than this distance in the complexes with the remaining triazoles, tetrazole and pentazole. The Cy-O' distances in these two complexes are long, and the H-Cy-O' angles are about 57°. These data suggest that Cy-H does not significantly interact with O'. This is consistent with the primary and secondary charge-transfer energies of these two complexes which are also reported in Table 2. Primary refers to the interaction associated with the tetrel bond. Figure 3 provides an orbital representation of the primary charge-transfer interaction in 1H-imid-23. The Nx lone pair donates charge to the σ antibonding C-O orbital in both 1H-imid-23 and 1H-124-tri-45. The σ antibonding C-O' orbital is the local in-plane C-O π* orbital of CO_2. The charge-transfer energies in these two complexes are greater than they are in the other planar tetrel-bonded complexes, except for 1H-pyra-12. The plot of Figure 4 shows that the primary charge-transfer energies correlate exponentially with the Nx-C distance, with a correlation coefficient of 0.978, and that these energies decrease as the number of nitrogen atoms in the ring increases. Thus, the charge-transfer energies reflect the relative strengths of the tetrel bonds in these complexes. The secondary charge-transfer energy in 1H-imid-23, which is also depicted in Figure 3, indicates that there is no Cy-H-O' interaction in this complex. Thus, it is reasonable to conclude that 1H-imid-23 and 1H-124tri-45 are stabilized solely by tetrel bonds.

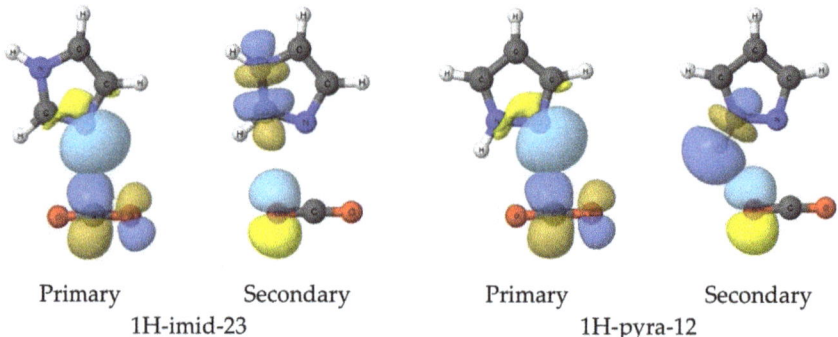

Figure 3. Orbital descriptions of the primary and secondary charge-transfer interactions in 1H-imid-23 and 1H-pyra-12.

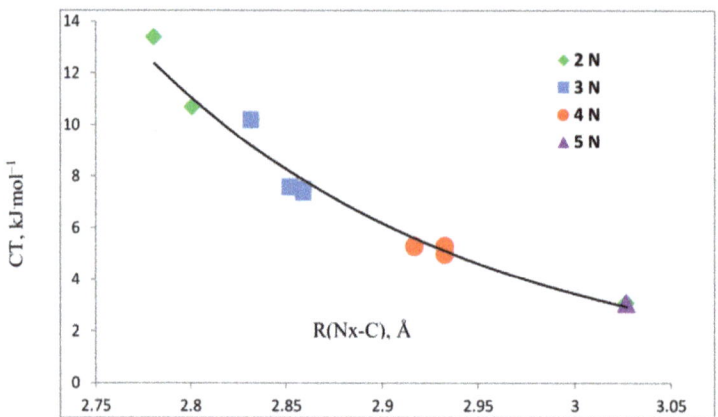

Figure 4. Primary $Nx_{lp} \rightarrow \sigma^*C\text{-}O$ charge-transfer energies versus the Nx-C distance. The legend indicates the number of nitrogen atoms in the azole ring.

The second group of planar complexes is also stabilized by Nx···C tetrel bonds and by a secondary interaction between Ny-H and O' and have Nx-C distances that are shorter than Ny-O' distances. This set is composed of complexes with pyrazole and the triazoles. As evident from Figure 2, they have the largest binding energies which a range from 20.3 to 22.7 kJ·mol^{-1} and Nx-C distances between 2.801 and 2.859 Å. The secondary interaction between Ny-H and O' may be described as a distorted hydrogen bond with H-Ny-O' angles of about 40°. Nevertheless, the interaction between Ny-H and O' must play a role in stabilizing these complexes, since the binding energies of the complexes with the triazoles are greater than the binding energy of 1H-imid-23, which has a much shorter Nx-C distance and no stabilizing secondary interaction. Figure 3 provides an orbital description of the primary and secondary charge-transfer interactions in 1H-pyra-12. The primary charge-transfer is $Nx_{lp} \rightarrow \sigma^*C\text{-}O$, with charge-transfer energies of 10.7 kJ·mol^{-1} for 1H-pyra-12 and about 7.5 kJ·mol^{-1}

for the complexes with the triazoles. These charge-transfer energies are greater than the energies of secondary back-donations of charge O'$_{lp}$→σ*Ny-H associated with the distorted hydrogen bond, which are 6 kJ·mol^{-1} for 1H-pyra-12 and about 2 kJ·mol^{-1} for the complexes with the triazoles. These data suggest that the tetrel bond is primarily responsible for the large binding energies of these complexes, with the Ny-H-O' interaction playing a secondary role.

The final group of planar complexes with tetrel bonds and Ny-H-O' interactions includes complexes of CO_2 with tetrazole and pentazole, which have Ny-O' distances that are shorter than Nx-C distances, as evident from Figure 2. These four complexes have the smallest binding energies, and the longest Nx-C distances. While the Nx-C and Ny-O' distances are similar for the tetrazole complexes, they are very different for 1H-pent-12. It may well be that the Ny-H-O' interaction in the latter complex is the stronger interaction. This interaction could be described as a distorted Ny-H···O' hydrogen bond, with H-Ny-O' angles of about 40°. The primary charge transfer associated with the tetrel bond arises from electron donation from Nx to the σ antibonding C-O orbital, while the secondary charge transfer involves back-donation of charge from O' to the antibonding σ Ny-H orbital. The secondary charge-transfer energy is greater than the primary charge transfer energy in 2H-tet-23 and 1H-pent-12. These data are consistent with the increased importance of the Ny-H-O' interaction in these complexes. In all of the planar complexes with Ny-H bonds interacting with O', the Nx-C and Ny-O' distances are the best compromise to produce a stable equilibrium complex.

Harmonic symmetric and asymmetric stretching and in-plane and out-of-plane bending frequencies of isolated CO_2 and of CO_2 in planar tetrel-bonded complexes are reported in Table S2 of the Supporting Information. The symmetric and asymmetric stretching frequencies and the out-of-plane bending frequencies of CO_2 change by less than 5 cm^{-1} upon complex formation. It is the in-plane CO_2 bending frequency that is most sensitive to complexation, since changes in this frequency most directly affect both the Nx···C tetrel bond and the Ny-H···O interaction. This frequency decreases by 9 to 31 cm^{-1} in the complexes, as evident from Table 3. Figure 5 illustrates that the in-plane bending frequency of the planar tetrel-bonded complexes decreases as the number of nitrogen atoms in the azole ring increases. The correlation coefficient of the exponential trend line is 0.951. It is interesting to note that the Ny-H stretching frequency is also changed by complex formation, thereby giving another property that supports the importance of the Ny-H-O' interaction in these complexes. This frequency is red-shifted by 22 to 30 cm^{-1} upon complexation, as evident from the plot of Figure S1 of the Supporting Information.

Table 3. CO_2 in-plane bending frequencies (ν), changes in these frequencies upon complexation (δν, cm^{-1}), and spin-spin coupling constants ^{1t}J(Nx-C) and J(Ny-O') (Hz) for planar CO_2:azole complexes stabilized by Nx···C tetrel bonds.

Azole	Complex	ν [a]	δν	^{1t}J(Nx-C)	J(Ny-O')
pyrazole	1H-pyra-12	634	−25.0	0.5	0.8
imidazole	1H-imid-23	628	−30.5	0.6	
triazoles	1H-123tri-12	640	−19.1	0.4	0.9
	2H-123tri-12	641	−18.4	0.3	1.0
	1H-124tri-12	640	−19.3	0.3	0.9
	1H-124tri-45	636	−23.4	0.4	
tetrazoles	1H-tet-12	646	−13.4	0.2	1.1
	2H-tet-23	646	−13.4	0.2	1.2
	2H-tet-12	646	−12.9	0.1	1.1
pentazoles	1H-pent-12	650	−8.6	0.0	1.5

[a] The degenerate bending frequency of isolated CO_2 is 659 cm^{-1}.

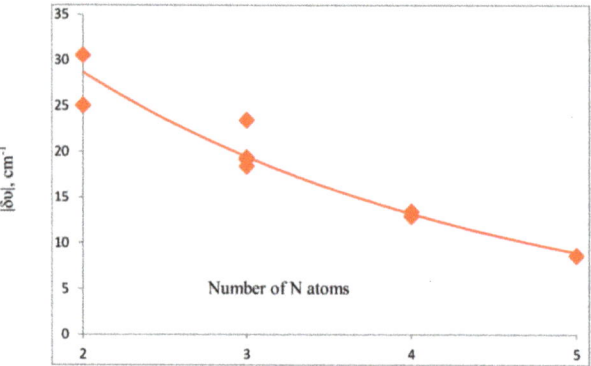

Figure 5. Absolute value of the change in the CO_2 in-plane bending frequency upon complex formation versus the number of nitrogen atoms in the azole ring.

Table 3 also reports the one-bond coupling constants $^{1t}J(Nx-C)$ across the tetrel bonds and $J(Ny-O')$ for the planar complexes. Only the FC term contributes to $^{1t}J(Nx-C)$, and this term has values between 0.0 and 0.6 Hz. These small values may be attributed to the nature of the FC term, which depends on s electron densities in the ground and excited states of the coupled nuclei. Since the tetrel bond basically forms through the π system of CO_2, there is little s-electron density at C in the direction of the Nx-C bond. Yet, despite the small values, $^{1t}J(Nx-C)$ exhibits a second-order correlation with the Nx-C distance, with a correlation coefficient of 0.978. A plot of $^{1t}J(Nx-C)$ versus the Nx-C distance is included as Figure S2 of the Supporting Information. Values of $J(Ny-O')$ are reported in Table 2 and are also small, ranging from 0.8 to 1.5 Hz. These are plotted against the Ny-O' distance in Figure S3 of the Supporting Information. The correlation coefficient of the second-order trend line is not as good, with a value of 0.877. However, what is most interesting is a comparison of Figures S2 and S3. It is evident that as the number of N atoms in the azole ring increases, $^{1t}J(Nx-C)$ decreases because the Nx-C distance increases, but $J(Ny-O')$ increases because the Ny-O' distance decreases.

3.3. Perpendicular Complexes Stabilized by Tetrel Bonds

Table S3 of the Supporting Information provides the structures, total energies and molecular graphs of the perpendicular tetrel-bonded complexes, and Figure 1 illustrates the structure of 4H-124tri-12p, which has C_{2v} symmetry. Table 4 reports the binding energies, charge-transfer energies and Nx-C and Ny-C distances of these complexes. The most stable 4H-124tri-12p complex has a binding energy of 18.1 kJ·mol^{-1} and N1-C and N2-C distances of 2.922 Å. The least stable complex is 1H-pent-23p, which has a binding energy of 11.2 kJ·mol^{-1} and N3-C and N2-C distances of 2.939 and 3.113 Å, respectively. Since there are two different N-C distances in five of the seven perpendicular complexes, the binding energies have been plotted against the average of the Nx-C and Ny-C distances. This plot has an exponential trend line with a correlation coefficient of 0.916. However, what is more informative is Figure 7, in which the binding energies of the complexes are plotted against the number of nitrogen atoms in the ring. This plot illustrates very well that the binding energies of perpendicular complexes decrease as the number of nitrogen atoms in the ring increases.

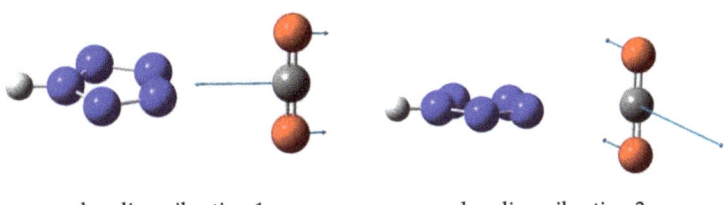

bending vibration 1   bending vibration 2

Figure 6. Representation of the two bending vibrations in the 1H-pent-34p complex.

Table 4. Binding energies ($-\Delta E$) and charge-transfer energies (CT, kJ·mol^{-1}), Nx-C and Ny-C distances (Å) and frequencies of bending vibration 1 (ν, cm^{-1}) in perpendicular azole:CO_2 complexes with tetrel bonds.

Azole	Complex	$-\Delta E$	R(Nx-C); R(Ny-C) [a]	CT [b]	ν [c,d]
triazoles	1H-123tri-23p	15.8	N2: 2.994; N3: 2.938	4.4	645
	4H-124tri-12p	18.1	N1: 2.922; N2: 2.922	5.6	644
tetrazoles	1H-tet-34p	15.1	N3: 2.959; N4: 2.981	4.9	648
	2H-tet-34p	13.7	N3: 3.021; N4: 2.967	4.2	647
	1H-tet-23p	13.3	N2: 3.088; N3: 2.954	4.1	647
pentazoles	1H-pent-34p	12.2	N3: 3.016; N4: 3.016	3.0	651
	1H-pent-23p	11.2	N2: 3.113; N3: 2.939	2.5	650

[a] Two electron-donor N atoms that form the tetrel bond. [b] The charge transfer is $(Nx_{lp} + Ny_{lp}) \rightarrow \pi^*$O-C-O. In the natural bond orbital (NBO) scheme, this is the sum of two charge-transfer interactions. [c] The degenerate bending vibrational frequency of isolated CO_2 is 659 cm^{-1}. [d] The bending vibrations are illustrated in Figure 6.

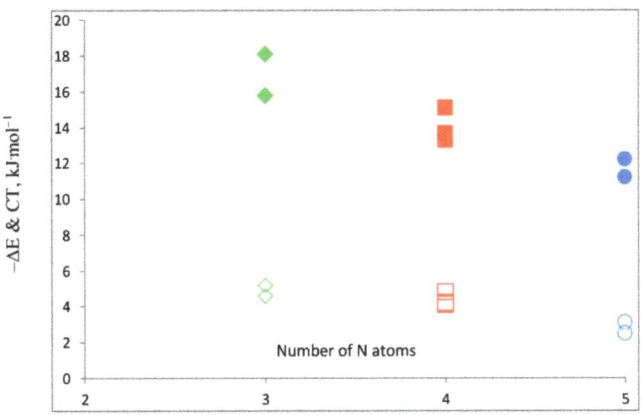

Figure 7. Binding energies (solid symbols) and charge-transfer energies (open symbols) versus the number of N atoms in the azole rings.

Charge-transfer energies for the perpendicular complexes are also reported in Table 4. These arise from electron donation from the lone pairs on Nx and Ny to the in-plane antibonding π^* O-C-O orbital of CO_2. The NBO method describes charge-transfer in these complexes as two charge-transfer interactions, and it is the sum that is reported in Table 3. The total charge-transfer energies vary from 2.5 kJ·mol^{-1} for 1H-pent-23p to 5.6 kJ·mol^{-1} for 4H-124tri-12p. Figure 7 presents a plot of the total charge-transfer energies versus the number of nitrogen atoms in the azole ring.

In the perpendicular complexes, there are four vibrational frequencies associated with the CO_2 molecule, two stretching and two bending vibrations, with the bending modes illustrated in Figure 6. Bending Vibration 1 may be roughly described as a bending motion that occurs in a plane perpendicular to the Nx-Ny bond, while Vibration 2 is a bending vibration in a plane that is parallel to the Nx-Ny bond. The stretching frequencies and bending frequency 2 are not very sensitive to complex formation, since the changes in these do not exceed 2.5 cm^{-1}, but bending vibration 1 is sensitive to complexation. As evident from Table 4, this frequency decreases by about 14 cm^{-1} in complexes with the triazoles, by 11 or 12 cm^{-1} with the tetrazoles and by 8 or 9 cm^{-1} with pentazole. The decrease in this frequency exhibits a linear dependence on the number of nitrogen atoms, with a correlation coefficient of 0.971.

Coupling constants $^{1t}J(Nx\text{-}C)$ and $^{1t}J(Ny\text{-}C)$ have been evaluated for these complexes. Only the FC term contributes to these coupling constants. However, because the carbon atom of CO_2 lies in the nodal plane, the computed FC terms are either 0.0 or 0.1 Hz.

3.4. Complexes Stabilized by Hydrogen Bonds

There are only four CO_2:azole complexes that are stabilized by Nz-H···O hydrogen bonds, 1H-pyrr, 1H-imid, 4H-124tri and 1H-pent. Table S4 of the Supporting Information provides their structures, total energies and molecular graphs, and 1H-imid is illustrated in Figure 1. Table 5 presents their binding energies, charge-transfer energies and Nz-O and Nz-H distances. The Nz-O distances are rather long, varying from 2.988 Å in 1H-pent to 3.175 Å in 1H-pyrr. The binding energies range from 10.1 kJ·mol^{-1} for 1H-pyrr to 17.6 kJ·mol^{-1} for 1H-pent, while the charge-transfer energies $O_{lp} \rightarrow \sigma^*$Nz-H range from 8.7 to 22.7 kJ·mol^{-1} in these same two complexes. Plots of these two properties versus the Nz-O distance are given in Figure 8. The correlation coefficients of the exponential trend lines are 0.987 for the binding energies and 0.999 for the charge-transfer energies. From Table 5 and Figure 8, it is apparent that the binding energies and charge-transfer energies increase as the number of N atoms in the ring increases, unlike the tetrel-bonded complexes for which the binding energies tend to decrease as the number of nitrogen atoms increases. As the number of nitrogen atoms in the ring increases, the azole molecule becomes more acidic and Nz-H becomes a better proton donor, while the azole molecule becomes a weaker base and a poorer electron-pair donor.

Table 5. Binding energies ($-\Delta E$) and charge-transfer energies (kJ·mol^{-1}), Nz-O and Nz-H distances (Å), changes in Nz-H stretching frequencies ($\delta\nu$, cm^{-1}) and spin-spin coupling constants $^{2h}J(Nz\text{-}O)$ (Hz) for CO_2:azole complexes stabilized by Nz-H···O hydrogen bonds.

Azole	Complex	$-\Delta E$	R(Nz-O)	$O_{lp} \rightarrow \sigma^*$Nz-H	$^{2h}J(Nz\text{-}O)$	R(Nz-H)	$\Delta\nu$(Nz-H)
pyrrole	1H-pyrr	10.1	3.175	8.7	2.1	1.006	−5.1
imidazole	1H-imid	11.2	3.152	10.1	2.2	1.007	−8.8
124-triazole	4H-124tri	12.9	3.112	12.3	2.6	1.008	−17.1
pentazole	1H-pent	17.6	2.988	22.7	4.6	1.014	−47.0

Table 5 also reports IR and NMR spectroscopic data for these complexes including the change in the Nz-H stretching frequency upon complex formation and the two-bond NMR coupling constant $^{2h}J(Nz\text{-}O)$ across the hydrogen bond. For a typical X-H···Y hydrogen bond, there is a red shift, that is a shift to lower energy of the X-H stretching frequency. The data of Table 5 show that this frequency decreases in the CO_2:azole complexes relative to the corresponding isolated azole. The red shifts range from 5 to 47 cm^{-1} and increase as the number of nitrogen atoms in the ring and the Nz-H distance in the complexes increase. The linear trend line that relates the Nz-H stretching frequency to the Nz-H distance has a correlation coefficient of 0.996. The second spectroscopic property of interest is the NMR coupling constant $^{2h}J(Nz\text{-}O)$ across the hydrogen bond. This coupling constant varies from 10.1 Hz in 1H-pyrr to 17.6 Hz in 1H-pent. The second-order trend line that relates $^{2h}J(Nz\text{-}O)$ to the Nz-O distance has a correlation coefficient of 0.9997.

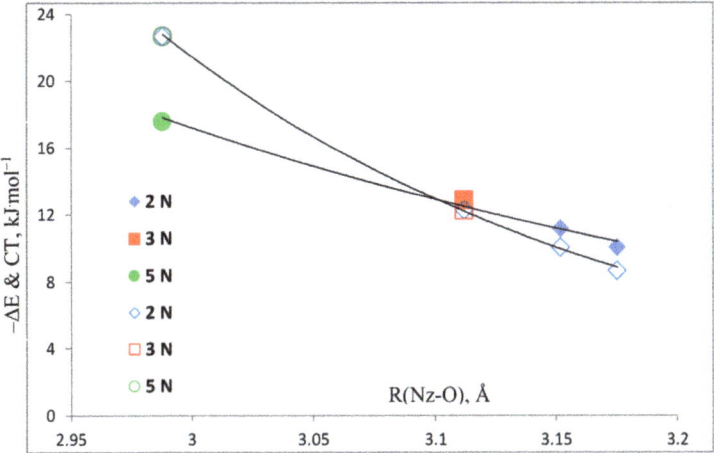

Figure 8. Binding energies (solid symbols) and charge-transfer energies (open symbols) versus the Nz-O distance.

As evident from Table 5, there are only four CO_2:azole complexes that are stabilized by hydrogen bonds. However, since there are ten different azoles, namely pyrrole, pyrazole, imidazole and pentazole, and two tautomers each of 1,2,3-triazole, 1,2,4-triazole and tetrazole, which differ in the position of the Nz-H bond, it would not be unreasonable to expect that there might be 10 hydrogen-bonded complexes. Why are there only four? Some insight into the answer to this question comes from the structures of the complexes 1H-pyrr, 1H-imid and 4H-124tri. In these, Nz-H is bonded to two C-H groups in the ring. Without an adjacent N atom with a lone pair of electrons available to form a tetrel bond, only an essentially linear Nz-H···O hydrogen bond can form. In contrast, pentazole does have N atoms with lone-pairs adjacent to Nz-H, and it does form one planar tetrel-bonded complex. However, the basicity of the azoles decreases and the acidity increases as the number of N atoms in the ring increases [52]. As a result, pentazole also forms a hydrogen-bonded complex with CO_2, and it is the most stable among these complexes.

4. Conclusions

Ab initio MP2/aug'-cc-pVTZ calculations have been performed to investigate complexes of CO_2 with the azoles pyrrole, pyrazole, imidazole, 1,2,3- and 1,2,4-triazole, tetrazole and pentazole. The results of these calculations support the following statements.

1. Three types of complexes have been found on the potential surfaces. These include ten complexes stabilized by tetrel bonds in which the azole molecule lies in the symmetry plane of the complex and seven complexes also stabilized by tetrel bonds, but which have the azole molecule perpendicular to the symmetry plane. In addition, there are four hydrogen-bonded complexes.
2. The ten complexes stabilized by tetrel bonds that have the azole molecule in the symmetry plane of the complex have some common characteristics.

    a. Those complexes that have an Ny-H group bonded to Nx are stabilized primarily by Nx···C tetrel bonds and by a secondary interaction between Ny-H and O', which assumes increased importance as the number of N atoms in the ring increases.

    b. The binding energies of the planar complexes do not correlate with the Nx-C distance, but the primary charge-transfer energies do.

c. The IR in-plane bending frequency of CO_2 is most sensitive to complex formation. The change in this frequency decreases as the number of N atoms increases. NMR spin-spin coupling constants $^{1t}J(Nx-C)$, which are FC dominated, are less than 1 Hz, since there is little s-electron density at C in the direction of the tetrel bond.

3. There are seven perpendicular tetrel-bonded complexes that arise when there are two adjacent N atoms in the ring, and each has a lone pair of electrons.

   a. The binding energies of perpendicular complexes decrease as the number of nitrogen atoms in the ring decreases.
   b. The IR bending mode of CO_2 that moves the C atom toward and away from the Nx-Ny bond is most sensitive to complex formation. The change in its frequency upon complex formation decreases in the order triazole > tetrazole > pentazole.
   c. The NMR coupling constants $^{1t}J(Nx-C)$ and $^{1t}J(Ny-C)$ are negligibly small since there is a node at C in the complex symmetry plane.

4. The four hydrogen-bonded complexes involve pyrrole, imidazole, 1,2,4-triazole and pentazole. Three of these form when the ring Nz-H is bonded to two C-H groups, thereby eliminating the possibility of tetrel-bond formation. The fourth forms with pentazole, which is the strongest acid.

   a. The binding energies of these complexes and their charge-transfer energies increase as the number of N atoms in the ring increases.
   b. Hydrogen bonding produces a red-shift of the IR Nz-H stretching band. The magnitude of the red shift increases as the number of N atoms in the ring increases.
   c. The NMR coupling constant $^{2h}J(Nz-O)$ across the hydrogen bond increases as the number of nitrogen atoms in the ring increases.

Supplementary Materials: Structures, total energies and molecular graphs of CO_2:azole complexes; bending and stretching frequencies of CO_2 in planar complexes; plots of stretching frequencies and coupling constants versus distances and/or the number of N atoms in the ring.

Acknowledgments: This work was carried out with financial support from the Ministerio de Economía, Industria y Competitividad (Project No. CTQ2015-63997-C2-2-P) and Comunidad Autónoma de Madrid (S2013/MIT2841, Fotocarbon). Thanks are also given to the Ohio Supercomputer Center and Centro Técnico de Informática (CTI-CSIC) for their continued computational support.

Author Contributions: I.A. and J.E.D.B. did the calculations. J.E.D.B., J.E. and I.A. contributed equally to the writing of this paper.

Conflicts of Interest: The authors declare no conflict of interest.

References

1. Supuran, C.T.; Scozzafava, A.; Conway, J. (Eds.) *Carbonic Anhydrase: Its Inhibitors and Activators*; CRC Press Book: Boca Raton, FL, USA, 2004.
2. Frost, S.C.; McKenna, R. (Eds.) *Carbonic Anhydrase: Mechanism, Regulation, Links to Disease, and Industrial Applications*; Springer: Dordrecht, The Netherlands, 2014.
3. Vedani, A.; Dunitz, J.D. Lone-Pair Directionality in Hydrogen Bond Potential Functions for Molecular Mechanics Calculations: The Inhibition of Human Carbonic Anhydrase II by Sulfonamides. *J. Am. Chem. Soc.* **1985**, *107*, 7653–7658. [CrossRef]
4. Eriksoon, A.E.; Jones, T.A.; Liljas, A. Refined Structure of Human Carbonic Anhydrase II at 2.0 Å Resolution. *Proteins* **1988**, *4*, 274–282. [CrossRef] [PubMed]
5. Liljas, A.; Hakansoon, K.; Jonsson, B.H.; Xue, Y. Inhibition and Catalysis of Carbonic Anhydrase. Recent Crystallographic Analyses. *Eur. J. Biochem.* **1994**, *219*, 1–10. [CrossRef] [PubMed]
6. Kiefer, L.L.; Paterno, S.A.; Fierke, C.A. Hydrogen Bond Network in the Metal Binding Site of Carbonic Anhydrase Enhances Zinc Affinity and Catalytic Efficiency. *J. Am. Chem. Soc.* **1995**, *117*, 6831–6837. [CrossRef]

7. Thoms, S. Hydrogen Bonds and the Catalytic Mechanism of Human Carbonic Anhydrase II. *J. Theor. Biol.* **2002**, *215*, 399–404. [CrossRef] [PubMed]
8. Supuran, C.T.; Vullo, D.; Manole, G.; Casini, A.; Scozzafava, A. Designing of Novel Carbonic Anhydrase Inhibitors and Activators. *Curr. Med. Chem.* **2004**, *2*, 49–68. [CrossRef]
9. Domsic, J.F.; Avvaru, B.S.; Kim, C.U.; Gruner, S.M.; Agbandje-McKenna, M.; Silverman, D.N.; McKenna, R. Entrapment of Carbon Dioxide in the Active Site of Carbonic Anhydrase II. *J. Biol. Chem.* **2008**, *283*, 30766–30771. [CrossRef] [PubMed]
10. Sjöblom, B.; Polentarutti, M.; Djinović-Carugo, K. Structural Study of X-ray Induced Activation of Carbonic Anhydrase. *PNAS* **2009**, *106*, 10609–10613. [CrossRef] [PubMed]
11. Avvaru, B.S.; Kim, C.U.; Sippel, K.H.; Gruner, S.M.; Agbandje-McKenna, M.; Silverman, D.N.; McKenna, R. A Short, Strong Hydrogen Bond in the Active Site of Human Carbonic Anhydrase II. *Biochemistry* **2010**, *49*, 249–251. [CrossRef] [PubMed]
12. Alberti, G.; Bertini, I.; Luchinat, C.; Scozzafava, A. A New Class of Inhibitors Capable of Binding Both the Acidic and Alkaline Forms of Carbonic Anhydrase. *Biochim. Biophys. Acta Protein Struct. Mol.* **1981**, *668*, 16–26. [CrossRef]
13. Ilies, M.; Banciu, M.D.; Ilies, M.A.; Scozzafava, A.; Caproiu, M.T.; Supuran, C.T. Carbonic Anhydrase Activators: Design of High Affinity Isozymes I, II, and IV Activators, Incorporating Tri-/Tetrasubstituted-Pyridinium-Azole Moieties. *J. Med. Chem.* **2002**, *45*, 504–510. [CrossRef] [PubMed]
14. Chen, D.; Oezguen, N.; Urvil, P.; Ferguson, C.; Dann, S.M.; Savidge, T.C. Regulation of Protein-Ligand Binding Affinity by Hydrogen Bond Pairing. *Sci. Adv.* **2016**, *2*, e1501240. [CrossRef] [PubMed]
15. Schofield, K.; Grimmett, M.R.; Keene, B.R.T. *The Azoles*; Cambridge University Press: Cambridge, UK, 1976.
16. Mangani, S.; Liljas, A. Crystal Structure of the Complex Between Human Carbonic Anhydrase II and the Aromatic Inhibitor 1,2,4-Triazole. *J. Mol. Biol.* **1993**, *232*, 9–14. [CrossRef] [PubMed]
17. Sumida, K.; Rogow, D.L.; Mason, J.A.; McDonald, T.M.; Bloch, E.D.; Herm, Z.R.; Bae, T.-H.; Long, J.R. Carbon Dioxide Capture in Metal–Organic Frameworks. *Chem. Rev.* **2012**, *112*, 724–781. [CrossRef] [PubMed]
18. Saha, S.; Chandra, S.; Garai, B.; Banerjee, R. Carbon Dioxide Capture by Metal Organic Frameworks. *Indian J. Chem.* **2012**, *51A*, 1223–1230.
19. Seth, S.; Savitha, G.; Moorthy, J.N. Carbon Dioxide Capture by a Metal–Organic Framework with Nitrogen-Rich Channels Based on Rationally Designed Triazole-Functionalized Tetraacid Organic Linker. *Inorg. Chem.* **2015**, *54*, 6829–6835. [CrossRef] [PubMed]
20. Li, P.-Z.; Wang, X.-J.; Liu, J.; Lim, J.S.; Zou, R.; Zhao, Y. A Triazole-Containing Metal–Organic Framework as a Highly Effective and Substrate Size-Dependent Catalyst for CO_2 Conversion. *J. Am. Chem. Soc.* **2016**, *138*, 2142–2145. [CrossRef] [PubMed]
21. Panda, T.; Pachfule, P.; Chen, Y.; Jiang, J.; Banerjee, R. Amino Functionalized Zeolitic Tetrazolate Framework (Ztf) with High Capacity for Storage of Carbon Dioxide. *Chem. Commun.* **2011**, *47*, 2011–2013. [CrossRef] [PubMed]
22. Du, N.; Park, H.B.; Robertson, G.P.; Dal-Cin, M.M.; Visser, T.; Scoles, L.; Guiver, M.D. Polymer Nanosieve Membranes for CO_2-Capture Applications. *Nat. Mater.* **2011**, *10*, 372–375. [CrossRef] [PubMed]
23. Kenarsari, S.D.; Yang, D.; Jiang, G.; Zhang, S.; Wang, J.; Russell, A.G.; Wei, Q.; Fan, M. Review of Recent Advances in Carbon Dioxide Separation and Capture. *RSC Adv.* **2013**, *3*, 22739–22773. [CrossRef]
24. Alkorta, I.; Elguero, J.; Del Bene, J.E. Azines as Electron-Pair Donors to CO_2 for N···C Tetrel Bonds. *J. Phys. Chem. A* **2017**, *121*, 8017–8025. [CrossRef] [PubMed]
25. Vaidhyanathan, R.; Iremonger, S.S.; Shimizu, G.K.H.; Boyd, P.G.; Alavi, S.; Woo, T.K. Direct Observation and Quantification of CO_2 Binding within an Amine-Functionalized Nanoporous Solid. *Science* **2010**, *330*, 650–653. [CrossRef] [PubMed]
26. Liao, P.-Q.; Zhang, W.-X.; Zhang, J.-P.; Chen, X.-M. Efficient Purification of Ethene by an Ethane-Trapping Metal-Organic Framework. *Nat. Commun.* **2015**, *6*, 8697. [CrossRef] [PubMed]
27. Liao, P.-Q.; Zhou, D.-D.; Zhu, A.-X.; Jiang, L.; Lin, R.-B.; Zhang, J.-P.; Chen, X.-M. Strong and Dynamic CO_2 Sorption in a Flexible Porous Framework Possessing Guest Chelating Claws. *J. Am. Chem. Soc.* **2012**, *134*, 17380–17383. [CrossRef] [PubMed]
28. Theuergarten, E.; Schlosser, J.; Schluns, D.; Freytag, M.; Daniliuc, C.G.; Jones, P.G.; Tamm, M. Fixation of Carbon Dioxide and Related Small Molecules by a Bifunctional Frustrated Pyrazolylborane Lewis Pair. *Dalton Trans.* **2012**, *41*, 9101–9110. [CrossRef] [PubMed]

29. Vogiatzis, K.D.; Mavrandonakis, A.; Klopper, W.; Froudakis, G.E. Ab Initio Study of the Interactions between CO_2 and N-Containing Organic Heterocycles. *ChemPhysChem* **2009**, *10*, 374–383. [CrossRef] [PubMed]
30. Prakash, M.; Mathivon, K.; Benoit, D.M.; Chambaud, G.; Hochlaf, M. Carbon Dioxide Interaction with Isolated Imidazole or Attached on Gold Clusters and Surface: Competition between σ H-Bond and π-Stacking Interaction. *PCCP* **2014**, *16*, 12503–12509. [CrossRef] [PubMed]
31. Hernández-Marín, E.; Lemus-Santana, A.A. Theoretical Study of the Formation of Complexes between CO_2 and Nitrogen Heterocycles. *J. Mex. Chem. Soc.* **2015**, *59*, 36–42.
32. Vidal-Vidal, Á.; Faza, O.N.; Silva López, C. CO_2 Complexes with Five-Membered Heterocycles: Structure, Topology, and Spectroscopic Characterization. *J. Phys. Chem. A* **2017**, *121*, 9118–9130. [CrossRef] [PubMed]
33. Pople, J.A.; Binkley, J.S.; Seeger, R. Theoretical Models Incorporating Electron Correlation. *Int. J. Quantum Chem. Quantum Chem. Symp.* **1976**, *10*, 1–19. [CrossRef]
34. Krishnan, R.; Pople, J.A. Approximate Fourth-Order Perturbation Theory of the Electron Correlation Energy. *Int. J. Quantum Chem.* **1978**, *14*, 91–100. [CrossRef]
35. Bartlett, R.J.; Silver, D.M. Many—Body Perturbation Theory Applied to Electron Pair Correlation Energies. I. Closed-Shell First Row Diatomic Hydrides. *J. Chem. Phys.* **1975**, *62*, 3258–3268. [CrossRef]
36. Bartlett, R.J.; Purvis, G.D. Many-Body Perturbation Theory, Coupled-Pair Many-Electron Theory, and the Importance of Quadruple Excitations for the Correlation Problem. *Int. J. Quantum Chem.* **1978**, *14*, 561–581. [CrossRef]
37. Del Bene, J.E. Proton Affinities of Ammonia, Water, and Hydrogen Fluoride and Their Anions: A Quest for the Basis-Set Limit Using the Dunning Augmented Correlation-Consistent Basis Sets. *J. Phys. Chem.* **1993**, *97*, 107–110. [CrossRef]
38. Dunning, T.H. Gaussian Basis Sets for Use in Correlated Molecular Calculations. I. The Atoms Boron through Neon and Hydrogen. *J. Chem. Phys.* **1989**, *90*, 1007–1023. [CrossRef]
39. Woon, D.E.; Dunning, T.H. Gaussian Basis Sets for Use in Correlated Molecular Calculations. V. Core-Valence Basis Sets for Boron Through Neon. *J. Chem. Phys.* **1995**, *103*, 4572–4585. [CrossRef]
40. Frisch, M.J.; Trucks, G.W.; Schlegel, H.B.; Scuseria, G.E.; Robb, M.A.; Cheeseman, J.R.; Scalmani, G.; Barone, V.; Mennucci, B.; Petersson, G.A.; et al. *Gaussian09*; revision D.01; Gaussian, Inc.: Wallingford, CT, USA, 2009.
41. Bader, R.F.W. A Quantum Theory of Molecular Structure and its Applications. *Chem. Rev.* **1991**, *91*, 893–928. [CrossRef]
42. Bader, R.F.W. *Atoms in Molecules: A Quantum Theory*; Oxford University Press: Oxford, UK, 1990.
43. Popelier, P.L.A. *Atoms in Molecules: An Introduction*; Prentice Hall: Harlow, UK, 2000.
44. Matta, C.F.; Boyd, R.J. *The Quantum Theory of Atoms in Molecules: From Solid State to DNA and Drug Design*; Wiley-VCH: Weinheim, Germany, 2007.
45. Keith, T.A. *AIMAll*; Version 17.11.14; TK Gristmill Software: Overland Park, KS, USA, 2017. Available online: aim.tkgristmill.com (accessed on 1 March 2018).
46. Reed, A.E.; Curtiss, L.A.; Weinhold, F. Intermolecular Interactions from a Natural Bond Orbital, Donor–Acceptor Viewpoint. *Chem. Rev.* **1988**, *88*, 899–926. [CrossRef]
47. Glendening, E.D.; Badenhoop, J.K.; Reed, A.E.; Carpenter, J.E.; Bohmann, J.A.; Morales, C.M.; Landis, C.R.; Weinhold, F. *NBO 6.0*; University of Wisconsin: Madison, WI, USA, 2013.
48. Perera, S.A.; Nooijen, M.; Bartlett, R.J. Electron Correlation Effects on the Theoretical Calculation of Nuclear Magnetic Resonance Spin-Spin Coupling Constants. *J. Chem. Phys.* **1996**, *104*, 3290–3305. [CrossRef]
49. Perera, S.A.; Sekino, H.; Bartlett, R.J. Coupled-Cluster Calculations of Indirect Nuclear Coupling Constants: The Importance of Non-Fermi Contact Contributions. *J. Chem. Phys.* **1994**, *101*, 2186–2196. [CrossRef]
50. Schäfer, A.; Horn, H.; Ahlrichs, R. Fully Optimized Contracted Gaussian Basis Sets for Atoms Li to Kr. *J. Chem. Phys.* **1992**, *97*, 2571–2577. [CrossRef]
51. Stanton, J.F.; Gauss, J.; Watts, J.D.; Nooijen, M.; Oliphant, N.; Perera, S.A.; Szalay, P.S.; Lauderdale, W.J.; Gwaltney, S.R.; Beck, S.; et al. *ACES II*; University of Florida: Gainesville, FL, USA, 1991.
52. Catalan, J.; Elguero, J. Basicity and Acidity of Azoles. *Adv. Heterocycl. Chem.* **1987**, *41*, 187–274.

© 2018 by the authors. Licensee MDPI, Basel, Switzerland. This article is an open access article distributed under the terms and conditions of the Creative Commons Attribution (CC BY) license (http://creativecommons.org/licenses/by/4.0/).

Article

Comparison between Tetrel Bonded Complexes Stabilized by σ and π Hole Interactions

Wiktor Zierkiewicz [1,*], Mariusz Michalczyk [1] and Steve Scheiner [2,*]

1. Faculty of Chemistry, Wrocław University of Science and Technology, Wybrzeże Wyspiańskiego 27, 50370 Wrocław, Poland; mariusz.michalczyk@pwr.edu.pl
2. Department of Chemistry and Biochemistry, Utah State University, Logan, UT 84322-0300, USA
* Correspondence: wiktor.zierkiewicz@pwr.edu.pl (W.Z.); steve.scheiner@usu.edu (S.S.); Tel.: +48-071-320-3455 (W.Z.); +1-435-797-7419 (S.S.)

Academic Editor: Steve Scheiner
Received: 30 May 2018; Accepted: 9 June 2018; Published: 11 June 2018

Abstract: The σ-hole tetrel bonds formed by a tetravalent molecule are compared with those involving a π-hole above the tetrel atom in a trivalent bonding situation. The former are modeled by TH_4, TH_3F, and TH_2F_2 (T = Si, Ge, Sn) and the latter by $TH_2=CH_2$, $THF=CH_2$, and $TF_2=CH_2$, all paired with NH_3 as Lewis base. The latter π-bonded complexes are considerably more strongly bound, despite the near equivalence of the σ and π-hole intensities. The larger binding energies of the π-dimers are attributed to greater electrostatic attraction and orbital interaction. Each progressive replacement of H by F increases the strength of the tetrel bond, whether σ or π. The magnitudes of the maxima of the molecular electrostatic potential in the two types of systems are not good indicators of either the interaction energy or even the full Coulombic energy. The geometry of the Lewis acid is significantly distorted by the formation of the dimer, more so in the case of the σ-bonded complexes, and this deformation intensifies the σ and π holes.

Keywords: MP2; DFT; NBO; MEP; AIM

1. Introduction

Recent years have seen a veritable explosion of research into noncovalent interactions that are analogous to the H-bond. The proton acting as a bridge between the two subunits in the H-bond can be replaced by any of a number of more electronegative atoms, without loss of binding strength. Depending upon the chemical family to which this bridging atom belongs, these noncovalent bonds have been denoted as halogen, chalcogen, and pnicogen bonds [1-10]. Data has also accumulated that this sort of bonding can also involve the inert gas atoms in aerogen bonds [11] and even the coinage metal atoms in so-called regium bonds [12]. All of these interactions have a number of features in common. Asymmetrical distribution of electron density around the bridging atom typically leads to one or more σ-hole [13–25] of positive electrostatic potential. Each such σ-hole is situated directly opposite a covalent bond involving the atom of interest, and can attract a nucleophile. To this Coulombic attraction is added other attractive forces identified with charge transfer, polarization, and dispersion.

Another of this set of noncovalent bonds which has begun to garner widespread attention is the tetrel bond, in which the bridging atom belongs to the C/Si Group 14 of the periodic table. Tetrel bonds play an essential role in numerous processes, as for instance the first stages of S_N2 reactions which are important in organic synthesis [26]. Other works include a study of the carbon bond as representative of tetrel bonds [27], acetonitrile complexes with tetrahalides [28], examples derived from crystal structures [29], steric crowding in FTR_3 (T = Si, Ge, Sn, Pb) complexes with various Lewis bases [30], factors controlling the strength of tetrel bonds [31], as well as a recent paper regarding the implications of deformation of the tetrel-containing molecule when paired with ammonia, pyrazine and nitrogen cyanide [32].

As study of these noncovalent bonds progressed it soon became apparent that σ-holes are not the only regions of positive potential that may be present. In cases where the bridging atom lies in a planar (or nearly planar) bonding environment, a positive area can develop above this molecular plane [33]. Like σ-holes, these π-holes serve as sites of attraction for an approaching nucleophile [34–42]. There have been a number of studies of noncovalent bonds of various sorts that have examined both σ and π-holes, and more interestingly, comparisons between the two [26,33,43–46]. Lastly, a valuable supplement to this matter concerning the molecular orbital theory-based description of σ, π and δ holes was introduced by Angarov and Kozuch [47]. It is stated there that many chalcogen and pnicogen bonds should be termed as hybrid σ/π hole interactions rather than simple σ-hole. However, these sorts of comparisons are largely absent in the context of tetrel bonds. Given the importance of tetrel bonds, and the preponderance of molecules in which both σ and π holes may be present, a thorough and comprehensive understanding of the forces that contribute to both, and how they compare with one another, is of paramount importance.

It is to this problem that this work is devoted. Systems are developed in which σ and π tetrel bonding may be directly compared with one another in a controlled fashion by quantum chemical calculations. The molecular electrostatic potential is evaluated for each Lewis acid, which reveals all plausible sites of attachment of a nucleophile, and geometry optimizations reveal which of these sites actually result in an equilibrium dimer. One is able to determine how accurate a measure of the binding strength are the intensities of the σ and π-holes. It is also possible to go one step further and assess whether the hole intensity in and of itself is an accurate indicator of the full Coulombic interaction between the two molecules. Beyond this, how does the latter electrostatic term compare with other attractive forces such as charge transfer and dispersion? Given the prior observation that tetrel bonds lead to sizable geometric deformations of the monomers [32,48], how do such distortions factor into the binding energy of the σ and π tetrel bonds? And as a bottom line, how do the strengths of σ and π tetrel bonds compare with one another?

2. Systems and Computational Methods

Tetravalent $TH_{4-n}F_n$ molecules, with T = Si, Ge, and Sn, were taken as systems which contain σ-holes of varying strength. As a point of comparison, $TH_{2-n}F_n=CH_2$ molecules place the T atom in a planar trivalent bonding situation, which can be expected to contain π-holes above the T atom. One can adjust the number n of electron-withdrawing F atoms in each molecule so as to modulate the strength of these two sorts of holes, and still facilitate a fair comparison. It is also possible to assess how sensitive the findings might be to the identity of the particular tetrel atom by comparing Si with Ge and Sn. NH_3 was taken as the universal electron donor, due first to the ready availability of its lone electron pair. The presence of only one such pair, coupled with the small size of this molecule, allows an unambiguous evaluation of the properties of the tetrel bond, minimizing any complicating secondary interactions.

All geometries were optimized at the MP2 level in conjunction with the aug-cc-pVDZ basis set [49,50]. For the Sn atom, the aug-cc-pVDZ-PP basis set from the EMSL library was applied for the purpose of including relativistic effects [51,52]. All complexes were characterized as minima by frequency analysis calculations. The interaction energies of the complexes were evaluated as the difference in energy between the dimer and the sum of the two monomers, frozen in the same geometry as in the dimer, then corrected for basis set superposition error (BSSE) by the standard counterpoise procedure [53]. The deformation energies of the two subunits were assessed as the difference in electronic energy between each unit within the geometry of the complex and that of the fully optimized isolated molecule. Computations were carried out with the Gaussian 09 suite of programs [54]. Energy decomposition analysis (EDA) was performed at the BLYP-D3(BJ)/ZORA/TZ2P level using the ADF program [55–57]. The molecular electrostatic potentials (MEPs) of the isolated monomers were evaluated on the electron density isosurface of $\varrho = 0.001$ au at the MP2/aug-cc-pVDZ level, and its extrema were determined using the WFA-SAS program [58]. MP2 electron densities were analyzed via AIM in order to identify the bond critical points (BCPs) [59] and to evaluate their properties. In order to incorporate electron correlation into the NBO analysis of interorbital electron transfer, the BLYP-D3(BJ) functional was applied within the context of the def2TZVPP basis set via the GenNBO program [60].

3. Results and Discussion

3.1. Electrostatic Potentials of Isolated Molecules

The molecular electrostatic potential (MEP) of the tetravalent, approximately tetrahedral TH_4, TH_3F and TH_2F_2 (T = Si, Ge or Sn) isolated molecules are displayed in Figure 1; analogous MEPs are shown in Figure 2 for the trivalent $TH_{2-n}F_n=CH_2$ analogues which are roughly planar.

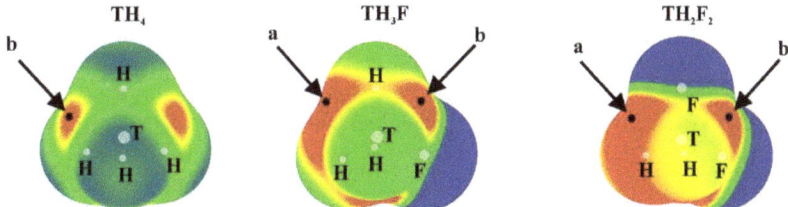

Figure 1. MEPs of TH_4, TH_3F and TH_2F_2 (T = Si, Ge or Sn) computed on the 0.001 au isodensity surface at the MP2/aug-cc-pVDZ-PP level. Colour ranges, in kcal/mol, are: red greater than 15, yellow between 8 and 15, green between 0 and 8, blue below 0 kcal/mol. The letters a and b mean different types of $V_{s,max}$.

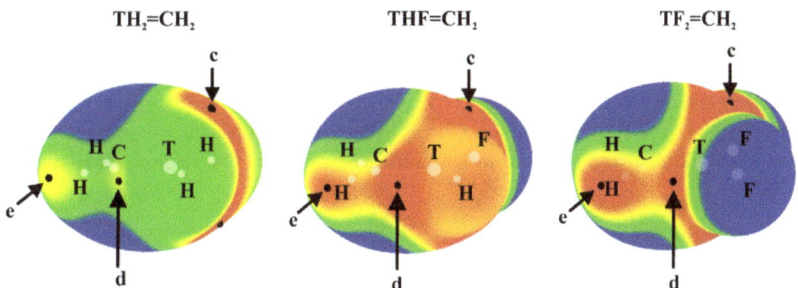

Figure 2. MEPs of $TH_{2-n}F_n=CH_2$ isolated molecules, computed on the 0.001 au isodensity surface at the MP2/aug-cc-pVDZ-PP level. Colour ranges, in kcal/mol, are: red greater than 15, yellow between 8 and 15, green between 0 and 8, blue below 0 kcal/mol. The letters c, d and e mean different types of $V_{s,\,max}$.

Positive values of the MEP are denoted in red, while blue represents negative regions. Each of the tetrahedral molecules in Figure 1 contains four σ-holes lying on the extension of each of the four covalent bonds. Due to its symmetry, all four of these MEP maxima are equivalent in TH_4. There are two types of maxima in the fluorosubstituted species: those opposite F are labeled a, and the b designation is applied to those opposite a H atom. The values of these maxima are collected in Table 1, where it is immediately obvious that a σ-holes opposite F atoms are more intense than their b analogues opposite the H atom. This pattern is consonant with the much greater electronegativity of F; the ratio of a/b values of $V_{s,max}$ varies between 1.4 and 1.8, and their numerical values are consistent with previous studies [24]. Another expected pattern evident in Table 1 is the increase in $V_{s,max}$ as progressively more F atoms are added to the molecule. One normally expects the hole to intensify as the tetrel atom is enlarged. While Sn certainly corresponds to the largest values of $V_{s,max}$, Si and Ge are less distinct from one another.

Table 1. Values of two maxima in the MEPs ($V_{s,max}$, kcal/mol) of tetravalent σ-hole donors at the MP2/aug-cc-pVDZ-PP level of theory.

T	$V_{s,max}$ [a]	TH_4	TH_3F	TH_2F_2
Si	a	-	41.8	43.9
	b	19.8	26.4	31.1
Ge	a	-	46.0	49.8
	b	18.4	25.3	30.4
Sn	a	-	54.8	59.6
	b	25.0	31.7	37.8

[a] a and b maxima lie respectively on the extensions of T–F and T–H bonds (see Figure 1).

The planar $TH_{2-n}F_n=CH_2$ molecules contain three primary types of MEP maximum as shown in Figure 2 (their values are given in Table 2). The first, and generally the most intense, is labeled c and occurs roughly above (and below) the T atom, skewed away from the C atom by a certain amount. Maximum d lies in the molecular plane, in a position corresponding roughly to the C=T bond midpoint, approximately on an extension of the T–H or T–F covalent bond.

In most cases, with the sole exceptions of $GeHF=CH_2$, and $SnHF=CH_2$, maximum c is considerably more positive than is d (see below for further discussion). The last maximum e is associated with the two CH_2 protons. This position would be pertinent to the formation of any possible CH···N H-bonds with an approaching NH_3 nucleophile. (Several other maxima appear in some of these molecules but are much weaker in intensity.) Focusing on maximum c, the site of the π-hole, one sees a clear intensification as H atoms are replaced by F. On the other hand, the expected trend of growing intensity with tetrel atom size is violated. Although Sn does indeed produce the largest π-holes, Si exceeds its larger Ge congener. The d patterns are more consistent with expectations, with the caveat that the addition of the second F atom reduces $V_{s,max}$. This lowering is sensible because the proximity of the very electronegative F atom to the hole would mitigate against its positive value. It might be noted here that several of the molecules in Figure 2 are not strictly planar. This point will be discussed in greater detail below.

Table 2. Values of maxima in the MEPs ($V_{s,max}$, kcal/mol) of $TR_2=CH_2$ π-hole donors, at the MP2/aug-cc-pVDZ-PP level of theory.

T	$V_{s,max}$ [a]	$TH_2=CH_2$	$THF=CH_2$	$TF_2=CH_2$
Si	c	21.0	32.4	48.8
	d	10.1	23.9	19.5
	e	12.1	16.3	18.8
Ge	c	19.4	29.8	44.8
	d	10.8	32.5	27.0
	e	12.2	19.2	24.0
Sn[b]	c	24.0	34.8	53.3
	d	14.8	43.6	37.3
	e	11.1	19.5	25.1

[a] Locations of the maxima are displayed in Figure 2. [b] In the $SnH_2=CH_2$ molecule there is another $V_{s,max}$ with a value of 18.1 kcal/mol located on the extension of the C=Sn bond (between two c maxima).

Finally, with respect to the ammonia molecule, the value of $V_{s,\,min}$ on the N atom at its lone pair position is -37.7 kcal/mol. Based on the positions and intensities of the various σ-holes, one would anticipate that a nucleophile such as NH_3 would be attracted to the a maximum, directly opposite the F atom if one is present, and that the strongest tetrel bonds would occur for T=Sn, followed by Ge and then by Si; TH_2F_2 ought to engage in a slightly stronger bond than would TH_3F.

3.2. σ-Hole Bonded Dimers

The optimized geometries in which NH_3 engages with the σ-holes of the tetravalent TH_4, TH_3F and TH_2F_2 molecules are illustrated in Figure 3. Consistent with the labeling in Figure 1a,b designate whether the N is located opposite the F or H atom, respectively. The interaction energies (E_{int}), corrected for BSSE, are collected in Table 3, along with the deformation energies (E_{def}) of the subunits as well as selected intermolecular geometrical parameters.

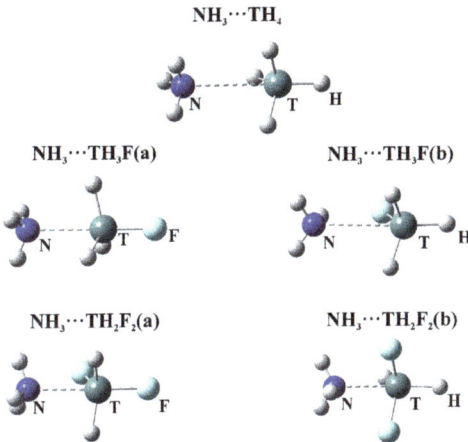

Figure 3. Optimized structures of σ-hole bonded tetrel complexes. (**a**) and (**b**) refer to σ-hole positions in Figure 1 (N—dark blue, T—green, H—white, F—Light blue).

Table 3. Interaction energy (E_{int}) corrected for BSSE, of indicated Lewis acid with NH_3 in σ-hole bonded complexes, along with deformation energy (E_{def}) of individual subunits, intermolecular distance and angle (energies in kcal/mol, distances in Å, angles in degrees). Data obtained at the MP2 level of theory.

Lewis Acid	E_{int}	E_{def} A [a]	E_{def} B [b]	R(N···T)	θ(R–T···N) [c]
SiH_4	−1.8	0.14	0	3.232	180
GeH_4	−1.59	0.11	0	3.332	179.6
SnH_4	−2.81	0.37	0	3.170	180
SiH_3F(a)	−7.43	1.93	0	2.557	180
SiH_3F(b)	−3.24	0.34	0	3.102	174
GeH_3F(a)	−7.34	1.49	0	2.630	179.9
GeH_3F(b)	−3.72	0.29	0	3.134	170.5
SnH_3F(a)	−10.29	1.78	0	2.667	180
SnH_3F(b)	−7.43	2.23	0.03	2.793	166.2
SiH_2F_2(a)	−10.42	5.07	0.02	2.390	177.6
SiH_2F_2(b)	−4.12	1.38	0	2.865	175.8
GeH_2F_2(a)	−10.84	3.97	0.02	2.458	174.3
GeH_2F_2(b)	−11.34	9.14	0.06	2.364	168.2
SnH_2F_2(a)	−15.29	3.77	0.04	2.521	169.2
SnH_2F_2(b)	−20.07	10.45	0.14	2.374	155.8

[a] Deformation energy of Lewis acid. [b] Deformation energy of Lewis base (NH_3). [c] R refers to F or H in complexes (**a**) and (**b**), respectively.

The presence of a tetrel bond is signaled first by the intermolecular R(N···T) distance which is smaller than the sum of the corresponding van der Waals radii. (This sum is equal to 3.85, 3.95 and 4.08 Å for Si, Ge and Sn, respectively.) The N atom lies very nearly directly opposite the F atom of the Lewis acid in the a dimers. The θ(R–T···N) angle in the last column of Table 3 is 180°, with the exception of TH$_2$F$_2$. Larger deviations of the θ(HT···N) angles from linearity are observed for the b complexes. These nonlinearities are due to attractive forces between the F and H atoms of the Lewis acid and base, respectively.

The interaction energies vary between less than 2 kcal/mol for the TH$_4$ molecules to as much as 20 kcal/mol for the difluorinated Lewis acids. The patterns match those of V$_{s,\,max}$ in Table 1, although imperfectly. In the first place, a dimers with the base opposite F are more strongly bound than b complexes opposite H, but this trend is reversed for GeH$_2$F$_2$ and SnH$_2$F$_2$. Whether a or b type, E$_{int}$ rises in the order Si~Ge < Sn, and also increases as more F atoms are added to the acid.

In order to more fully understand the nature of the tetrel bond, and the effects that factor into it, one must first recognize that the formation of such a bond relies on a certain amount of distortion of the monomer geometry. The crowded nature of the tetravalent bonding surrounding the tetrel atom impedes the approach of a nucleophile. Three of the substituents must be peeled back away from this nucleophile to facilitate its approach, which in turn produces a certain amount of deformation energy within the molecule. The magnitude of this deformation energy is listed in Table 3 as E$_{def}$ A for the acid. The NH$_3$ molecule need undergo only very little internal deformation so E$_{def}$ B is quite small. E$_{def}$ A is very small for the unsubstituted TH$_4$ molecules, not surprising in view of the long intermolecular separations of more than 3 Å. Monofluorination brings the N in much closer, to about 2.6 Å for the a dimers, and the deformation energies are thus larger, nearly 2 kcal/mol. The intermolecular distance is shorter after difluorination and E$_{def}$ A is correspondingly larger, 4–5 kcal/mol. Note that some of the b dimers have an even closer approach, and thus a correspondingly higher deformation energy.

These energies can be correlated to the geometrical changes within the monomers. Summation of the three θ(R$_1$TR$_2$) angles of the R substituents that come into contact with the nucleophile offers a convenient measure of these distortions. On one extreme, in a fully tetrahedral environment, this sum would be equal to three times 109.5° or 328.5°, which would change to 360° if these three substituents peel back to lie in a plane in a bipyramidal arrangement. This measure of the geometry is listed in Table 4 along with the amount it changes as a result of complexation with NH$_3$. Note that there is a very strong correlation between the latter change and deformation energy E$_{def}$ A in Table 3. In fact, the correlation coefficient is 0.999. In either case, the quantity is larger for b than for the a complex for Ge and Sn.

One would expect that the MEPs of these molecules would likewise be altered by the geometrical distortions accompanying dimerization. The effect of the deformation upon the value of V$_{s,\,max}$ is reported in the last three columns of Table 4 where it may be seen that the partial planarization yields fairly large increases in the MEP maximum, as much as 35 kcal/mol. On a percentage basis, these increases vary from 28% to a near doubling. Note also that the deformation-induced V$_{s,\,max}$ increase is especially large for the b dimers of Ge and Sn. And it is in just these complexes that one sees an anomalously large interaction energy. On the other hand, it is not just the b geometries for which V$_{s,\,max}$ grows upon deformation.

Table 4. Planarity measure and MEP maximum of TH$_2$F$_2$ molecule in its geometry within the monomer and within its complex with NH$_3$.

	Σθ(R1TR2), degs			V$_{s,\,max}$, kcal/mol		
	Monomer	Complex	Change	Monomer	Complex	Change
Si a	332.3	350.0	17.7	43.9	60.9	17.0
Si b	324.1	334.3	10.2	31.1	42.9	11.8
Ge a	335.1	351.0	15.9	49.8	63.7	13.9
Ge b	320.7	346.9	26.2	30.4	58.9	28.5
Sn a	337.1	352.9	15.8	59.6	74.0	14.4
Sn b	318.4	347.5	29.1	37.8	72.8	35.0

The MEP maximum rises also in the a structures, albeit by not as much in the Ge and Sn cases. As a net result, $V_{s,max}$ is larger for a than for b in all of the complexes in Table 4, so one cannot explain the larger interaction energies for the latter solely in terms of MEP. There are of course other aspects of the interaction besides electrostatic attraction. Table 5 presents other components based on an EDA analysis, viz. orbital interaction E_{oi} and dispersion E_{disp}. E_{elec} contributes a fairly consistent 52–65% of the total attractive force, differing little between a and b structures. Dispersion makes a smaller contribution, especially in the more strongly bound dimers where it amounts to only about 5%. The orbital interaction term is perhaps more interesting, particularly for the TH_2F_2 systems. Parallel to the full E_{def}, E_{oi} is larger for the b dimers than for a for both T=Ge and Sn, but the reverse is true for T=Si. It would thus appear that a large part of this pattern can be traced to orbital interactions.

Table 5. EDA/BLYP-D3(BJ)/ZORA/TZ2P decomposition of the interaction energy of σ-hole bonded complexes into Pauli repulsion (E_{Pauli}), electrostatic (E_{elstat}), orbital interaction (E_{oi}) and dispersion (E_{disp}) terms. All energies in kcal mol^{-1}. The relative values in percent express the contribution of each to the sum of all attractive energy terms.

Lewis Acid	ΔE	E_{Pauli}	E_{elec}	%	E_{oi}	%	E_{disp}	%
SiH_4	−2.12	5.8	−4.2	53	−1.98	25	−1.73	22
GeH_4	−1.69	5.03	−3.52	52	−1.53	23	−1.67	25
SnH_4	−3.04	10.55	−8.12	60	−3.18	23	−2.29	17
SiH_3F(a)	−8.64	29.45	−22.44	59	−12.88	34	−2.77	7
SiH_3F(b)	−3.66	8.78	−7.43	60	−2.85	23	−2.15	17
GeH_3F(a)	−7.29	27.28	−21.29	62	−10.59	31	−2.7	8
GeH_3F(b)	−3.95	9.52	−8.24	61	−2.99	22	−2.24	17
SnH_3F(a)	−9.92	33.5	−27.86	64	−12.54	29	−3.01	7
SnH_3F(b)	−7.54	28.65	−23.26	64	−9.85	27	−3.07	8
SiH_2F_2(a)	−11.22	48.26	−36	61	−20.34	34	−3.11	5
SiH_2F_2(b)	−4.8	16.04	−13.24	64	−4.99	24	−2.6	12
GeH_2F_2(a)	−10	45.54	−34.9	63	−17.59	32	−3.05	5
GeH_2F_2(b)	−10.53	62.38	−46.24	63	−23.32	32	−3.34	5
SnH_2F_2(a)	−14.16	50.53	−41.9	65	−19.42	30	−3.36	5
SnH_2F_2(b)	−18.91	75.93	−61.53	65	−29.58	31	−3.73	4

This supposition is confirmed by NBO analysis of the charge transfer. Table S1 demonstrates that two measures of charge transfer conform to the trends listed above. The total intermolecular charge transfer CT is computed as the sum of atomic charge on either monomer. ΣE(2) represents the energetic consequence of transfers from particular molecular orbitals, in this case from the N lone pair to the four antibonding σ*(T–R) orbitals. Both of these parameters are larger for the b than for the a dimer for Ge and Sn, but smaller for Si. And furthermore, they are also larger for a than for b for all the monofluorinated TH_3F molecules, as was the case for the full interaction energy.

An alternate means of analyzing the molecular interactions derives from AIM treatment of the topology of the total electron density. Diagrams of the various dimers are provided in Figure S1 for the illustrative Ge set of dimers where small green dots indicate the position of bond critical points. The density, density Laplacian, and total electron energy at the intermolecular bond critical points are collected in Table S2. It might first be noted that there are certain anomalies in this data. In addition to the expected T···N bond paths, there are a number of bond paths placed by AIM between N and certain F atoms of the Lewis acid. Such bonds are reported only for the b type dimers, but not in all cases. The presence of a true N···F bond would contribute to the stability of these geometries. In one case, SiH_2F_2(b), a bond path connects N with one of the H atoms of the Lewis acid. Indeed in this case, AIM does not provide evidence of a T···N tetrel bond at all. Dispensing with these anomalies, there are patterns in the AIM data that are consistent with the full energetics. The AIM measures of the Ge···N and Sn···N tetrel bonds in TH_2F_2(b) are larger than those for the a analogue, while the opposite may be said for all three TH_3F dimers.

In summary, the σ-hole directly opposite the F atom is consistently much more positive than one opposite H. Nonetheless, due to a combination of factors, that include deformation-induced intensification, and a greater degree of charge transfer, the latter position becomes competitive with the former as a site for tetrel bonding, and can even surpass the location opposite F as a preferred binding site in certain cases.

3.3. π-Hole Bonded Complexes

As indicated in Figure 2, the MEPs of the planar $TH_{2-n}F_n=CH_2$ molecules have maxima (c) above the molecular plane, in the plane near the C=T midpoint (d), and (e) associated with the CH_2 protons. The c regions represent the π-hole above the T atom so are the focus of the calculations. The structures of the relevant complexes with NH_3 are illustrated in Figure 4, and their energetics and geometric details reported in Table 6.

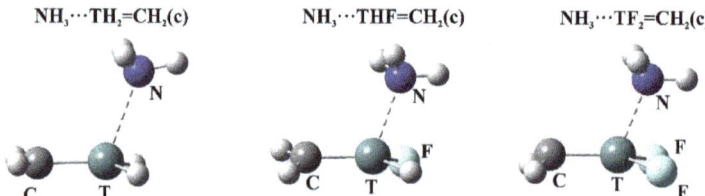

Figure 4. Optimized structures of π-hole bonded tetrel complexes (N—dark blue, T—green, H—white, F—light blue).

Table 6. Interaction energy (E_{int}) corrected for BSSE, subunit deformation energy (E_{def}), and intermolecular geometrical parameters (energies in kcal/mol, distances in Å, angles in degrees) in π-hole bonded complexes with NH_3. Data obtained at the MP2 level of theory.

Lewis Acid	E_{int}	E_{int} (Planar) [a]	E_{def} A	E_{def} B	R(N···T)	θ(R–T···N)
$SiH_2=CH_2$	−7.82	−3.57	2.08	0.10	2.176	113.3
$GeH_2=CH_2$	−3.72	−2.79	0.88	0.04	2.460	112.0
$SnH_2=CH_2$	−5.79	−4.80	0.63	0.05	2.582	104.6
$SiHF=CH_2$	−19.64	−8.10	6.29	0.15	2.052	111.9
$GeHF=CH_2$	−14.13	−6.71	4.84	0.14	2.184	110.8
$SnHF=CH_2$	−19.37	−10.49	6.41	0.18	2.356	100.2
$SiF_2=CH_2$	−28.30	−15.70	5.81	0.15	2.003	116.2
$GeF_2=CH_2$	−27.26	−14.90	8.01	0.16	2.094	111.5
$SnF_2=CH_2$	−29.02	−19.17	6.75	0.21	2.296	106.4

[a] Lewis acid molecule restrained to planarity.

As in the σ-hole complexes, all T···N distances are shorter than the sum of the van der Waals radii of the corresponding atoms. The θ(R–T···N) angles are all greater than 90°, reflecting the position of the π-hole maximum. E_{int} varies from a minimum value of 3.7 kcal/mol all the way up to nearly 30 kcal/mol. Just as in the case of $V_{s,max}$ for these π-holes, E_{int} increases steadily as H atoms are replaced by F, with large increments in both quantities associated with each such substitution.

As in the case of the tetravalent σ-hole complexes described above, formation of the π-hole dimers also impose a certain geometric distortion into the monomers. The deformation energies listed in Table 6 are not insignificant, particularly for the mono and difluorinated species for which E_{def} A varies between 5 and 8 kcal/mol. In this same vein, the various $TH_{2-n}F_n=CH_2$ monomers are not all fully planar and become even less so upon formation of the π-hole dimers. It is a matter of some interest how the interactions might be affected if these molecules were forced to be fully planar within the context of the dimer. Comparison of the second and third columns of Table 6 reveals that such a restriction would severely diminish the interaction energy. This reduction varies from only 1 kcal/mol for GeH$_2$=CH$_2$ and SnH$_2$=CH$_2$, but can be as large as 12 kcal/mol for some of the fluorinated species. As a rule of thumb, the various fluorinated Lewis acids lose roughly half of their interaction energy if forced into a planar conformation. But at the same time, it should be stressed that even these reduced interaction energies, in the framework of enforced planarity, still exceed those of the σ-hole dimers in Table 3. The EDA interaction energy contributions of the π-dimers are listed in Table 7.

Table 7. EDA/BLYP-D3(BJ)/ZORA/TZ2P decomposition of the interaction energy of π-hole bonded complexes into Pauli repulsion (E$_{Pauli}$), electrostatic (E$_{elstat}$), orbital interaction (E$_{oi}$) and dispersion (E$_{disp}$) terms. All energies in kcal/mol. The relative values in percent express the contribution of each to the sum of all attractive energy terms.

Lewis Acid	E$_{int}$	E$_{Pauli}$	E$_{elec}$	%	E$_{oi}$	%	E$_{disp}$	%
SiH$_2$=CH$_2$	−9.15	101.19	−64.42	58	−43.04	39	−2.88	3
GeH$_2$=CH$_2$	−4.24	54.47	−35.18	60	−20.63	35	−2.91	5
SnH$_2$=CH$_2$	−6.73	54.61	−38.58	63	−19.77	32	−3.00	5
SiHF=CH$_2$	−19.87	128.97	−86.78	58	−58.74	40	−3.17	2
GeHF=CH$_2$	−12.33	107.09	−72.17	60	−44.06	37	−3.19	3
SnHF=CH$_2$	−19.09	84.72	−65.89	63	−34.47	33	−3.46	3
SiF$_2$=CH$_2$	−27.53	139.3	−97.5	58	−66.02	40	−3.31	2
GeF$_2$=CH$_2$	−24.73	122.68	−88.6	60	−55.41	38	−3.39	2
SnF$_2$=CH$_2$	−26.98	88.69	−74.07	64	−38.05	33	−3.55	3

As in the σ-hole dimers, electrostatics contribute roughly 58–64% of the total attractive interaction. Dispersion is considerably smaller in the π complexes, less than 5%. Orbital interactions account for the difference, making up some 32–40%, as compared to roughly 30% for the σ-dimers. Perhaps more revealing are the absolute values of these components. Both the electrostatic and orbital interaction energies are much larger in magnitude for the π-dimers in Table 7 than for the σ-complexes in Table 5. For example, E$_{elec}$ for the three TH$_4$ complexes vary between 3.5 and 8.1 kcal/mol, whereas the analogous values for the corresponding TH$_2$=CH$_2$ systems lie in the 35.2–64.4 kcal/mol range. The monofluorinated σ dimers cover the 21.3–27.9 range, which is greatly exceeded by the 65.9–86.8 kcal/mol range for the corresponding π-dimers. The same sort of enlargement of the π vs σ complexes is observed in the orbital interaction energies. It is only the dispersion component which is quite similar for the two types of complexes. (This similarity may be due to the use of the Grimme empirical correction, which is not sensitive to the variation of the wave function [61].)

The enlarged contribution from orbital interactions is verified by NBO analysis. As reported in Table S3, the total charge transfer is quite substantial, varying between 113 and 197 me, larger than the same quantities observed for the σ-hole dimers in Table S1. The same amplification applies to the sum of E(2) interorbital transfers, which reach up to nearly 80 kcal/mol in some cases. The magnitudes of these quantities do not closely match the interaction energies. For example, the charge transfers are greatest for Si, as compared to Ge and Sn although the dimers involving Si are not the most strongly bound.

Unlike the σ-hole dimers, the AIM molecular diagrams indicate only a single intermolecular bond path, which corresponds to the T···N tetrel bond, as illustrated in Figure S2. The numerical values of the properties of each bond critical point are displayed in Table S5. Like the interaction energies in Table 6, each successive replacement of H by F adds an increment. The comparisons between the three tetrel atoms are, however, not as clear. Taking the three THF=CH$_2$ acids as an example, Ge presents the weakest dimer, whereas it shows the largest ϱ$_{BCP}$ and H. Comparisons show that the AIM indicators of tetrel bond strength are considerably larger for the π than for the σ-hole tetrel bonds, consistent with the energetic data.

It was pointed out above that the tetravalent TR$_4$ molecules undergo significant distortion upon complexation with NH$_3$, which in turn enlarges their σ-hole. Table 8 compiles the same sort of data for the π-bonding $TR_2=CH_2$ molecules where the deformation from planarity about both the C and the T atoms are measured by the deviation from 360° of the sum of the three bond angles in which they engage. As may be seen from the second column in Table 8 this nonplanarity only occurs for the difluorinated $GeF_2=CH_2$ and $SnF_2=CH_2$ monomers, and is more exaggerated for the C atom. However, all species become significantly nonplanar in the π-bonded dimers. These deformations about the C atom are fairly small, and only occur for fluorinated species, obeying the T = Si < Ge < Sn pattern. Perhaps more to the point of the interaction of NH_3 with the T atom, these nonplanar deformations are fairly small, less than 10°.

Table 8. Planarity measure of $TR_2=CH_2$ molecule in its geometry within the monomer and its π-bonded c complex with NH_3.

	$\Sigma\theta(R_1CR_2)$, degs			$\Sigma\theta(R_1TR_2)$, degs		
	Monomer	Complex	Change	Monomer	Complex	Change
$SiH_2=CH_2$	360	359.9	−0.1	359.9	353.9	−6.0
$GeH_2=CH_2$	360	359.6	−0.4	359.9	356.8	−3.1
$SnH_2=CH_2$	360	359.6	−0.4	360	359	−1.0
$SiHF=CH_2$	359.9	359.6	−0.3	360	353.3	−6.7
$GeHF=CH_2$	360	359	−1.0	360	355.6	−4.4
$SnHF=CH_2$	359.8	343.4	−16.4	359.9	359.8	−0.1
$SiF_2=CH_2$	360	359.5	−0.5	360	352.5	−7.5
$GeF_2=CH_2$	353.7	345.6	−8.1	357.9	357	−0.9
$SnF_2=CH_2$	337.5	327.2	−10.3	351.6	359.8	+8.2

Contrary to the C deformations, the T nonplanarities follow the opposite Si > Ge > Sn pattern. (It is interesting that the $SnF_2=CH_2$ molecule actually becomes more planar about the Sn atom upon complexation.) In summary, the geometrical distortions induced by π-tetrel bonding are less severe than in the σ-bonded cases, where the deformation measures ranged all the way up to nearly 30°. As in the case of the σ-bonded systems, the deformations of the π-bonding $TR_2=CH_2$ molecules also raise the value of $V_{s,max}$, as is evident in Table 9.

Table 9. Magnitude of $V_{s,max}$ (kcal/mol) on T atom of isolated $TR_2=CH_2$ molecule and its value when the molecule is distorted to that within the π-bonded c complex.

	Monomer	Complex	Change
$SiH_2=CH_2$	21.0	23.0	2.0
$GeH_2=CH_2$	19.4	20.6	1.2
$SnH_2=CH_2$	24.0	24.4	0.4
$SiHF=CH_2$	32.4	39.2	6.8
$GeHF=CH_2$	29.8	34.7	4.9
$SnHF=CH_2$	34.8	49.1	14.3
$SiF_2=CH_2$	48.8	54.3	5.5
$GeF_2=CH_2$	44.8	58.4	13.6
$SnF_2=CH_2$	53.3	78.5	25.2

This increase is quite small for $TH_2=CH_2$ but grows as F substituents are added. Just as the trivalent molecules undergo larger geometrical perturbations than do their tetravalent sisters, so too are the π-hole enhancements smaller than those observed in the σ-holes.

3.4. Other Geometries

In addition to the c maximum in the MEP of the planar Lewis acids, there is also a d maximum located in the approximate molecular plane, in the vicinity of the T=C midpoint, as detailed in Figure 2. However, optimization of the dimer geometry does not necessarily lead to a minimum in the potential energy surface with the NH_3 in this position. It is only for the monosubstituted $THF=CH_2$ molecule that such a configuration represents a minimum. In some sense this structure resembles a σ-hole dimer, with N situated directly opposite the F atom, rather than a π-dimer. The AIM molecular diagram confirms this to be a T···N tetrel bond for Ge and Sn although the bond path for the former is much more curved than is usually the case, as illustrated in Figure S3. But it must be added that this tetrel bond vanishes for the Si system in Figure S3a, leaving only two weak H···N interactions, whose ϱ and $\nabla^2 \varrho$ values just barely meet the criteria of hydrogen bonds.

As may be seen in Table S5, these d dimers are also more weakly bound than the c π-dimers: the former span an E_{int} range between 2.5 and 8.9 kcal/mol, in comparison to the 14.1–19.4 kcal/mol range of the latter. This comparative weakness is in contrast to the values of $V_{s,max}$ in Table 2, for which the d maxima are comparable to, and even exceed the c values. The weaker nature of the d minima extends beyond energetics, encompassing also longer N···T distances, and lower E(2) energies, charge transfer, and electronic properties of the BCPs as well, with details contained in Tables S6 and S7. Given the values of $V_{s,\,max}$ in Table 2, it is perhaps not entirely surprising that it is only the $THF=CH_2$ unit that engages in this d bonding. More specifically, the c maximum is much larger than d for both $TH_2=CH_2$ and $TF_2=CH_2$; it is only the monofluorinated species for which the two maxima have comparable values. The EDA results obtained for d complexes are provided in Table S8. As in their analogous c complexes, electrostatic energy contributes about 52–64% of the total attractive interaction, while dispersion is considerably larger, from 9 to 27% in $SnHF=CH_2$ and $SiHF=CH_2$, respectively. Therefore, the largest contribution of E_{disp} is for the least stable d complex. Orbital interactions in these complexes account for about 23% (average value) which is smaller than those in their stronger c cousins (average value of 37%).

In addition to the dimer geometries described above there were a number of secondary minima, all quite a bit weaker than those described above, none with E_{int} larger than 2 kcal/mol. These weak secondary minima are displayed in Table S9 for the tetrahedral TH_4, TH_3F and TH_2F_2 molecules, along with their calculated properties. Most dimers are held together by weak H-bonds, and none show any evidence of containing any sort of tetrel bond. Table S10 contains the analogous secondary minima for the planar Lewis acids. Again the primary attractive forces are weak H-bonds and the total interaction energies are rather small.

3.5. Discussion

Although the tetrel bond has not been studied as intensively as some of its cousins, e.g., the H-bond or halogen bond, there are nevertheless some prior data that offer points of comparison and context with our own results. The study of complexes of TH_4 and its mono, tri, and tetrafluorinated derivatives with ammonia (T=Si, Ge, Sn) [31] led to similar conclusions for this different subset of systems. Comparison between intensities of σ-holes exhibits strong similarities and the same trends as those examined here. This earlier work had shown how incorporating monomer deformation energies into the full energetics can lead to somewhat different patterns than the interactions between pre-deformed subunits. Another recent study [62] places the same σ-hole donors in complexes with various π-electron systems acting as Lewis bases. The same Si < Ge < Sn pattern was found there as for the weaker b complexes above, somewhat different than for the more strongly bound a complexes. This work also noted that geometry deformation of the Lewis acid can be negligible, but becomes important for the stronger complexes. Decomposition of interaction energies revealed that the complexes are electrostatically driven and dispersion becomes significant only when the complexes are exceptionally weakly bonded. The vital role of the Pauli repulsion which exceeded the absolute value of the electrostatic component was also noticed. Our results are consistent with these observations. One factor driving the small values of dispersion energy may be the small size of the base, including only a single non-hydrogen atom.

There have been a number of prior studies comparing σ- and π-hole bonded systems. Li's group [35] paired $F_2C=CFTF_3$ with three Lewis bases including formaldehyde, water, and ammonia, and found π-hole bonded complexes were generally preferred for T=C but the opposite for Si and Ge. With particular respect to NH$_3$, the bonding grew in strength as the tetrel atom became larger for both σ- and π-hole complexes, in partial agreement with our results which showed some deviations from this pattern. The interaction energies correlated with the σ-hole intensity of T which was, in turn, strongly associated with the hybridization of C atoms in the order sp^3 < sp^2 < sp. A recent [26] perspective article indicates the dominating influence of electrostatic and dispersive terms in both weak σ- and π-hole dimers, in complexes whose deformation energies are close to 0, which was confirmed by Xu et al. [33] based on TH$_3$F (T=C and Si) complexes with pyrazine and 1,4-dicyanobenzene. As in the current work, the π-complexes were more stable than their σ counterparts in terms of larger interaction energy, also exhibiting shorter binding distance, greater electron density at BCPs, and larger CT. Also consistent with the data reported above was the distribution of attractive and repulsive components of the interaction, and the consistency with the magnitudes of MEP maxima. Distinctions arise on shifting from tetrel to aerogen atoms. In our own earlier study of aerogen bonds formed between AeOF$_2$ (Ae = Kr, Xe) and diazines [43], the σ-hole bonded complexes were considerably stronger than their π-hole analogues.

In the context of the replacement of H atoms by the much more electronegative F, it is typically observed that the interaction grows stronger with each such substitution. For example, early work suggested that tetrafluorosilane was bound to ammonia more tightly than unsubstituted silane [63]. This conclusion was confirmed in later calculations confined to Si [64,65] as well as in the other works that extended to complexes containing heavier tetrel atoms [31,66], and is consistent with our own findings above.

It has been shown in the literature that there are systems where the intensities of the MEP maxima or minima are not necessarily well correlated with interaction energies [32,67,68]. For instance, in the tetrel-bonded complexes of formamidine with TH$_3$F (T = C, Si, Ge, and Sn) the interaction energy increases in the order C < Ge < Si < Sn, inconsistent with the magnitude of the σ-hole on the T atom [68]. A similar pattern was found in our current work for the σ-hole bonded (a) dimers. In a recent work [32], a series of complexes pairing Lewis acids TF$_4$ or ZF$_5$ (T = Si, Ge, Sn and Z = P, As, Sb) with Lewis bases NH$_3$, pyrazine, and HCN, the tetrel molecules TF$_4$ have a considerably larger (more than 10 kcal/mol) value of $V_{s,max}$ than their corresponding pnicogen ZF$_5$ cousin, but nonetheless smaller interaction energy. Moreover, another inconsistency was observed with respect to $V_{s,min}$ which is more negative for NCH than for pyrazine, but the latter complexes investigated were more strongly bound. Similar discrepancies arise in halogen bonded complexes involving chlorinated and methylated amines [67].

The issue of geometrical deformations of the monomers and their impact on tetrel-bonded complexes has been described recently in a few papers [47,69]. In our own latest work devoted to implications of monomer deformation upon tetrel and pnicogen bonds [32] it was shown that complexation can cause monomer deformation which results in a multifold increase in the intensity of $V_{s,max}$, which in turn amplifies the magnitude of the interaction energy.

4. Conclusions

In conclusion, the π-complexes formed above the plane of the TR$_2$=CH$_2$ molecules are more strongly bound than are their quasi-tetrahedral TR$_4$ σ congeners, given the same degree of fluorosubstitution. Starting with the unsubstituted species, the interaction energies of TH$_2$=CH$_2$ vary between 3.7 and 7.8 kcal/mol, considerably larger in magnitude than the 1.6–2.8 kcal/mol range for the σ-bonded TH$_4$ species. In the difluorinated sets, the ranges of binding energies of TF$_2$=CH$_2$ and TF$_2$H$_2$ are respectively 27.3–29.3 and 10.4–15.3 kcal/mol. This distinction cannot be attributed to the intensity of the π and σ-holes in the MEPs, as they are roughly comparable, and indeed the σ-holes tend to be a bit more intense. In fact, the latter σ-holes grow even larger when the TR$_4$ molecules deform into the geometries they adopt within their complex with NH$_3$. Contrary to the general similarity between the intensities of the σ and π-holes, the full evaluation of the electrostatic interaction reveals a much greater Coulombic attraction for the π-dimers, coupled with an enlarged orbital interaction energy. It should be emphasized that the stronger binding in the π-complexes cannot be attributed to any geometrical distortions undergone by these pseudoplanar molecules. In the first place, their geometrical deformation upon dimerization is less than that of their tetravalent analogues. And even when these $TR_2=CH_2$ Lewis acid molecules are forced into a fully planar internal geometry, their interaction energy with NH_3 remains larger than their σ-hole TR_4 counterparts, even if the latter are permitted to deform within the dimer.

Supplementary Materials: The following are available online, Figure S1: AIM diagrams showing the bond critical points (green dots) in Ge-containing complexes stabilized by σ-hole tetrel bonds, Figure S2: Bond critical points (green dots) in several Ge-containing complexes stabilized by π-hole tetrel bond, Figure S3: AIM molecular diagram of $THF=CH_2/NH_3$ d dimers wherein the base occupies the d maximum of the MEP of the acid, Table S1: NBO values of sum of the E(2) for LP(N)→σ*(T-X), (T= Si, Ge or Sn and X=H or F) orbital interaction and total charge transfer (CT) from NH_3 to $TH_{2-n}F_n$ in σ-hole bonded complexes obtained at the BLYP-D3(BJ)/def2-TVZPP level, Table S2: AIM data for σ-hole bonded complexes. Bond critical point (BCP) properties: electron density ρ, Laplacian of electron density $\nabla^2\rho$ (both in atomic units) and total electron energy (H, kcal mol^{-1}). Calculations were performed at the MP2/aug-cc-pVDZ-PP level, Table S3: NBO values of sum of the E(2) for LP(N)→σ*(T-X), (T= Si, Ge or Sn and X=H or F) orbital interaction and total charge transfer (CT) from NH_3 to $TH_{2-n}F_n=CH_2$ in π-hole bonded complexes obtained at the BLYP-D3(BJ)/def2-TVZPP level, Table S4: AIM data for π-hole bonded complexes. Bond critical point (BCP) properties: electron density ρ, Laplacian of electron density $\nabla^2\rho$ (both in atomic units) and total electron energy (H, kcal mol^{-1}). Calculations were performed at the MP2/aug-cc-pVDZ level, Table S5: Geometry and energetics for d complexes, Table S6: NBO properties of d complexes, Table S7: AIM parameters of d complexes, Table S8: EDA/BLYP-D3(BJ)/ZORA/TZ2P decomposition of the interaction energy of π-hole bonded complexes d into Pauli repulsion (E_{Pauli}), electrostatic (E_{elec}), orbital interaction (E_{oi}) and dispersion (E_{disp}) terms. All energies in kcal/mol. The relative values in percent express the contribution of each to the sum of all attractive terms, Table S9: Secondary minima for dimers of NH_3 with σ-hole donors. Data obtained at the MP2/aug-cc-pVDZ-PP level of theory. E_{int} corrected for BSSE (in kcal/mol). Distances are in Å, Table S10: Secondary minima for dimers of NH_3 with π-hole donors. Data obtained at the MP2/aug-cc-pVDZ-PP level of theory. E_{int} corrected for BSSE (in kcal/mol). Distances are in Å.

Author Contributions: Conceptualization, S.S. and W.Z.; Data curation, W.Z. and M.M.; Supervision, S.S.; Visualization, W.Z. and M.M.; Writing—original draft, W.Z. and M.M.; Writing—review & editing, S.S.

Acknowledgments: This work was financed in part by a statutory activity subsidy from the Polish Ministry of Science and Higher Education for the Faculty of Chemistry of Wroclaw University of Science and Technology. A generous computer time from the Wroclaw Supercomputer and Networking Center is acknowledged.

Conflicts of Interest: The authors declare no conflict of interest.

References

1. Desiraju, G.R.; Ho, P.S.; Kloo, L.; Legon, A.C.; Marquardt, R.; Metrangolo, P.; Politzer, P.; Resnati, G.; Rissanen, K. Definition of the halogen bond (IUPAC recommendations 2013). *Pure Appl. Chem.* **2013**, *85*, 1711–1713. [CrossRef]
2. Cavallo, G.; Metrangolo, P.; Milani, R.; Pilati, T.; Priimagi, A.; Resnati, G.; Terraneo, G. The halogen bond. *Chem. Rev.* **2016**, *116*, 2478–2601. [CrossRef] [PubMed]
3. García-Llinás, X.; Bauzá, A.; Seth, S.K.; Frontera, A. Importance of R-CF3···O tetrel bonding interactions in biological systems. *J. Phys. Chem. A* **2017**, *121*, 5371–5376. [CrossRef] [PubMed]
4. Esrafili, M.D.; Mohammadian-Sabet, F. Homonuclear chalcogen-chalcogen bond interactions in complexes pairing YO3 and YHX molecules (Y=S, Se, X=H, Cl, Br, CCH, NC, OH, OCH3): Influence of substitution and cooperativity. *Int. J. Quantum Chem.* **2016**, *116*, 529–536. [CrossRef]
5. Bauzá, A.; Mooibroek, T.J.; Frontera, A. σ-Hole opposite to a lone pair: unconventional pnicogen bonding interactions between ZF3(Z=N, P, As, and Sb) compounds and several donors. *Chem. Phys. Chem.* **2016**, *17*, 1608–1614. [CrossRef] [PubMed]
6. Bauza, A.; Alkorta, I.; Frontera, A.; Elguero, J. On the reliability of pure and hybrid DFT methods for the evaluation of halogen, chalcogen, and pnicogen bonds involving anionic and neutral electron donors. *J. Chem. Theory Comput.* **2013**, *9*, 5201–5210. [CrossRef] [PubMed]
7. Iwaoka, M.; Komatsu, H.; Katsuda, T.; Tomoda, S. Quantitative evaluation of weak nonbonded Se···F interactions and their remarkable nature as orbital interactions. *J. Am. Chem. Soc.* **2002**, *124*, 1902–1909. [CrossRef] [PubMed]

8. Azofra, L.M.; Alkorta, I.; Scheiner, S. Chalcogen bonds in complexes of SOXY (X, y = F, Cl) with nitrogen bases. *J. Phys. Chem. A* **2015**, *119*, 535–541. [CrossRef] [PubMed]
9. Scheiner, S. Effects of multiple substitution upon the P···N noncovalent interaction. *Chem. Phys.* **2011**, *387*, 79–84. [CrossRef]
10. Kolar, M.H.; Hobza, P. Computer modeling of halogen bonds and other σ-hole interactions. *Chem. Rev.* **2016**, *116*, 5155–5187. [CrossRef] [PubMed]
11. Bauza, A.; Frontera, A. Aerogen bonding interaction: a new supramolecular force? *Angew. Chem. Int. Ed.* **2015**, *54*, 7340–7343. [CrossRef] [PubMed]
12. Stenlid, J.H.; Johansson, A.J.; Brinck, T. σ-Holes and σ-lumps direct the Lewis basic and acidic interactions of noble metal nanoparticles: Introducing regium bonds. *Phys. Chem. Chem. Phys.* **2018**, *20*, 2676–2692. [CrossRef] [PubMed]
13. Murray, J.S.; Lane, P.; Politzer, P. Expansion of the σ-hole concept. *J. Mol. Model.* **2009**, *15*, 723–729. [CrossRef] [PubMed]
14. Murray, J.S.; Lane, P.; Clark, T.; Riley, K.E.; Politzer, P. σ-Holes, π-holes and electrostatically-driven interactions. *J. Mol. Model.* **2012**, *18*, 541–548. [CrossRef] [PubMed]
15. Bundhun, A.; Ramasami, P.; Murray, J.S.; Politzer, P. Trends in σ-hole strengths and interactions of F3MX molecules (M = C, Si, Ge and X = F, Cl, Br, I). *J. Mol. Model.* **2013**, *19*, 2739–2746. [CrossRef] [PubMed]
16. Bauza, A.; Mooibroek, T.J.; Frontera, A. Tetrel-bonding interaction: Rediscovered supramolecular force? *Angew. Chem. Int. Ed.* **2013**, *52*, 12317–12321. [CrossRef] [PubMed]
17. Politzer, P.; Murray, J.S.; Clark, T. Halogen bonding and other σ-hole interactions: A perspective. *Phys. Chem. Chem. Phys.* **2013**, *15*, 11178–11189. [CrossRef] [PubMed]
18. Clark, T.; Hennemann, M.; Murray, J.S.; Politzer, P. Halogen bonding: The σ-hole: Proceedings of "Modeling interactions in biomolecules II", Prague, September 5th-9th, 2005. *J. Mol. Model.* **2007**, *13*, 291–296. [CrossRef] [PubMed]
19. Auffinger, P.; Hays, F.A.; Westhof, E.; Ho, P.S. Halogen bonds in biological molecules. *Proc. Natl. Acad. Sci. USA* **2004**, *101*, 16789–16794. [CrossRef] [PubMed]
20. Clark, T. σ-Holes. *WIREs Comput. Mol. Sci.* **2013**, *3*, 13–20. [CrossRef]
21. Politzer, P.; Lane, P.; Concha, M.C.; Ma, Y.; Murray, J.S. An overview of halogen bonding. *J. Mol. Model.* **2007**, *13*, 305–311. [CrossRef] [PubMed]
22. Stone, A.J. Are Halogen Bonded Structures Electrostatically Driven? *J. Am. Chem. Soc.* **2013**, *135*, 7005–7009. [CrossRef] [PubMed]
23. Politzer, P.; Murray, J.S. Halogen bonding: An interim discussion. *Chem. Phys. Chem.* **2013**, *14*, 278–294. [CrossRef] [PubMed]
24. Eramian, H.; Tian, Y.-H.; Fox, Z.; Beneberu, H.Z.; Kertesz, M. On the anisotropy of van der waals atomic radii of O, S, Se, F, Cl, Br, and I. *J. Phys. Chem. A* **2013**, *117*, 14184–14190. [CrossRef] [PubMed]
25. Politzer, P.; Murray, J.S.; Clark, T. Halogen bonding: An electrostatically-driven highly directional noncovalent interaction. *Phys. Chem. Chem. Phys.* **2010**, *12*, 7748–7757. [CrossRef] [PubMed]
26. Grabowski, S.J. Hydrogen bonds, and σ-hole and π-hole bonds-mechanisms protecting doublet and octet electron structures. *Phys. Chem. Chem. Phys.* **2017**, *19*, 29742–29759. [CrossRef] [PubMed]
27. Mani, D.; Arunan, E. The X-C···Y (X = O/F, y = O/S/F/Cl/Br/N/P) 'carbon bond' and hydrophobic interactions. *Phys. Chem. Chem. Phys.* **2013**, *15*, 14377–14383. [CrossRef] [PubMed]
28. Helminiak, H.M.; Knauf, R.R.; Danforth, S.J.; Phillips, J.A. Structural and energetic properties of acetonitrile-group IV (A & B) halide complexes. *J. Phys. Chem. A* **2014**, *118*, 4266–4277. [PubMed]
29. George, J.; Dronskowski, R. Tetrel bonds in infinite molecular chains by electronic structure theory and their role for crystal stabilization. *J. Phys. Chem. A* **2017**, *121*, 1381–1387. [CrossRef] [PubMed]
30. Scheiner, S. Steric Crowding in Tetrel Bonds. *J. Phys. Chem. A* **2018**, *122*, 2550–2562. [CrossRef] [PubMed]
31. Scheiner, S. Systematic elucidation of factors that influence the strength of tetrel bonds. *J. Phys. Chem. A* **2017**, *121*, 5561–5568. [CrossRef] [PubMed]
32. Zierkiewicz, W.; Michalczyk, M.; Scheiner, S. Implications of monomer deformation for tetrel and pnicogen bonds. *Phys. Chem. Chem. Phys.* **2018**, *20*, 8832–8841. [CrossRef] [PubMed]
33. Xu, H.; Cheng, J.; Yang, X.; Liu, Z.; Li, W.; Li, Q. Comparison of σ-hole and π-hole tetrel bonds formed by pyrazine and 1,4-dicyanobenzene: The interplay between anion-π and tetrel bonds. *Chem. Phys. Chem.* **2017**, *18*, 2442–2450. [CrossRef] [PubMed]

34. Politzer, P.; Murray, J.S. A unified view of halogen bonding, hydrogen bonding and other σ-hole interactions. In *Noncovalent Forces*; Scheiner, S., Ed.; Springer: Dordrecht, The Netherlands, 2015; Volume 19, pp. 291–322.
35. Wenbo, D.; Xin, Y.; Jianbo, C.; Wenzuo, L.; Qingzhong, L. Comparison for σ-hole and π-hole tetrel-bonded complexes involving F_2CCFTF_3 (TC, Si, and Ge): Substitution, hybridization, and solvation effects. *J. Fluorine Chem.* **2018**, *207*, 38–44.
36. Politzer, P.; Murray, J.S.; Clark, T. σ-Hole bonding: A physical interpretation. *Top. Curr. Chem.* **2015**, *358*, 19–42. [PubMed]
37. Azofra, L.M.; Alkorta, I.; Scheiner, S. Noncovalent interactions in dimers and trimers of SO_3 and CO. *Theor. Chem. Acc.* **2014**, *133*, 1586. [CrossRef]
38. Bauza, A.; Ramis, R.; Frontera, A. A combined theoretical and cambridge structural database study of π-hole pnicogen bonding complexes between electron rich molecules and both nitro compounds and inorganic bromides (YO_2Br, Y = N, P, and As). *J. Phys. Chem. A* **2014**, *118*, 2827–2834. [CrossRef] [PubMed]
39. Bauza, A.; Mooibroek, T.J.; Frontera, A. Directionality of π-holes in nitro compounds. *Chem. Commun.* **2015**, *51*, 1491–1493. [CrossRef] [PubMed]
40. Sanchez-Sanz, G.; Trujillo, C.; Solimannejad, M.; Alkorta, I.; Elguero, J. Orthogonal interactions between nitryl derivatives and electron donors: Pnictogen bonds. *Phys. Chem. Chem. Phys.* **2013**, *15*, 14310–14318. [CrossRef] [PubMed]
41. Murray, J.S.; Lane, P.; Clark, T.; Riley, K.E.; Politzer, P. Σ-holes, π-holes and electrostatically-driven interactions. *J. Mol. Model.* **2012**, *18*, 541–548. [CrossRef] [PubMed]
42. Del Bene, J.E.; Alkorta, I.; Elguero, J. Characterizing complexes with pnicogen bonds involving sp2 hybridized phosphorus atoms: $(H_2C=PX)_2$ with X = F, Cl, OH, CN, NC, CCH, H, CH_3, and BH_2. *J. Phys. Chem. A* **2013**, *117*, 6893–6903. [CrossRef] [PubMed]
43. Zierkiewicz, W.; Michalczyk, M.; Scheiner, S. Aerogen bonds formed between $AeOF_2$ (Ae = Kr, Xe) and diazines: comparisons between σ-hole and π-hole complexes. *Phys. Chem. Chem. Phys.* **2018**, *20*, 4676–4687. [CrossRef] [PubMed]
44. Politzer, P.; Murray, J.S. σ-holes and π-holes: Similarities and differences. *J. Comput. Chem.* **2018**, *39*, 464–471. [CrossRef] [PubMed]
45. Zhang, Y.; Wang, D.; Wang, W. Beyond the σ-hole and π-hole: The origin of the very large electrophilic regions of fullerenes and carbon nanotubes. *Comp. Theor. Chem.* **2018**, *1128*, 56–59. [CrossRef]
46. Wei, Y.; Li, Q. Comparison for σ-hole and π-hole tetrel-bonded complexes involving cyanoacetaldehyde. *Mol. Phys.* **2018**, *116*, 222–230. [CrossRef]
47. Angarov, V.; Kozuch, S. On the σ, π and δ hole interactions: A molecular orbital overview. *New J. Chem.* **2018**, *42*, 1413. [CrossRef]
48. Scilabra, G.; Terraneo, G. Resnati, Fluorinated elements of Group 15 as pnictogen bond donor sites. *J. Fluorine Chem.* **2017**, *203*, 62–74. [CrossRef]
49. Møller, C.; Plesset, M.S. Note on an approximation treatment for many-electron systems. *Phys. Rev.* **1934**, *46*, 618.
50. Dunning, T.H., Jr. Gaussian basis sets for use in correlated molecular calculations. I. The atoms boron through neon and hydrogen. *J. Chem. Phys.* **1989**, *90*, 1007. [CrossRef]
51. Peterson, K.A. Systematically convergent basis sets with relativistic pseudopotentials. I. Correlation consistent basis sets for the post-*d* group 13–15 elements. *J. Chem. Phys.* **2003**, *119*, 11099. [CrossRef]
52. Schuchardt, K.L.; Didier, B.T.; Elsethagen, T.; Sun, L.; Gurumoorthi, V.; Chase, J.; Li, J.; Windus, T.L. Basis Set Exchange: A community database for computational sciences. *J. Chem. Inf. Model.* **2007**, *47*, 1045–1052. [CrossRef] [PubMed]
53. Boys, S.F.; Bernardi, F. The calculation of small molecular interactions by the differences of separate total energies. Some procedures with reduced errors. *Mol. Phys.* **1970**, *19*, 553–566. [CrossRef]
54. Frisch, M.J.; Trucks, G.W.; Schlegel, H.B.; Scuseria, G.E.; Robb, M.A.; Cheeseman, J.R.; Scalmani, G.; Barone, V.; Mennucci, B.; Petersson, G.A.; et al. *Gaussian 09, Revision E.01*; Gaussian, Inc.: Wallingford, CT, USA, 2009.
55. Te Velde, G.; Bickelhaupt, F.M.; Baerends, E.J.; Fonseca Guerra, C.; van Gisbergen, S.J.A.; Snijders, J.G.; Ziegler, T. Chemistry with ADF. *J. Comput. Chem.* **2001**, *22*, 931. [CrossRef]
56. Fonseca Guerra, C.; Snijders, J.G.; te Velde, G.; Baerends, E.J. Towards an order-N DFT method. *Theor. Chem. Acc.* **1998**, *99*, 391. [CrossRef]
57. *ADF 2014*; Vrije Universiteit: Amsterdam, The Netherlands, 2014.

58. Bulat, F.; Toro-Labbe, A.; Brinck, T.; Murray, J.S.; Politzer, P. Quantitative analysis of molecular surfaces: Areas, volumes, electrostatic potentials and average local ionization energies. *J. Mol. Model.* **2010**, *16*, 1679. [CrossRef] [PubMed]
59. Todd, A. *Keith AIMAll Version 14.11.23*, TK Gristmill Software: Overland Park, KS, USA, 2014.
60. Glendening, E.D.; Landis, C.R.; Weinhold, F. NBO 6.0: Natural bond orbital analysis program. *J. Comput. Chem.* **2013**, *34*, 1429–1437. [CrossRef] [PubMed]
61. Grimme, S.; Ehrlich, S.; Goerigk, L. Effect of the damping function in dispersion corrected density functional theory. *J. Comp. Chem.* **2011**, *32*, 1456–1465. [CrossRef] [PubMed]
62. Grabowski, S.J. Tetrel bonds with π-electrons acting as lewis bases—theoretical results and experimental evidences. *Molecules* **2018**, *23*, 1183. [CrossRef] [PubMed]
63. Rossi, A.R.; Jasinski, J.M. Theoretical studies of neutral silane-ammonia adducts. *J. M. Chem. Phys. Lett.* **1990**, *169*, 399–404. [CrossRef]
64. Alkorta, I.; Rozas, I.; Elguero, J. Molecular complexes between silicon derivatives and electron-rich groups. *J. Phys. Chem. A* **2001**, *105*, 743–749. [CrossRef]
65. Schoeller, W.W.; Rozhenko, A. Pentacoordination at fluoro-substituted silanes by weak lewis donor addition. *Eur. J. Inorg. Chem.* **2000**, *2000*, 375–381. [CrossRef]
66. Grabowski, S.J. Tetrel bond–σ-hole bond as a preliminary stage of the S_N2 reaction. *Phys. Chem. Chem. Phys.* **2014**, *16*, 1824–1834. [CrossRef] [PubMed]
67. Xu, H.; Cheng, J.; Yu, X.; Li, Q. Abnormal tetrel bonds between formamidine and th_3f: substituent effects. *Chem. Sel.* **2018**, *3*, 2842–2849.
68. Zierkiewicz, W.; Michalczyk, M. On the opposite trends of correlations between interaction energies and electrostatic potentials of chlorinated and methylated amine complexes stabilized by halogen bond. *Theor. Chem. Acc.* **2017**, *136*, 125. [CrossRef]
69. Fanfrlık, J.; Zierkiewicz, W.; Svec, P.; Rezac, J.; Michalczyk, M.; Ruzickova, Z.; Ruzicka, A.; Michalska, D.; Hobza, P. Pnictogen bonding in pyrazine•PnX_5 (Pn = P, As, Sb and X = F, Cl, Br) complexes. *J. Mol. Model.* **2017**, *23*, 128. [CrossRef] [PubMed]

© 2018 by the authors. Licensee MDPI, Basel, Switzerland. This article is an open access article distributed under the terms and conditions of the Creative Commons Attribution (CC BY) license (http://creativecommons.org/licenses/by/4.0/).

Article

Tetrel Bonds with π-Electrons Acting as Lewis Bases—Theoretical Results and Experimental Evidences

Sławomir J. Grabowski [1,2]

[1] Faculty of Chemistry, University of the Basque Country and Donostia International Physics Center (DIPC), P.K. 1072, 20080 Donostia, Spain; s.grabowski@ikerbasque.org; Tel.: +34-943-01-5477
[2] IKERBASQUE, Basque Foundation for Science, 48011 Bilbao, Spain

Academic Editor: Steve Scheiner
Received: 30 April 2018; Accepted: 11 May 2018; Published: 15 May 2018

Abstract: MP2/aug-cc-pVTZ calculations were carried out for the ZFH_3-B complexes (Z = C, Si, Ge, Sn and Pb; B = C_2H_2, C_2H_4, C_6H_6 and $C_5H_5^-$; relativistic effects were taken into account for Ge, Sn and Pb elements). These calculations are supported by other approaches; the decomposition of the energy of interaction, Quantum Theory of Atoms in Molecules (QTAIM) and Natural Bond Orbital (NBO) method. The results show that tetrel bonds with π-electrons as Lewis bases are classified as Z⋯C links between single centers (C is an atom of the π-electron system) or as Z⋯π interactions where F-Z bond is directed to the mid-point (or nearly so) of the CC bond of the Lewis base. The analogous systems with Z⋯C/π interactions were found in the Cambridge Structural Database (CSD). It was found that the strength of interaction increases with the increase of the atomic number of the tetrel element and that for heavier tetrel elements the ZFH_3 tetrahedral structure is more deformed towards the structure with the planar ZH_3 fragment. The results of calculations show that the tetrel bond is sometimes accompanied by the Z-H⋯C hydrogen bond or even sometimes the ZFH_3-B complexes are linked only by the hydrogen bond interaction.

Keywords: electron charge shifts; tetrel bond; hydrogen bond; π-electrons as Lewis bases; σ-hole

1. Introduction

The tetrel bond is a Lewis acid—Lewis base interaction that may play an important role in some chemical and biological processes [1]; for example, it may be considered as a preliminary stage of the S_N2 reaction [2]. This interaction was classified as the σ-hole bond by Politzer and coworkers since it may be defined as an interaction between the 14th Group element acting as the Lewis acid centre through its σ-hole and a region that is rich of the electron density by a lone electron pair, π-electron system etc. [3,4]. The σ-hole is usually located in this case in the elongation of the covalent bond to the tetrel centre and it is often characterized by the positive electrostatic potential (EP) [3,4]. It seems that first time the tetrel bond was analyzed in terms of the σ-hole concept in the SiF_4 complexes with amines [5]. In spite of the fact that the term "tetrel bond" appeared recently [6] and also this interaction was classified as the σ-hole bond in the last decade [3–5] it was analyzed in earlier studies. For example, the $SiF_4 \cdots NH_3$ and $SiF_4 \cdots (NH_3)_2$ complexes were analyzed theoretically since ab initio MO calculations were performed with the use of STO-3G and STO-6G basis sets [7], another study that is more complex concerns a large sample of complexes of silicon derivatives with electron-rich groups [8]. Both latter studies were performed before the proposition of the σ-hole concept [9,10] and the introduction of the tetrel bond term [6].

One can mention other studies on tetrel bonds as for example that one where this interaction was analyzed as a preliminary stage of the S_N2 reaction [11], the theoretical analysis of structural and energetic properties of acetonitrile complexes with the 14 Group tetrahalides [12]; a study on the Lewis acid carbon center, the corresponding interaction was labeled as the carbon bond and it was compared with the hydrogen bond [13]; it is worth mentioning that the carbon bond is a sub-class of the tetrel bond interactions [2,11]. The tetrel-hydride interaction is another sub-class of tetrel bonds where the negatively charged H-atom plays the role of Lewis base center [14].

There are other, more recent studies on this kind of interaction; only few are mentioned here; the analysis of factors which influence the strength of tetrel bonds [15], the analysis of mechanisms of S_N2 reactions, among them at the C centre [16], the role of tetrel bonds in the crystal structures' stabilization [17], the theoretical analysis of the H-Si···N and F-Si···N linear or nearly so arrangements [18], comparison of neutral and charge assisted tetrel bonds [19], the geometry deformations of monomers linked by tetrel bond [20] or the balance between the attractive forces of tetrel interactions and the steric repulsions in crystal structures [21].

One may cite numerous other examples since the number of studies on this kind of interaction has increased rapidly. However, it seems that there are no systematic and extensive studies on the tetrel bonds with π-electrons playing a role of Lewis bases or at most they are very rare and they are not a main goal of investigations. For example, very recently, various σ-hole bonds were analyzed with the use of few theoretical approaches: halogen, chalcogen, pnicogen and tetrel bonds were compared [22]; three different types of Lewis bases were considered there, neutral species (NH_3), anion (Cl^-) and the π-electron system (C_2H_2). Hence two tetrel bonded complexes with acetylene molecule playing a role of the Lewis base were analyzed there among various other complexes. These are the $SiFH_3···C_2H_2$ and $GeFH_3···C_2H_2$ complexes [22].

The other issue that is not analyzed so frequently concerns the tetrel bond interactions with the heavier tetrel elements, such analyses are rather rare and mainly concern germanium species. There are more experimental studies on tetrel bonds with heavier tetrel elements playing a role of the Lewis acid centers, however only sometimes such experimental analyses are supported by theoretical results [23]. One of examples where heavier tetrel elements were considered in tetrel bonds is a study on the SnF_4 and PbF_4 complexes with NH_3 and HCN that play a role of Lewis bases through the nitrogen centre [24]. A theoretical study analogous to the latter one was performed on the lighter tetrel species since the complexes of CF_4, SiF_4 and GeF_4 Lewis acid units with NH_3 and AsH_3 Lewis bases were analyzed [25].

Returning to the π-electron species—numerous theoretical and experimental studies on interactions where such systems play a role of Lewis bases may be mentioned. These are mainly those studies that concern hydrogen bonded systems [26]. However other Lewis acid—Lewis base interactions with π-electron donors were analyzed very often [27–30]. One can even mention the triel bonds between the boron or aluminium Lewis acid center and acetylene or ethylene [31,32] or the recent study where the multivalent halogen centers act as the Lewis acids [33].

The aim of this study is an analysis of the tetrel bonds in complexes of ZFH_3 species, where Z labels the following centers; C, Si, Ge, Sn and Pb, thus light and heavy tetrels are taken into account; the acetylene, ethylene, benzene and cyclopentadienyl anion were chosen as the π-electron moieties acting as the Lewis bases. Different theoretical techniques are applied here to deepen the understanding of the nature of these tetrel bond interactions; i.e., the Quantum Theory of Atoms in Molecules (QTAIM) [34], Natural Bond Orbital (NBO) approach [35], the decomposition of the energy of interaction [36,37] as well as the analysis of the electrostatic potential (EP) distribution [38]. The short descriptions of the theoretical approaches applied here are included in the section that concerns the computational details.

2. Results and Discussion

2.1. Energetic and Geometric Parameters

Figure 1 presents examples of complexes analyzed here. All kinds of Lewis bases that are considered are shown in selected examples of the figure. The molecular graphs are presented since they reflect geometry of species analyzed. However these graphs are discussed further here in the section on QTAIM results.

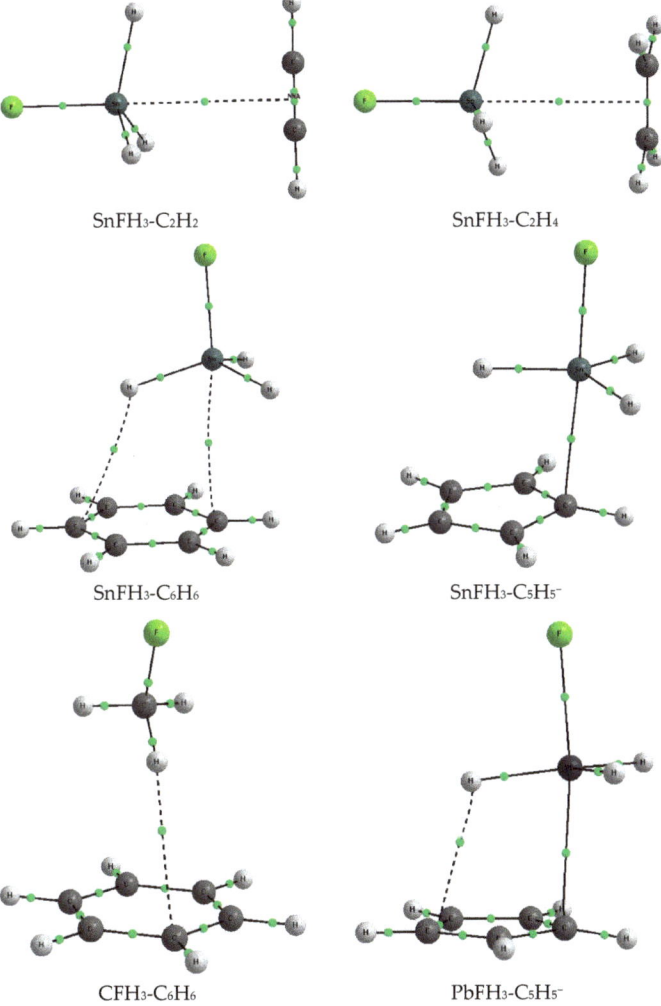

Figure 1. The molecular graphs of the selected complexes analyzed here; big circles—attractors, small green circles—BCPs, the nonnuclear attractor (NNA) is located (small red circle) between two BCPs of the CC bond in a case of the SnFH$_3$-C$_2$H$_2$ complex.

The energetic parameters of analyzed complexes, among them, the binding and interaction energies corrected for BSSE, E_{bin} and E_{int}, respectively, are included in Table 1. One can see much stronger interactions, i.e., greater $-E_{int}$ and $-E_{bin}$ values, for complexes with the cyclopentadienyl anion, than for complexes with the other Lewis base units. This may be explained since the complexes of $C_5H_5^-$ anion that are linked through the tetrel bonds are assisted by negative charge; the latter anion is much stronger base than the remaining species chosen here. The following tendencies are also observed here, and it does not depend on the choice of $-E_{int}$ or $-E_{bin}$ value for the discussion; the strength of interaction for the same Lewis base increases in the following order of the tetrel center C < Si < Ge < Sn ≅ Pb. It was observed earlier for tetrel bond interactions [3,4] and it was explained by the increase of the electrostatic part of the energy of interaction since the electrostatic potential (EP) at the tetrel σ-hole increases with the increase of the atomic number [3,4]. The calculations performed here show the EP of tetrel σ-hole of the ZFH$_3$ species equal to 0.033; 0.062; 0.068; 0.081; 0.080 au for C, Si, Ge, Sn and Pb centers, respectively (0.001 au electron density surfaces were chosen). The EP values for tin and lead centers are almost equal one to each other. This is why for complexes analyzed here similar Lewis acid properties are observed for these centers; however the strongest interaction is observed for the SnFH$_3$-C$_5$H$_5^-$ complex if the interaction energy is considered while if the binding energy is taken into account thus it is the PbFH$_3$-C$_5$H$_5^-$ complex.

Table 1. The energetic parameters of complexes analyzed (in kcal/mol); interaction energy, E_{int}, binding energy, E_{bin}, deformation energy, E_{def} and BSSE correction. The distance between Lewis base and Lewis acid units is included—the shortest Z···C distance was chosen, the values in parentheses show if this distance is greater than the corresponding sum of van der Waals radii (positive values) or if it is lower (negative ones), distances in Å.

Complex	Distance	E_{int}	E_{bin}	E_{def}	BSSE
CFH$_3$-C$_2$H$_2$	3.428 (+0.53)	−1.3	−1.2	0.0	0.3
CFH$_3$-C$_2$H$_4$	3.458 (+0.56)	−1.3	−1.3	0.0	0.4
CFH$_3$-C$_6$H$_6$	3.398 (+0.50)	−2.7	−2.7	0.0	1.0
CFH$_3$-C$_5$H$_5^-$	3.241 (+0.34)	−10.8	−10.5	0.3	1.2
SiFH$_3$-C$_2$H$_2$	3.344 (+0.04)	−2.4	−2.3	0.0	0.6
SiFH$_3$-C$_2$H$_4$	3.315 (−0.02)	−2.7	−2.6	0.1	0.8
SiFH$_3$-C$_6$H$_6$	3.253 (−0.05)	−3.7	−3.7	0.1	1.3
SiFH$_3$-C$_5$H$_5^-$	2.477 (−0.82)	−27.5	−18.1	9.4	2.0
GeFH$_3$-C$_2$H$_2$	3.285 (−0.02)	−2.6	−2.6	0.1	1.1
GeFH$_3$-C$_2$H$_4$	3.253 (−0.05)	−2.9	−2.9	0.1	1.5
GeFH$_3$-C$_6$H$_6$	3.203 (−0.10)	−4.3	−4.2	0.1	2.7
GeFH$_3$-C$_5$H$_5^-$	2.525 (−0.78)	−29.3	−21.0	8.3	4.4
SnFH$_3$-C$_2$H$_2$	3.325 (−0.13)	−3.4	−3.3	0.1	1.3
SnFH$_3$-C$_2$H$_4$	3.280 (−0.17)	−3.8	−3.7	0.2	1.8
SnFH$_3$-C$_6$H$_6$	3.183 (−0.27)	−5.5	−5.2	0.3	3.2
SnFH$_3$-C$_5$H$_5^-$	2.519 (−0.93)	−41.9	−30.6	11.4	5.2
PbFH$_3$-C$_2$H$_2$	3.323 (−0.18)	−3.5	−3.4	0.1	2.2
PbFH$_3$-C$_2$H$_4$	3.267 (−0.23)	−3.8	−3.7	0.1	3.1
PbFH$_3$-C$_6$H$_6$	3.148 (−0.35)	−5.8	−5.6	0.3	5.7
PbFH$_3$-C$_5$H$_5^-$	2.624 (−0.88)	−39.3	−31.8	7.6	8.3

If the Lewis acid unit is the same thus the interaction strength increases in the following order $C_2H_2 < C_2H_4 < C_6H_6 < C_5H_5^-$. One can also see that the $-E_{int}$ or $-E_{bin}$ values do not exceed 6 kcal/mol for all complexes of acetylene, ethylene and benzene while they are much greater in a case of complexes with cyclopentadienyl, especially large values are observed for the above-mentioned tin and lead complexes.

The BSSE corrections are greater for stronger interactions, especially large values are observed for interactions in cyclopentadienyl complexes. The deformation energy, E_{def}, is a parameter that is related to geometrical changes of the interacting systems. For example, in a case of the strong A-H···B hydrogen bonds, the complexation often leads to the meaningful elongation of the A-H proton donating bond that results in the greater E_{def} values [39]. Steric effects are very important for tetrel bonded species [11,21] since the tetrel center, often characterized by the sp^3 hybridization and surrounded by four substituents (like for the systems considered here) is hardly available for the Lewis base (nucleophilic attack). Thus the tetrel-base link should cause greater deformations connected with the increase of the availability of the tetrel center. In other words the ZFH$_3$ tetrahedral system should be closer to the trigonal bipyramid in the ZFH$_3$-B complex with the ZH$_3$ part being closer to planarity. For complexes of acetylene, ethylene and benzene E$_{def}$ does not exceed 0.3 kcal/mol indicating negligible changes of geometry resulting from complexation. However for the C$_5$H$_5^-$ complexes, if one excludes the CFH$_3$-C$_5$H$_5^-$ complex with this energy amounting only 0.3 kcal/mol, E$_{def}$ is close to 10 kcal/mol or even exceeds this value. This is in agreement with changes of geometry; one can see (Figure 1) the ZH$_3$ part close to planarity and the F-Z···C arrangement close to linearity for two complexes presented; SnFH$_3$-C$_5$H$_5^-$ and PbFH$_3$-C$_5$H$_5^-$.

The above-presented EP values at the Z-tetrel centre concern the σ-hole that occurs in the extension of F-Z bonds. For the ZFH$_3$ species analyzed here, similarly as for other sp^3 hybridized tetrel centers four σ-holes located in extensions of covalent bonds to Z-center occur. However the electronegative F-substituent enhances F-Z σ-hole [3,4] that results in greater positive EP values than those of other H-Z σ-holes. For example, for the SnFH$_3$ molecule the EP value at the F-Sn σ-hole is equal to +0.081 au while this value for the H-Sn σ-hole amounts +0.048 au. For the clarity of the results' presentation only interactions of the F-Z σ-hole are considered here; it means that the F-Z σ-hole is directed to the π-electron system in the configurations analyzed.

Table 1 presents the Lewis acid—Lewis base distances, for each of complexes the shortest Z···C contact was chosen. One can see that these distances are usually greater than 3 Å, only for the C$_5$H$_5^-$ complexes where stronger interactions are observed such distances amount ~2.5 Å (except of the CFH$_3$-C$_5$H$_5^-$ complex where this distance is equal to ~3.2 Å). It was pointed out in numerous studies that the distance between interacting units is roughly related to the strength of interaction, this was observed for the hydrogen bonded complexes [40] but it seems that such dependence occurs also for other types of interactions [2]. It is often stated in various studies that the sum of van der Waals radii of two atoms being in contact roughly indicates at which distance a significant so-called noncovalent interaction begins [41]. Table 1 presents how the Z···C distances are related to the corresponding sum of Z and C van der Waals radii. These distances are greater than the corresponding sum for carbon complexes (Z = C) and for the SiFH$_3$-C$_2$H$_2$ complex while for the remaining ones these distances are lower than the van der Waals sum; the following van der Waals radii were applied here, H—1.2 Å [42], C—1.7 Å, Si—2.1 Å, Ge—2.1 Å, Sn—2.25 Å and Pb—2.3 Å [43]. One can see that for the strongest interactions in the C$_5$H$_5^-$ complexes the Z···C distance is almost by 1 Å lower than the sum of van der Waals radii. If one considers only the Z···C distances as a measure of the strength of interaction thus the interactions in the CFH$_3$ complexes are very weak and one may contest even their stabilizing nature.

Table 2 presents geometrical parameters related to the changes resulting from complexation; one of them is a percentage elongation of the Z-F bond related to the corresponding isolated ZFH$_3$ species that is not involved in the tetrel bond. One can see that these elongations correspond to the deformation energies, the greatest values are observed for the cyclopentadienyl complexes. For the species analyzed here three F-Z-H angles in the Lewis acid unit are very close one to each other, however for each complex considered its average value is considered in the further discussions; the latter angle is defined in Figure 2. For the isolated ZFH$_3$ species this tetrahedral angle, labeled as α_{iso}, amounts from 101.4° for PbFH$_3$ to 108.8° for CFH$_3$. It decreases in the complex to α_{comp} (up to 90° corresponding to the planar ZH$_3$ system in the trigonal bipyramid structure). The angle decrease values, [(α_{iso} − α_{comp})/α_{iso}] × 100%, are presented in Table 2. They correspond to the deformation energies discussed earlier here as well as to the elongations of the Z-F bonds since the greatest decreases are observed for the strongest interactions. The Z-F bond elongations result from the $\pi_{CC} \to \sigma_{ZF}^*$ and $\sigma_{CH} \to \sigma_{ZF}^*$ overlaps; Table 2 shows the orbital-orbital NBO energies corresponding to these overlaps; E$_{NBO}^1$ is a sum of all energies of such interactions for the F−C bond considered. Similarly the E$_{NBO}^2$ energy summarizes all $\pi_{CC} \to \sigma_{ZH}^*$ and $\sigma_{CH} \to \sigma_{ZH}^*$ overlaps, however in this case the whole ZFH$_3$ species is considered for one E$_{NBO}^2$ value (i.e., three C-H bonds). One can see (Table 2) that both E$_{NBO}^1$ and E$_{NBO}^2$ values are much greater for the C$_5$H$_5^-$ complexes than for other ones.

Table 2. The characteristics of complexes analyzed; ZF% and Angle% are the percentage increase of the Z-F distance and the percentage decrease of the F-Z-H angle, respectively; E_{NBO}^1 and E_{NBO}^2 are the NBO energies defined in the text (in kcal/mol); El-trans (au) is the electron charge transfer from the Lewis base to the Lewis acid while Z-charge is the charge of the Z-center in the complex considered (both charges in au calculated within NBO approach).

Complex	ZF%	Angle%	E_{NBO}^1	E_{NBO}^2	El-Trans	Z-Charge [a]
CFH_3-C_2H_2	0.14	0.00	0.5	0.0	−0.001	−0.158
CFH_3-C_2H_4	0.14	0.00	0.6	0.0	−0.002	−0.158
CFH_3-C_6H_6	0.29	0.09	0.4	0.8	−0.004	−0.156
CFH_3-$C_5H_5^-$	1.95	0.28	1.5	2.4	−0.030	−0.378
$SiFH_3$-C_2H_2	0.25	0.65	1.7	0.3	−0.014	1.227
$SiFH_3$-C_2H_4	0.31	0.74	2.3	0.4	−0.021	1.218
$SiFH_3$-C_6H_6	0.31	0.74	1.8	1.0	−0.019	1.221
$SiFH_3$-$C_5H_5^-$	4.46	9.06	24.5	20.9	−0.250	1.086
$GeFH_3$-C_2H_2	0.40	0.85	3.1	0.6	−0.019	1.025
$GeFH_3$-C_2H_4	0.52	0.85	4.1	0.7	−0.027	1.015
$GeFH_3$-C_6H_6	0.52	0.85	3.2	1.6	−0.025	1.021
$GeFH_3$-$C_5H_5^-$	5.18	8.95	33.4	22.1	−0.250	0.906
$SnFH_3$-C_2H_2	0.46	1.34	4.1	1.2	−0.024	1.246
$SnFH_3$-C_2H_4	0.57	1.43	5.4	1.9	−0.035	1.232
$SnFH_3$-C_6H_6	0.62	1.72	4.8	2.9	−0.035	1.240
$SnFH_3$-$C_5H_5^-$	5.11	11.45	37.9	40.5	−0.289	1.148
$PbFH_3$-C_2H_2	0.59	0.99	4.9	1.0	−0.027	1.085
$PbFH_3$-C_2H_4	0.78	1.28	7.0	2.4	−0.039	1.070
$PbFH_3$-C_6H_6	0.83	1.68	6.1	3.9	−0.044	1.080
$PbFH_3$-$C_5H_5^-$	5.44	8.78	36.5	31.0	−0.289	1.013

[a] Z-charge for CFH_3: −0.157, $SiFH_3$: +1.240, $GeFH_3$: +1.042, $SnFH_3$: +1.267, $PbFH_3$: +1.106 (all in au).

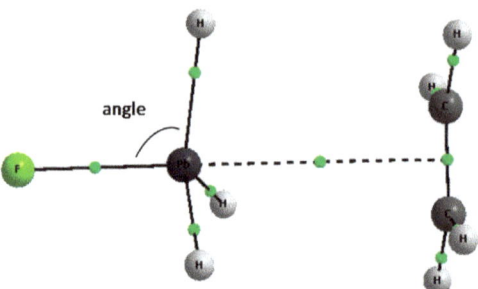

Figure 2. The definition of the angle expressing the change of the tetrahedral structure into the structure being closer to the trigonal bipyramid.

Table 2 shows the electron charge transfer values from the Lewis base to the Lewis acid unit, these transfers are especially great for the ZFH_3-$C_5H_5^-$ complexes, except of the CFH_3-$C_5H_5^-$ one. Such electron charge redistributions resulting from complexation are usually great for those complexes where the geometry deformations are important [2]; this is observed also here. The charge of the Z-central tetrel atom is shown in Table 2; one can see that this charge decreases (is "more negative") in complexes in comparison with the corresponding isolated ZFH_3 species. This is in opposite to the A-H···B hydrogen bonded systems where the complexation usually results in the increase of the positive charge of the central H-atom [35].

2.2. Nature of Interactions—Decomposition of Interaction Energy

Table 3 presents terms of the energy of interaction resulting from the Ziegler and Rauk decomposition scheme [36,37] (see the Computational Details section).

Table 3. The terms of the energy of interaction (kcal/mol); Pauli repulsion, ΔE_{Pauli}, electrostatic, ΔE_{elstat}, orbital, ΔE_{orb}, dispersion, ΔE_{disp}, and the total interaction energy, ΔE_{int}.

Complex	ΔE_{Pauli}	ΔE_{elstat}	ΔE_{orb}	ΔE_{disp}	ΔE_{int}
CFH_3-C_2H_2	2.4	−1.3	−0.8	−1.4	−1.1
CFH_3-C_2H_4	2.7	−1.4	−0.9	−1.6	−1.2
CFH_3-C_6H_6	4.7	−2.4	−1.5	−3.6	−2.8
CFH_3-$C_5H_5^-$	8.9	−11.0	−5.6	−4.2	−11.9
$SiFH_3$-C_2H_2	5.6	−3.4	−2.7	−2.2	−2.6
$SiFH_3$-C_2H_4	7.3	−4.3	−3.5	−2.8	−3.2
$SiFH_3$-C_6H_6	8.2	−4.1	−3.3	−4.5	−3.8
$SiFH_3$-$C_5H_5^-$	54.2	−43.4	−34.9	−5.6	−29.8
$GeFH_3$-C_2H_2	7.1	−4.4	−3.2	−2.7	−3.2
$GeFH_3$-C_2H_4	9.1	−5.5	−4.1	−3.4	−3.8
$GeFH_3$-C_6H_6	10.6	−5.6	−4.1	−5.8	−5.0
$GeFH_3$-$C_5H_5^-$	56.1	−47.7	−32.8	−6.3	−30.7
$SnFH_3$-C_2H_2	8.7	−5.8	−3.8	−3.0	−3.9
$SnFH_3$-C_2H_4	11.6	−7.3	−5.0	−3.9	−4.6
$SnFH_3$-C_6H_6	14.0	−7.8	−5.6	−6.9	−6.2
$SnFH_3$-$C_5H_5^-$	75.3	−68.1	−41.8	−6.6	−41.2
$PbFH_3$-C_2H_2	9.2	−6.3	−3.8	−2.8	−3.6
$PbFH_3$-C_2H_4	12.9	−8.3	−5.1	−3.7	−4.2
$PbFH_3$-C_6H_6	17.9	−10.0	−6.6	−7.4	−6.2
$PbFH_3$-$C_5H_5^-$	71.4	−65.9	−37.2	−6.6	−38.4

One can see that only for the CFH_3 complexes with acetylene, ethylene and benzene the dispersion term, ΔE_{disp}, is the most important attractive one. This is typical for weak van der Waals interactions where attractive interaction energy terms related to charge distributions and to electron charge shifts, ΔE_{elstat} and ΔE_{orb}, are less important [2]. For the above-mentioned three complexes, $-\Delta E_{int}$ does not exceed 3 kcal/mol, and in two cases it is close to 1 kcal/mol. The electron charge shifts for these complexes (Table 2) do not exceed 4 millielectrons! The latter is connected with the practically unchanged carbon charge in the CFH_3 unit in those complexes in comparison with the isolated CFH_3 molecule. For the remaining complexes electrostatic interaction energy is the most important attractive term, only in a case of the $SiFH_3$-C_6H_6 and $GeFH_3$-C_6H_6 complexes the dispersive term is slightly "less negative" than the electrostatic one. If one excludes the above-mentioned three CFH_3 complexes, thus for majority of remaining complexes the orbital interaction, ΔE_{orb}, is the next most important attractive term, after electrostatic interaction.

It was discussed in recent studies on hydrogen bonds and on other σ-hole bonds that these interactions are accompanied by effects that are a response for the Pauli repulsion [2,44]. The latter was also discussed for halogen bonds where multivalent halogen center plays a role of the Lewis acid while the π-electrons are the Lewis base [33]. For such interactions correlations were found between the repulsion interaction energy and different terms of the attractive interaction. It was found in earlier studies than the orbital interaction energy (if one refers to the decomposition scheme applied here) well correlates with the repulsion term, correlations for other interaction energy terms are not so good. However, in general, the sum of all attractive terms correlates with the Pauli repulsion term [2,33,44]. Figure 3 presents such a correlation for the complexes analyzed here. Thus the attractive interaction which is related to various effects related to complexation, among them to the electron charge redistribution, is a response for the Pauli repulsion.

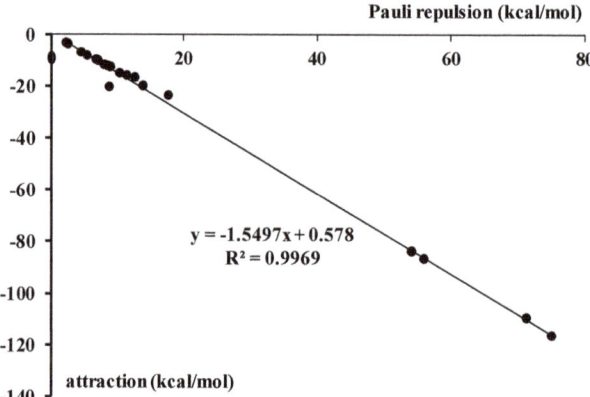

Figure 3. The linear correlation between the repulsion interaction energy and the sum of attractive terms (both in kcal/mol) for the ZFH$_3$-B complexes analyzed here.

The orbital interaction reflecting electron shifts corresponds to energy terms which are named in the other way in other decomposition schemes; most often they are labeled as the delocalization interaction energy, induction, charge transfer, polarization and others [2]. Figure 4 shows, for the complexes analyzed here, the correlation between the orbital energy, ΔE_{orb}, and the electron charge shift resulting from complexation.

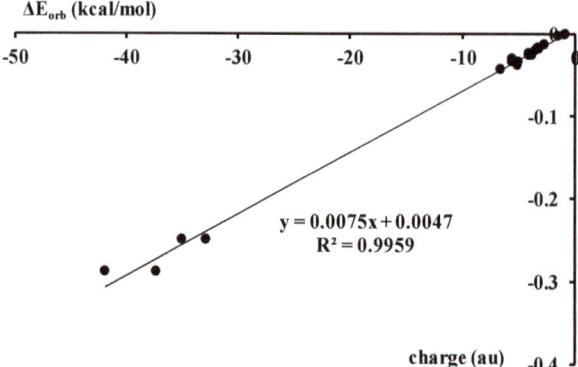

Figure 4. The linear correlation between the orbital interaction energy, ΔE_{orb}, and the electron charge shift from the Lewis base unit to the Lewis acid (au).

2.3. Quantum Theory of Atoms in Molecules Parameters

Table 4 presents characteristics of the bond critical point (BCP) of the bond path that connects the Lewis acid and Lewis base units of the complex. It is a link between the tetrel center (tetrel attractor) or the hydrogen center (hydrogen attractor) and the critical point of the Lewis base species. This critical point may correspond to the carbon atom attractor, to the bond critical point (BCP) of the CC bond or to the non-nuclear attractor (NNA) located on the CC bond path. Hence one can see that there are various topologies of complexes analyzed here.

Table 4. The QTAIM parameters (in au) of BCP of the Lewis acid—Lewis base bond path; electron density at BCP, ρ_{BCP}, its laplacian, $\nabla^2\rho_{BCP}$, and the total electron energy density at BCP, H_{BCP}. The bond path type is also indicated.

Complex	ρ_{BCP}	$\nabla^2\rho_{BCP}$	H_{BCP}	BP-Type
CFH_3-C_2H_2	0.005	0.020	0.001	C⋯C
CFH_3-C_2H_4	0.005	0.018	0.001	C⋯C
CFH_3-C_6H_6	0.007	0.022	0.001	(C)H⋯C
CFH_3-$C_5H_5^-$	0.011	0.038	0.001	(C)H⋯C
$SiFH_3$-C_2H_2	0.007	0.022	0.001	Si⋯C
$SiFH_3$-C_2H_4	0.008	0.023	0.001	(Si)H⋯C
$SiFH_3$-C_6H_6	0.008	0.025	0.001	(Si)H⋯C
$SiFH_3$-$C_5H_5^-$	0.036	0.012	−0.012	Si⋯C
$GeFH_3$-C_2H_2	0.009	0.027	0.001	Ge⋯NNA(CC)
$GeFH_3$-C_2H_4	0.010	0.028	0.001	Ge⋯BCP(CC)
$GeFH_3$-C_6H_6	0.010	0.028	0.001	Ge⋯C
-	0.007	0.024	0.001	(Ge)H⋯C
$GeFH_3$-$C_5H_5^-$	0.037	0.045	−0.008	Ge⋯C
$SnFH_3$-C_2H_2	0.010	0.025	0.001	Sn⋯NNA(CC)
$SnFH_3$-C_2H_4	0.011	0.029	0.001	Sn⋯BCP(CC)
$SnFH_3$-C_6H_6	0.012	0.029	0.001	Sn⋯C
-	0.007	0.022	0.001	(Sn)H⋯C
$SnFH_3$-$C_5H_5^-$	0.045	0.066	−0.011	Sn⋯C
$PbFH_3$-C_2H_2	0.011	0.034	0.001	Pb⋯NNA(CC)
$PbFH_3$-C_2H_4	0.013	0.036	0.001	Pb⋯BCP(CC)
$PbFH_3$-C_6H_6	0.014	0.039	0.001	Pb⋯C
-	0.009	0.026	0.001	(Pb)H⋯C
$PbFH_3$-$C_5H_5^-$	0.042	0.074	−0.008	Pb⋯C
-	0.015	0.041	0.001	(Pb)H⋯C

The above-mentioned bond path may concern the tetrel bond if the Z-center of the Lewis acid unit is linked with the Lewis base critical point or it may concern the hydrogen bond if the H-atom attractor of the Lewis acid is linked with the Lewis base critical point. One may expect the (Z)H⋯C bond paths show some "artificial interactions", especially since the meaning of the bond path and its usefulness to analyze interactions is often a subject of controversies [45] and disputes [46,47]. The presented here preliminary results on tetrel bonds where π-electrons play a role of Lewis bases need additional extended studies. However few arguments that the accompanying (Z)H⋯C bond paths observed for some benzene and cyclopentadienyl complexes may correspond to weak hydrogen bonds are listed here. The electron densities at the (Z)H⋯C bond critical points (BCPs) are not meaningless and they are comparable sometimes with such values for the Z⋯C BCPs; see the $GeFH_3$-C_6H_6 complex for example (Table 4). The $PbFH_3$-$C_5H_5^-$ complex is an example where the greatest electron density at the H⋯C BCP is observed since it amounts 0.015 au; note that for the medium in strength hydrogen bond in the water dimer the electron density at the H⋯O BCP is equal to 0.023 au (MP2/6-311++G(d,p) results [48]). Additionally the H⋯C intermolecular contacts correspond to the attractive electrostatic interactions since the carbon centers of the C_6H_6 and $C_5H_5^-$ moieties are characterized by the negative electrostatic potentials (EPs) while the H-centers of the ZFH_3 species by the positive EPs.

Particularly the following cases of bond paths are observed for complexes analyzed here. For the CFH_3-C_2H_2 and CFH_3-C_2H_4 complexes the irregular and nonlinear carbon-carbon bond paths are observed that may result from weak interactions (Figure 5); formally according to the QTAIM approach, they may be attributed to the tetrel bonds. For the CFH_3-C_6H_6 (Figure 1) and CFH_3-$C_5H_5^-$ complexes the H⋯C intermolecular bond paths are observed which may be attributed to the C-H⋯C hydrogen bonds! For the $SiFH_3$-C_2H_2, $SiFH_3$-C_2H_4 and $SiFH_3$-C_6H_6 complexes the non-linear bond paths are detected, similarly as for the CFH_3-C_2H_2 and CFH_3-C_2H_4 complexes, which are attributed to the Si⋯C or H⋯C intermolecular links (Table 4).

In a case of the SiFH$_3$-C$_5$H$_5^-$ complex the clear almost linear Si···C bond path corresponding to the strong tetrel bond is observed, similarly as for the other ZFH$_3$-C$_5$H$_5^-$ complexes for Z = Ge, Sn and Pb. In a case of the PbFH$_3$-C$_5$H$_5^-$ complex the additional H···C bond path corresponding to the Pb-H···C hydrogen bond is observed (Figure 1). For the ZFH$_3$-C$_6$H$_6$ complexes (Z = Ge, Sn, Pb) the tetrel and hydrogen bonds are observed with the corresponding bond paths, the SnFH$_3$-C$_6$H$_6$ complex representing such a situation is presented in Figure 1. Similarly the SnFH$_3$-C$_2$H$_2$, SnFH$_3$-C$_2$H$_4$ complexes in Figure 1 reflect the same situation in analogues tin and lead complexes; in the case of acetylene Lewis base the Z···NNA bond path is observed while in the case of ethylene Lewis base this is the Z···BCP bond path.

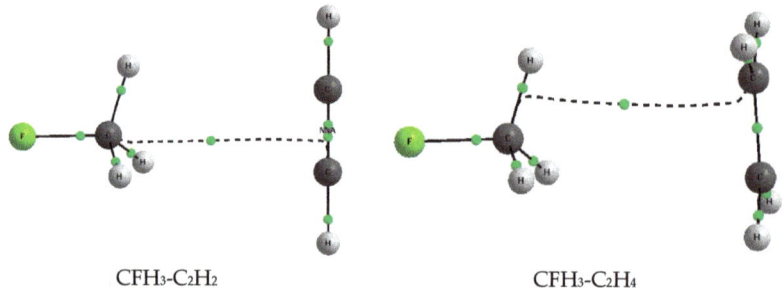

Figure 5. The molecular graphs of the CFH$_3$-C$_2$H$_2$ and CFH$_3$-C$_2$H$_4$ complexes; big circles—attractors, small green circles—BCPs, the nonnuclear attractor (NNA) is located (small red circle) between two BCPs in a case of the CFH$_3$-C$_2$H$_2$ complex.

The characteristics of bond critical points presented in Table 4 reflect the strength of interaction. It was discussed in various studies that these characteristics may be often treated as measures of the strength of interaction [49]; especially for homogeneous samples of complexes. Numerous relationships were found between characteristics of the H···B BCP and the strength of interaction for the A-H···B hydrogen bonds. For complexes analyzed here greater values of the electron density at the bond critical point, ρ_{BCP}'s, are observed for the C$_5$H$_5^-$ complexes. The Laplacian of the electron density at BCP, $\nabla^2\rho_{BCP}$, is positive for all complexes analyzed which may show these are not covalent interactions; the H$_{BCP}$ values are positive and close to zero for all complexes of acetylene, ethylene and benzene as well as for the CFH$_3$-C$_5$H$_5^-$ complex. For the remaining complexes of the C$_5$H$_5^-$ anion the negative H$_{BCP}$ values are observed that may indicate these are partly covalent in nature interactions.

One may ask what is the difference between the Z···π and Z···C tetrel bonds that are presented here. These "two kinds" of connections correspond to the types of bond paths. For the majority of acetylene and ethylene complexes former connections are observed while for the benzene and cyclopentadienyl complexes the latter ones. The Z···π bond path is a link between Z-attractor that corresponds to the nucleus and BCP or NNA located at the CC bond of acetylene or ethylene (see Figure 1). The Z···C bond path is a link between Z and C attractors corresponding to nuclei. This difference occurs within the Quantum Theory of Atoms in Molecules (QTAIM) scheme but it seems it is not observed in other approaches; for example in both cases the same orbital-orbital overlaps occur that correspond to the Z···π interactions; i.e., $\pi_{CC} \rightarrow \sigma_{ZF}^*$ ones. All other accompanying overlaps specified earlier here are the same in both cases of contacts. The similar situations were observed earlier for the hydrogen bonded complexes with the π-electron systems playing a role of Lewis bases [50].

3. Computational Details

The calculations were performed with the Gaussian16 set of codes [51] using the second-order Møller-Plesset perturbation theory method (MP2) [52], and the aug-cc-pVTZ basis set [53].

The relativistic effects for the heavier Ge, Sn and Pb atoms were taken into account. The calculations for these elements were done with quasi-relativistic small-core effective core potentials: ECP10MDF, ECP28MDF and ECP60MDF, for Ge, Sn and Pb, respectively [54]. For the latter elements the basis sets corresponding to aug-cc-pVTZ were applied, i.e., ECP10MDF_AVTZ, ECP28MDF_AVTZ and ECP60MDF_AVTZ, respectively [55]. Frequency calculations were performed for the complexes analyzed and their monomers to confirm that the optimized structures correspond to energetic minima. The binding energy, E_{bin}, was calculated as difference between the energy of the complex and the sum of energies of monomers optimized separately while the interaction energy, E_{int}, is a difference between the energy of the complex and the sum of energies of monomers which geometries come from the geometry of the complex considered [56]. The binding and interaction energies are negative but their difference—the deformation energy, $E_{def} = E_{bin} - E_{int}$, is positive and it is connected with the change of geometries of monomers resulting from the complexation [39]. The Counterpoise (CP) correction was applied to calculate the basis set superposition error BSSE [57]; hence the E_{bin} and E_{int} values corrected for BSSE are analyzed in this study.

The Quantum Theory of 'Atoms in Molecules' (QTAIM) was also applied to characterize critical points (BCPs) in terms of the electron density (ρ_{BCP}), its Laplacian ($\nabla^2 \rho_{BCP}$) and the total electron energy density at BCP (H_{BCP}) which is the sum of the potential electron energy density (V_{BCP}) and the kinetic electron energy density (G_{BCP}) [34]. The AIMAll program was used to carry out the QTAIM calculations [58].

The Natural Bond Orbital (NBO) method [35] was applied to calculate atomic charges, the electron charge shifts from the Lewis bases to the Lewis acids as well as the orbital-orbital interactions. The $n_B \rightarrow \sigma_{AH}^*$ orbital-orbital interaction is characteristic for the A-H···B hydrogen bond; n_B labels the lone electron pair of the B Lewis base center and σ_{AH}^* is the antibonding orbital of the A-H Lewis acid bond [35]. In a case of the hydrogen bonds where π-electrons and σ-electrons play a role of the Lewis bases, A-H···π and A-H···σ systems, the $\pi_B \rightarrow \sigma_{AH}^*$ and $\sigma_B \rightarrow \sigma_{AH}^*$ overlaps, respectively, are the most important orbital-orbital interactions [59]. The similar situation occurs for the tetrel bonds analyzed here, they may be classified as the Z···π or Z···C interactions (Z labels the tetrel centre). The $\pi_{CC} \rightarrow \sigma_{ZF}^*$ and $\pi_{CC} \rightarrow \sigma_{ZH}^*$ overlaps are observed here as the most important interactions; besides the $\sigma_{CH} \rightarrow \sigma_{ZF}^*$ and $\sigma_{CH} \rightarrow \sigma_{ZH}^*$ overlaps are also detected but they are characterized by lower energies than the former interactions. For example, the $\pi_{CC} \rightarrow \sigma_{ZF}^*$ interaction is calculated as the second-order perturbation theory energy (Equation (1)):

$$\Delta E (\pi_{CC} \rightarrow \sigma_{ZF}^*) = -2 \langle \pi_{CC} | F | \sigma_{ZF}^* \rangle^2 / (\varepsilon(\sigma_{ZF}^*) - \varepsilon(\pi_{CC})), \quad (1)$$

$\langle \pi_{CC} | F | \sigma_{ZF}^* \rangle$ designates the Fock matrix element and $(\varepsilon(\sigma_{ZF}^*) - \varepsilon(\pi_{CC}))$ is the orbital energy difference. The similar equations (to Equation (1)) for the remaining above-mentioned orbital-orbital interactions may be given.

The energy decomposition analysis (EDA) [36,37] was carried out with the BP86 functional [60,61] in conjunction with the Grimme dispersion corrections (BP86-D3) [62] using uncontracted Slater-type orbitals (STOs) as basis functions for all elements with triple-ζ quality (ADF-basis set TZP). The energy decomposition analysis (EDA) was performed with the use of the ADF2013.01 program [63] for all complexes analyzed here and characterized by geometries resulting from the MP2/aug-cc-pVTZ optimizations. The EDA method follows the energy partition of Morokuma [36,37]. The interaction energy, ΔE_{int}, between two fragments (A and B) in the A-B link, in the particular electronic reference state and in the frozen geometry of AB is considered in this approach. The ΔE_{int} interaction energy is divided into three components and the additional dispersion term, ΔE_{disp} (Equation (2)):

$$\Delta E_{int} = \Delta E_{elstat} + \Delta E_{Pauli} + \Delta E_{orb} + \Delta E_{disp}, \quad (2)$$

The ΔE_{elstat} term corresponds to the electrostatic interaction between the unperturbed charge distributions of atoms and is usually attractive. The Pauli repulsion, ΔE_{Pauli}, is the energy change associated with the transformation from the superposition of the unperturbed electron densities of the isolated fragments to the wavefunction which properly obeys the Pauli principle through explicit antisymmetrization and renormalization of the product wavefunction; it comprises the destabilizing interactions between electrons of the same spin on either fragment. The orbital interaction, ΔE_{orb}, accounts for charge transfer and polarization effects.

Figure 6 presents the correlation between the interaction energy calculated within the MP2/aug-cc-pVTZ approach (Table 1), thus at the level corresponding to the systems' optimizations, and ΔE_{int} DFT energy calculated with the use of ADF codes. The excellent correlation observed here partly justifies the use of DFT calculations for the previously optimized MP2 geometries.

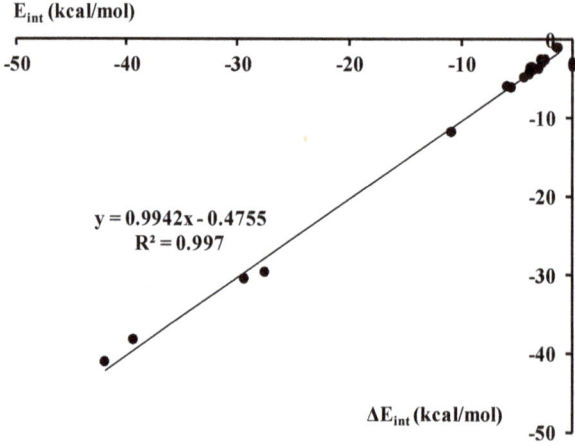

Figure 6. The linear correlation between the MP2 E_{int} interaction energy and the ΔE_{int} energy calculated within the DFT approach; both energies in kcal/mol.

4. Conclusions and Perspectives

The tetrel bonds in complexes where the π-electron system plays a role of the Lewis base were analyzed here. Practically for all complexes considered the NBO approach shows the existence of the $\pi_{CC} \rightarrow \sigma_{FZ}^*$ overlaps, however in a case of complexes of CFH$_3$ with acetylene, ethylene and benzene the corresponding energies are negligible thus the existence of tetrel bonds is problematic. On the other hand for the remaining complexes the above-mentioned interactions are significant that may indicate the existence of the tetrel bonds. The QTAIM approach often shows the complicated topology, sometimes the additional bond paths corresponding to the hydrogen bonds are observed, or like for the CFH$_3$-C$_5$H$_5^-$ complex, only C-H···C intermolecular link is observed that may indicate the existence of the hydrogen bond and not of the tetrel bond. However for the other cyclopentadienyl complexes the interactions are very strong and the Z···C bond paths exist there. Hence there is no doubt that these complexes are linked by the tetrel bonds; all theoretical approaches applied in this study support the existence of such interactions in these complexes.

Only for some of acetylene and ethylene complexes one may observe the link between tetrel center and the site corresponding to π-electrons, NNA or BCP of the CC bond of the Lewis base unit (see the SnFH$_3$-C$_2$H$_2$ and SnFH$_3$-C$_2$H$_4$ complexes in Figure 1 as examples). In a case of the C$_6$H$_6$ and C$_5$H$_5^-$ aromatic systems, the Z···C bond path is observed that suggest the one-atom Lewis base center and not the π-electron system. It means that the existence of two types of tetrel bonds may be considered within the QTAIM approach, Z···π and Z···C. However other approaches applied here do not distinguish rather between these types. Such a situation was earlier observed for the hydrogen bonded complexes [50].

The question arises if the interactions analyzed theoretically here really exist. This is why the Cambridge Structural Database (CSD) [64] search was performed. The following search criteria were taken into account; non-disordered structures, R less than 10%, 3D coordinated determined, non polymeric structures, single crystal structures and no errors (CSD updates up to February of 2018 were taken into account). The additional condition was that the Z tetrel center (C, Si, Ge, Sn and Pb) has to form two intermolecular Z···C contacts within corresponding sum of van der Waals radii. Two Z···C contacts were required since one may expect that in a case of double and triple CC bonds, or if CC bond concerns delocalized aromatic system; at least two Z···C distances within the van der Waals sum should be observed. 218 systems of crystal structures fulfilling those requirements were found in CSD. However only in 10 cases the clear tetrel-CC bond contacts with the tetrahedral (sp^3 hybridized) tetrel center were observed which suggest the existence of the tetrel···π-electrons interactions. Figure 7 shows an example where one can observe the F-Si···CC contact (CC bond of the aromatic phenyl ring). This issue requires additional studies on the experimental crystal structures however. It seems that the search criteria could be also improved. More detailed study on experimental crystal structures' results is in the progress.

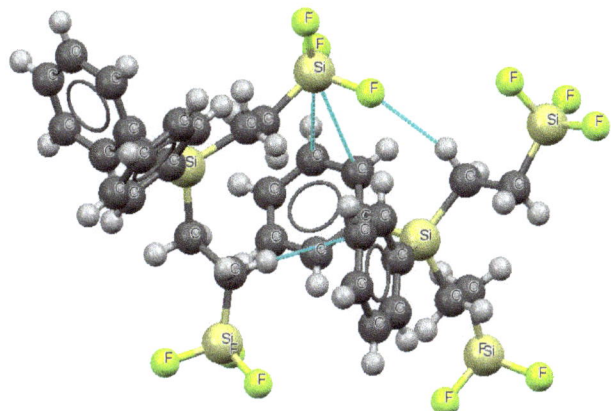

Figure 7. The fragment of the crystal structure (refcode: BAKZOF) where the Si···CC tetrel bond is observed.

Funding: Financial support comes from Eusko Jaurlaritza (GIC IT-588-13) and the Spanish Government MINECO/FEDER (CTQ2016-80955).

Acknowledgments: Technical and human support provided by Informatikako Zerbitzu Orokora—Servicio General de Informática de la Universidad del País Vasco (SGI/IZO-SGIker UPV/EHU), Ministerio de Ciencia e Innovación (MICINN), Gobierno Vasco Eusko Jaurlanitza (GV/EJ), European Social Fund (ESF) is gratefully acknowledged.

Conflicts of Interest: The author declares no conflict of interest.

References

1. García-LLinás, X.; Bauzá, A.; Seth, S.K.; Frontera, A. Importance of R-CF$_3$···O Tetrel Bonding Interactions in Biological Systems. *J. Phys. Chem. A* **2017**, *121*, 5371–5376. [CrossRef] [PubMed]
2. Grabowski, S.J. Hydrogen bonds, and σ-hole and π-hole bonds–mechanisms protecting doublet and octet electron structures. *Phys. Chem. Chem. Phys.* **2017**, *19*, 29742–29759. [CrossRef] [PubMed]
3. Politzer, P.; Murray, J.S.; Clark, T. Halogen bonding and other σ-hole interactions: A perspective. *Phys. Chem. Chem. Phys.* **2013**, *15*, 11178–11189. [CrossRef] [PubMed]
4. Bundhun, A.; Ramasami, P.; Murray, J.S.; Politzer, P. Trends in σ-hole strengths and interactions of F$_3$MX molecules (M = C, Si, Ge and X = F, Cl, Br, I). *J. Mol. Model.* **2013**, *19*, 2739–2746. [CrossRef] [PubMed]

5. Politzer, P.; Murray, J.S.; Lane, P.; Concha, M.C. Electrostatically Driven Complexes of SiF$_4$ with Amines. *Int. J. Quantum Chem.* **2009**, *109*, 3773–3780. [CrossRef]
6. Bauzá, A.; Mooibroek, T.J.; Frontera, A. Tetrel-Bonding Interaction: Rediscovered Supramolecular Force? *Angew. Chem. Int. Ed.* **2013**, *52*, 12317–12321. [CrossRef] [PubMed]
7. Chehayber, J.M.; Nagy, S.T.; Lin, C.S. Ab initio studies of complexes between SiF$_4$ and ammonia. *Can. J. Chem.* **1984**, *62*, 27–31. [CrossRef]
8. Alkorta, I.; Rozas, I.; Elguero, J. Molecular Complexes between Silicon Derivatives and Electron-Rich Groups. *J. Phys. Chem. A* **2001**, *105*, 743–749. [CrossRef]
9. Clark, T.; Hennemann, M.; Murray, J.S.; Politzer, P. Halogen bonding: The σ-hole. *J. Mol. Model.* **2007**, *13*, 291–296. [CrossRef] [PubMed]
10. Politzer, P.; Lane, P.; Concha, M.C.; Ma, Y.; Murray, J.S. An overview of halogen bonding. *J. Mol. Model.* **2007**, *13*, 305–311. [CrossRef] [PubMed]
11. Grabowski, S.J. Tetrel bond-σ-hole bond as a preliminary stage of the S$_N$2 reaction. *Phys. Chem. Chem. Phys.* **2014**, *16*, 1824–1834. [CrossRef] [PubMed]
12. Helminiak, H.M.; Knauf, R.R.; Danforth, S.J.; Phillips, J.A. Structural and Energetic Properties of Acetonitrile–Group IV (A & B) Halide Complexes. *J. Phys. Chem. A* **2014**, *118*, 4266–4277. [PubMed]
13. Mani, D.; Arunan, E. The X-C···Y (X = O/F, Y = O/S/F/Cl/Br/N/P) 'carbon bond' and hydrophobic interactions. *Phys. Chem. Chem. Phys.* **2013**, *15*, 14377–14383. [CrossRef] [PubMed]
14. Li, Q.-Z.; Zhuo, H.-Y.; Li, H.-B.; Liu, Z.-B.; Li, W.-Z.; Cheng, J.-B. Tetrel-Hydride Interaction between XH$_3$F (X = C, Si, Ge, Sn) and HM (M = Li, Na, BeH, MgH). *J. Phys. Chem. A* **2015**, *119*, 2217–2224. [CrossRef] [PubMed]
15. Scheiner, S. Systematic Elucidation of Factors That Influence the Strength of Tetrel Bonds. *J. Phys. Chem. A* **2017**, *121*, 5561–5568. [CrossRef] [PubMed]
16. Kubelka, J.; Bickelhaupt, F.M. Activation Strain Analysis of S$_N$2 Reactions at C, N, O, and F centers. *J. Phys. Chem. A* **2017**, *121*, 885–891. [CrossRef] [PubMed]
17. George, J.; Dronskowski, R. Tetrel Bonds in Infinite Molecular Chains by Electronic Structure Theory and Their Role for Crystal Stabilization. *J. Phys. Chem. A* **2017**, *121*, 1381–1387. [CrossRef] [PubMed]
18. Marín-Luna, M.; Alkorta, I.; Elguero, J. A theoretical study of the H$_n$F$_{4-n}$Si:N-base (n = 1–4) tetrel-bonded complexes. *Theor. Chem. Acc.* **2017**, *136*, 41. [CrossRef]
19. Liu, M.; Li, Q.; Scheiner, S. Comparison of tetrel bonds in neutral and protonated complexes of pyridine TF$_3$ and furan TF$_3$ (T = C, Si, and Ge) with NH$_3$. *Phys. Chem. Chem. Phys.* **2017**, *19*, 5550–5559. [CrossRef] [PubMed]
20. Zierkiewicz, W.; Michalczyk, M.; Scheiner, S. Implications of monomer deformations for tetrel and pnicogen bonds. *Phys. Chem. Chem. Phys.* **2018**, *20*, 8832–8841. [CrossRef] [PubMed]
21. Scheiner, S. Steric Crowding in tetrel Bonds. *J. Phys. Chem. A* **2018**, *122*, 2550–2562. [CrossRef] [PubMed]
22. Grabowski, S.J.; Sokalski, W.A. Are Various σ-Hole Bonds Steered by the Same Mechanisms? *ChemPhysChem* **2017**, *18*, 1569–1577. [CrossRef] [PubMed]
23. Roy, S.; Drew, M.G.B.; Bauzá, A.; Frontera, A.; Chattopadhyay, S. Non-covalent tetrel bonding interactions in hemidirectional lead (II) complexes with nickel(II)–salen type metalloligands. *New J. Chem.* **2018**, *42*, 6062–6076.
24. Grabowski, S.J. Tetrel bonds, penta- and hexa-coordinated tin and lead centres. *Appl. Organomet. Chem.* **2017**, *31*, e3727. [CrossRef]
25. Grabowski, S.J. Lewis Acid Properties of Tetrel tetrafluorides—The Coincidence of the σ-Hole Concept with the QTAIM Approach. *Crystals* **2017**, *7*, 43. [CrossRef]
26. Nishio, M.; Hirota, M.; Umezawa, Y. *The CH/π Interaction, Evidence, Nature, and Consequences*; Wiley-VCH: New York, NY, USA, 1998.
27. Vasilyev, A.V.; Lindeman, S.V.; Kochi, J.K. Noncovalent binding of the halogens to aromatic donors. Discrete structures of labile Br$_2$ complexes with benzene and toluene. *Chem. Commun.* **2001**, *10*, 909–910. [CrossRef]
28. Duarte, D.J.R.; de las Vallejos, M.M.; Peruchena, N.M. Topological analysis of aromatic halogen/hydrogen bonds by electron charge density and electrostatic potentials. *J. Mol. Model.* **2010**, *16*, 737–748. [CrossRef] [PubMed]
29. Zhuo, H.; Li, Q.; Li, W.; Cheng, J. Is π halogen bonding or lone pair···π interaction formed between borazine and some halogenated compounds. *Phys. Chem. Chem. Phys.* **2014**, *16*, 159–165. [CrossRef] [PubMed]

30. Liu, C.; Zeng, Y.; Li, X.; Meng, L.; Zhang, X. A comprehensive analysis of P···π pnicogen bonds: Substitution effects and comparison with Br···π halogen bonds. *J. Mol. Model.* **2015**, *21*, 143. [CrossRef] [PubMed]
31. Fau, S.; Frenking, G. Theoretical investigation of the weakly bonded donor-acceptor complexes X_3B-H_2, $X_3B-C_2H_4$, and $X_3B-C_2H_2$ (X = H, F, Cl). *Mol. Phys.* **1999**, *96*, 519–527.
32. Grabowski, S.J. Triel Bonds, π-Hole-π-Electrons Interactions in Complexes of Boron and Aluminium Trihalides and Trihydrides with Acetylene and Ethylene. *Molecules* **2015**, *20*, 11297–11316. [CrossRef] [PubMed]
33. Grabowski, S.J. New Type of Halogen Bond: Multivalent Halogen Interacting with π- and σ-Electrons. *Molecules* **2017**, *22*, 2150. [CrossRef] [PubMed]
34. Bader, R.F.W. *Atoms in Molecules, a Quantum Theory*; Oxford University Press: Oxford, UK, 1990.
35. Weinhold, F.; Landis, C. *Valency and Bonding, a Natural Bond Orbital Donor—Acceptor Perspective*; Cambridge University Press: Cambridge, UK, 2005.
36. Morokuma, K. Molecular Orbital Studies of Hydrogen Bonds. III. C=O···H-O Hydrogen Bond in $H_2CO···H_2O$ and $H_2CO···2H_2O$. *J. Chem. Phys.* **1971**, *55*, 1236–1244. [CrossRef]
37. Ziegler, T.; Rauk, A. On the calculation of bonding energies by the Hartree–Fock Slater method. *Theor. Chim. Acta* **1977**, *46*, 1–10. [CrossRef]
38. Murray, J.S.; Politzer, P. Molecular electrostatic potentials and noncovalent interactions. *WIREs Comput. Mol. Sci.* **2017**, *7*, e1326. [CrossRef]
39. Grabowski, S.J.; Sokalski, W.A. Different types of hydrogen bonds: Correlation analysis of interaction energy components. *J. Phys. Org. Chem.* **2005**, *18*, 779–784. [CrossRef]
40. Gilli, P.; Bertolasi, V.; Ferretti, V.; Gilli, G. Evidence for resonance-assisted hydrogen bonding. 4. Covalent nature of the strong homonuclear hydrogen bond. Study of the O-H···O system by crystal structure correlation methods. *J. Am. Chem. Soc.* **1994**, *116*, 909–915. [CrossRef]
41. Murray, J.S.; Politzer, P. Molecular Surfaces, van der Waals Radii and Electrostatic Potentials in Relation to Noncovalent Interactions. *Croat. Chem. Acta* **2009**, *82*, 267–275.
42. Pauling, L. *The Nature of the Chemical Bond*, 3rd ed.; Cornell University Press: New York, NY, USA, 1960.
43. Batsanov, S.S. Van der Waals Radii of Elements. *Inorg. Mater.* **2001**, *37*, 871–885. [CrossRef]
44. Grabowski, S.J. Hydrogen bonds and other interactions as a response to protect doublet/octet electron structure. *J. Mol. Model.* **2018**, *24*, 38. [CrossRef] [PubMed]
45. Poater, J.; Solà, M.; Bickelhaupt, F.M. Hydrogen-Hydrogen Bonding in Planar Biphenyl, Predicted by Atoms-In-Molecules Theory, Does Not Exist. *Chem. Eur. J.* **2006**, *12*, 2889–2895. [CrossRef] [PubMed]
46. Bader, R.F.W. Pauli Repulsions Exist Only in the Eye of the Beholder. *Chem. Eur. J.* **2006**, *12*, 2896–2901. [CrossRef] [PubMed]
47. Poater, J.; Solà, M.; Bickelhaupt, F.M. A Model of the Chemical Bond Must Be Rooted in Quantum Mechanics, Provide Insight, and Possess Predictive Power. *Chem. Eur. J.* **2006**, *12*, 2902–2905. [CrossRef] [PubMed]
48. Grabowski, S.J. Ab Initio Calculations on Conventional and Unconventional Hydrogen Bonds–Study of the Hydrogen Bond Strength. *J. Phys. Chem. A* **2001**, *105*, 10739–10746. [CrossRef]
49. Espinosa, E.; Molins, E.; Lecomte, C. Hydrogen bond strengths revealed by topological analyses of experimentally observed electron densities. *Chem. Phys. Lett.* **1998**, *285*, 170–173. [CrossRef]
50. Grabowski, S.J.; Ugalde, J.M. Bond Paths Show Preferable Interactions: Ab Initio and QTAIM Studies on the X-H···π Hydrogen Bond. *J. Phys. Chem. A* **2010**, *114*, 7223–7229. [CrossRef] [PubMed]
51. Frisch, M.J.; Trucks, G.W.; Schlegel, H.B.; Scuseria, G.E.; Robb, M.A.; Cheeseman, J.R.; Scalmani, G.; Barone, V.; Petersson, G.A.; Nakatsuji, H.; et al. *Gaussian 16, Revision A.03*, Gaussian, Inc.: Wallingford, CT, USA, 2016.
52. Møller, C.; Plesset, M.S. Note on an Approximation Treatment for Many-Electron Systems. *Phys. Rev.* **1934**, *46*, 618–622. [CrossRef]
53. Woon, D.E.; Dunning, T.H., Jr. Gaussian Basis Sets for Use in Correlated Molecular Calculations. III. The second row atoms, Al-Ar. *J. Chem. Phys.* **1993**, *98*, 1358–1371. [CrossRef]
54. Metz, B.; Stoll, H.; Dolg, M. Small-core multiconfiguration-Dirac-Hartree-Fock-adjusted pseudopotentials for post-d main group elements: Application to PbH and PbO. *J. Chem. Phys.* **2000**, *113*, 2563–2569. [CrossRef]
55. Peterson, K.A. Systematically convergent basis sets with relativistic pseudopotentials. I. Correlation consistent basis sets for the post-d group 13-15 elements. *J. Chem. Phys.* **2003**, *119*, 11099–11112. [CrossRef]
56. Piela, L. *Ideas of Quantum Chemistry*; Elsevier Science Publishers: Amsterdam, The Netherlands, 2007; pp. 684–691.

57. Boys, S.F.; Bernardi, F. The calculation of small molecular interactions by the differences of separate total energies. Some procedures with reduced errors. *Mol. Phys.* **1970**, *19*, 553–561. [CrossRef]
58. Keith, T.A. *AIMAll (Version 11.08.23)*; TK Gristmill Software: Overland Park, KS, USA, 2011. Available online: aim.tkgristmill.com.
59. Grabowski, S.J. Dihydrogen bond and X-H···σ interaction as sub-classes of hydrogen bond. *J. Phys. Org. Chem.* **2013**, *26*, 452–459. [CrossRef]
60. Becke, A.D. Density-functional exchange-energy approximation with correct asymptotic behavious. *Phys. Rev. A* **1988**, *38*, 3098–3100. [CrossRef]
61. Perdew, J.P. Density-functional approximation for the correlation energy of the inhomogeneous electron gas. *Phys. Rev. B* **1986**, *33*, 8822–8824. [CrossRef]
62. Grimme, S.; Antony, J.; Ehrlich, S.; Krieg, H. A consistent and accurate ab initio parametrization of density functional dispersion correction (DFT-D) for the 94 elements H-Pu. *J. Chem. Phys.* **2010**, *132*, 154104. [CrossRef] [PubMed]
63. Velde, G.T.E.; Bickelhaupt, F.M.; Baerends, E.J.; Guerra, C.F.; van Gisbergen, S.J.A.; Snijders, J.G.; Ziegler, T. Chemistry with ADF. *J. Comput. Chem.* **2001**, *22*, 931–967. [CrossRef]
64. Wong, R.; Allen, F.H.; Willett, P. The scientific impact of the Cambridge Structural Database: A citation-based study. *J. Appl. Crystallogr.* **2010**, *43*, 811–824. [CrossRef]

Sample Availability: Samples of the compounds are not available from the author.

© 2018 by the author. Licensee MDPI, Basel, Switzerland. This article is an open access article distributed under the terms and conditions of the Creative Commons Attribution (CC BY) license (http://creativecommons.org/licenses/by/4.0/).

MDPI
St. Alban-Anlage 66
4052 Basel
Switzerland
Tel. +41 61 683 77 34
Fax +41 61 302 89 18
www.mdpi.com

Molecules Editorial Office
E-mail: molecules@mdpi.com
www.mdpi.com/journal/molecules

www.ingramcontent.com/pod-product-compliance
Lightning Source LLC
LaVergne TN
LVHW070443100526
838202LV00014B/1657